Practical Nanotechnology
for Petroleum Engineers

Practical Nanotechnology for Petroleum Engineers

Chun Huh, Hugh Daigle, Valentina Prigiobbe, and Maša Prodanović

CRC Press
Taylor & Francis Group
Boca Raton London New York

CRC Press is an imprint of the
Taylor & Francis Group, an **informa** business

CRC Press
Taylor & Francis Group
6000 Broken Sound Parkway NW, Suite 300
Boca Raton, FL 33487-2742

First issued in paperback 2020

ISBN 13: 978-0-367-65648-5 (pbk)
ISBN 13: 978-0-8153-8149-5 (hbk)

Library of Congress Cataloging-in-Publication Data

Names: Huh, Chun, author. | Hugh, Daigle, author. | Maša, Prodanović, author. | Prigiobbe, Valentina, author.
Title: Practical nanotechnology for petroleum engineers / Chun Huh, Hugh Daigle, Maša Prodanović, and Valentina Prigiobbe.
Description: Boca Raton : Taylor & Francis a CRC title, part of the Taylor & Francis imprint, a member of the Taylor & Francis Group, the academic division of T&F Informa, plc, 2019. | Includes bibliographical references and index.
Identifiers: LCCN 2018051256| ISBN 9780815381495 (hardback : acid-free paper) | ISBN 9781351210362 (ebook)
Subjects: LCSH: Oil field chemicals. | Petroleum engineering--Materials. | Oil well drilling--Materials. | Nanostructured materials--Indiustrial applications. | Petroleum industry and trade--Technological innovations.
Classification: LCC TN871.27 .H84 2019 | DDC 622/.3380284--dc23
LC record available at https://lccn.loc.gov/2018051256

Visit the Taylor & Francis Web site at
http://www.taylorandfrancis.com

and the CRC Press Web site at
http://www.crcpress.com

Contents

Preface

With the investment of billions of dollars for research and development, nanotechnology is now an important contributor to materials engineering, electronics, medical, and other industries. In the upstream oil and gas industry, however, there has been very little recognition of its potential, and accordingly until the last ten years or so, no effort has been made to adapt and utilize the tremendous nanotechnology developments made in other industries for our benefit. We, the four co-authors of this book, had the fortunate opportunity to actively participate in the oil industry's new effort to utilize nanotechnology for improved oil recovery and oilfield operations; and this book is our modest but sincere effort to spread the word among the upstream oil industry professionals that, when carefully adopted and appropriately modified to meet the unique requirements for oilfield application, nanotechnology can indeed provide significant benefits to the upstream oil industry.

As expounded in this book, nanotechnology is in simple terms the manufacture and manipulation of nanoparticles, i.e., particles smaller than 100 nm. In addition to the well-recognized properties of nanoparticles that make them uniquely useful in many industries, their two obvious attributes that are particularly important for the oil industry are their smallness and the very large surface area per unit mass. That is, properly designed nanoparticles can go through tiny and tortuous pore pathways of oil and gas reservoirs for a long distance without being retained. And their large surface area can serve as secure substrates to carry functional chemicals in significant quantities; or as in-situ reaction sites to convert a substantial amount of material, deep down in the reservoir. Such capabilities open up new possibilities of carrying out a wide variety of chemical reactions, physical functions, and engineering maneuvers that are well developed in many scientific and engineering disciplines but hitherto were impossible to bring deep down to the subsurface formations. This exciting potential motivated us to prepare this current-status report to encourage other upstream researchers and developers to participate.

In publishing this book, we would like to express our sincere gratitude, in particular, to Prof. Steven Bryant (now with the University of Calgary) who was the main driver to push the nanotechnology applications development at the University of Texas at Austin. We also thank Profs. David DiCarlo, Kishore Mohanty, Gary Pope, Matthew Balhoff (all with Petroleum and Geosystems Engineering), Keith Johnston (Chemical Engineering), Thomas Milner (Biomedical Engineering), and the graduate students, postdoctoral fellows, and undergraduate assistants at the University of Texas at Austin who worked on the nanoparticle applications development with enthusiasm and dedication, with the realization that we all are participating in a new venture to help transform the upstream oil industry for its continued sustainability. Our particular appreciation goes to Joanna Castillo, without whose meticulous and tireless editing help, the publication of this book would have been impossible.

Chun Huh, Hugh Daigle, Valentina Prigiobbe, and Maša Prodanović

Authors

Dr. Chun Huh received his B.S. in chemical engineering from Seoul National University, and Ph.D. also in chemical engineering from University of Minnesota. Before joining the Department of Petroleum & Geosystems Engineering, University of Texas at Austin as a Research Professor in January, 2004, Dr. Huh worked as an Engineering Advisor at ExxonMobil Upstream Research Company in Houston, Texas. He is one of the leading experts on surfactant- and polymer-based improved oil recovery (IOR) processes. "Chun Huh equation," which predicts ultralow interfacial tension from microemulsion solubilization, is widely used for the design of surfactant-based IOR processes. He is also the formulator of the "Huh-Scriven paradox," which first demonstrated the singularity generation when fluid-dynamics solutions are attempted for spreading/wetting of fluids on solid. At UT-Austin, Dr. Huh has started research on use of nanoparticles for a wide variety of upstream oil industry applications, co-authoring more than 60 publications on the subject. Dr. Huh received the Society of Petroleum Engineers' IOR Pioneer Award in 2012; and is a SPE Distinguished Member and a member of the US National Academy of Engineering.

Dr. Hugh Daigle is an Assistant Professor in the Hildebrand Department of Petroleum and Geosystems Engineering at the University of Texas at Austin and holds the Anadarko Petroleum Corporation Centennial Fellowship #2 in Petroleum Engineering. He obtained an A.B. *magna cum laude* in Earth and Planetary Sciences from Harvard University in 2004 and a Ph.D. in Earth Science from Rice University in 2011. Before starting his current position in 2013, he worked for 5 years in the oil and gas industry as a wireline logger and petrophysicist, both before and after graduate school. Dr. Daigle is director of the Nanoparticles for Subsurface Engineering Industrial Affiliates Program at the University of Texas at Austin, and is author or co-author of over 80 technical papers on topics ranging from nanotechnology applications in the upstream oil and gas industry to natural gas hydrates in marine sediments and percolation theory. Outside his university duties, he is an avid birdwatcher and classical musician.

Dr. Valentina Prigiobbe is an Assistant Professor in the Department of Civil, Environmental, and Ocean Engineering at Stevens Institute of Technology. She obtained a Laurea with magna cum laude in Engineering for Environment and Territory from the University of Rome Tor Vergata in 2000 and two Ph.D. degrees, one in Environmental Engineering from the same university in 2005 and one in Mechanical and Process Engineering from ETH Zurich in 2010. Before starting her current position in 2015, she was postdoctoral researcher in the Hildebrand Department of Petroleum and Geosystems Engineering at the University of Texas at Austin. Dr. Prigiobbe is author or co-author of more than 40 technical papers on topics ranging from flow and transport in porous media to nanotechnology and separation processes with applications to water treatment, carbon capture utilization and storage, and contaminated site remediation. Outside her university duties, she is a swimmer and painter.

Maša Prodanović has been an associate professor at the Department of Petroleum and Geosystems Engineering, The University of Texas at Austin since September 2016. Prior to her current post she has held an assistant professor position 2010-2016, a Research Associate position in the Center for Petroleum and Geosystems Engineering (UT Austin) 2007-2010, and prestigious J. T. Oden Postdoctoral Fellowship at the Institute of Computational Engineering and Sciences 2005-2007. She holds a Bachelor of Science in Applied Mathematics from the University of Zagreb, Croatia and a PhD in Applied Mathematics and Statistics from Stony Brook University, New York, USA. Her research interests include multiphase flow and image-based porous media characterization especially applied to heterogeneous porous media, ferrohydrodynamics, pore network models, shale gas flow, particulate flow and formation damage, sediment mechanics and fracturing. She received NSF CAREER award in 2013, Interpore Procter & Gamble Research Award for Porous Media Research in 2014, SPE Faculty Innovative Teaching Award in 2014, Texas 10 (top faculty) and Stony Brook 40 Under Forty awards in 2017.

1

Introduction

Nanotechnology is a truly huge area with billions of dollars of research and development funding, and publications of thousands of papers and patents, every year. Consequently, novel nanoscale structured materials, in the form of solid composites, complex fluids, and functional nanoparticle-fluid combinations, and a tremendous variety of applications built on those materials, are bringing major technological advances in many industries. A few examples are the extraordinary material strength, elasticity, and thermal conductivity of nano-based metal and polymer composites; targeted and programmed delivery of drugs and enhanced imaging of human organs in medicine; and highly localized measurements of chemical/physical properties using nanosensors. These and many other advances are due to the orders-of-magnitude increase in interfacial area and associated excess stress and chemical potential for the nano-structured materials, as well as some chemical and physical properties that are unique to nanoscale. In the petroleum engineering discipline, until around 2010, research and applications development of nanotechnology have been virtually non-existent. This is because the oil and gas reservoirs are generally deep, remote from surface, and in harsh environments that exhibit extreme heterogeneities over lengthscales from microns to thousands of meters. Downhole technologies and treatments are therefore difficult to direct and control. Also, huge volumes of material are required for most treatments so that the cost of producing the nanoscale materials to the proper standards of size and desired functionality must be kept low.

Despite these difficulties, the tremendous benefits that could be gained from the immense nanotechnology advances made by other industries are beginning to be recognized. In particular, the recent, explosive advances in development of *functional nanoparticles* and their novel use in a wide variety of medical, biological, electronics, and engineering applications, offer truly exciting and unique opportunities for the upstream oil industry. This is because the key processes for oil exploration and production largely occur in porous rocks deep underground, and the nano-size particles that are engineered to carry out specific tasks in remote oil reservoirs can now be made to flow long-distance through rock pores, reach the target location, and perform the required function(s). When synthesized in a specific size range and with a design coating, nanoparticles exhibit unique properties because, while almost of molecular size, they still retain many useful colloidal characteristics. Because of these reasons, there is now a flurry of activities to develop novel upstream applications of nanoparticles. One of the main objectives of this book is to introduce these recent developments, as well as the nanotechnology concepts/applications that have been developed by other industries and are relevant to the oil industry.

Spurred in part by the Society of Petroleum Engineers Workshops and Forums on the nanotechnology application development for exploration and production, held in 2008 (Dubai, UAE), 2009 (Kuala Lumpur, Malaysia), 2010 (Cairo, Egypt), 2011 (Farro, Portugal), 2012 (Noordwijk, the Netherlands), and 2013 (Kyoto, Japan), there is now a widespread interest in the upstream oil industry on the nanotechnology application developments. A number of major oil companies and service companies are vigorously working on developing practical applications; and so are several universities. In view of the extensive advances made in other industries, the near-term applications will be the adaptation of what is already developed in those industries, i.e., harvesting the "low-hanging" fruits. A number of overview papers describing the nanotechnology's potential for upstream applications have recently been published (Amanullah and Ramasamy 2014; Bennetzen and Mogensen 2014; El-Diasty and Ragab 2013; Kapusta et al. 2011; Kong and Ohadi 2010; Krishnamoorti 2006; Matteo et al. 2012; Mokhatab et al. 2006; Pourafshary et al. 2009; Saggaf 2008).

While the above are largely "opinion" papers on upstream oil industry's nanotechnology needs and prospects, there are some reviews on very recent progress on applications development, mainly in enhanced oil recovery (EOR) areas (*e.g.*, Agista et al. 2018; Bera and Belhaj 2016; Kazemzadeh et al. 2018; Negin et al. 2016; ShamsiJazeyi et al. 2014).

There is an immense literature on nanotechnology, especially on nanoparticles, and for those who are seriously interested in the comprehensive overview, Klabunde and Richards (2009), Schaefer (2010), Schmid et al. (2006), Skandon and Singhal (2004), are good reference sources. For environmental nanotechnology, monographs by Wiesner and Bottero (2007) and Zhang et al. (2009) are good introductory resources. A comprehensive handbook on various nanoparticle applications by Hosokawa et al. (2012) is also an excellent resource.

1.1 Why Are Nanoparticles So Exciting?

Simply put, nanotechnology is the technology of using *nanoparticles* in their individual form, in the form of their 1-D, 2-D, 3-D assembly, or as a composite among any of the above and/or with a bulk material. The nanoparticle is a solid particle and its size at least in one dimension is smaller than 100 nm. With this broad definition, nanoparticles include sphere and cube and their variations (*e.g.*, spheroids, multi-edged or star-shaped grains); wire (carbon nano-wires); tube (single-walled or multi-walled carbon nanotubes); sheets (graphene, graphene oxide); hollow particle (fullerene); core-shell particle (metal oxide-silica nanospheres); and other sophisticated geometries. The main reason that the nanoparticles carry such unique properties is their extremely high surface-to-volume ratio due to their small size. A well-known illustration, shown in Figure 1.1, is the chopping of a cube with 1 cm sides into cubes with 10 nm sides. For the same mass of material, by chopping, the total surface area can be increased by 10^{12} times, which means that the ratio of surface molecule-to-bulk molecule also increased drastically. As the surface molecules are responsible for excess surface free energy, many of the material properties are consequently changed.

The extremely large surface area per mass is a major reason that nanoparticles are so useful in many industries. By attaching functional chemicals or biological entities to the surface of nanoparticles, and then by ushering them to the target locations, the chemical's or biological entity's desired functions can be carried out for specific, selective, and/or localized purposes. While ushering the chemicals (or biological entities) to the target by themselves is not feasible, the use of nanoparticles allows such specific placement, while the nanoparticles' large surface area also allows the use of a sufficient amount of chemicals (or biological entities). This unique capability

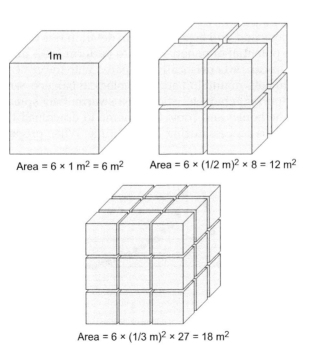

FIGURE 1.1
By decreasing the size of the particles, the surface area per mass of the material can be drastically increased.

of nanoparticles opens up many application development opportunities for upstream oil industry. Another key reason that nanoparticles are so exciting for oil and gas reservoir application is, simply, its being nano size. This is because nanoparticles with above unique capabilities can flow freely in reservoir rock without retention by straining. Figure 1.2 shows that, with an appropriate surface coating, 96% of a highly concentrated dispersion (18.7 wt%) of 5-nm silica nanoparticles that had been injected into a very tight (10 md) limestone core came out in the effluent water (Roberts et al. 2012). This means that a broad spectrum of applications of colloidal particles developed for more than 100 years, which was hitherto impossible to utilize for oil/gas reservoirs because of their almost immediate filtration, can now be utilized. For example, the solid core of the nanoparticles can be superparamagnetic, electrically conductive, piezoelectric, or catalytic—the properties that are highly useful for many potential oilfield applications, as will be described in detail in this book.

Still another important reason that nanoparticles are so potentially important for upstream oil industry is that they can be employed as stabilizers for foams and emulsions ("Pickering emulsion"; Binks and Horozov 2006) having some unique advantages over surfactants. Again, this means that a broad spectrum of applications of Pickering emulsions and foams actively developed during the past 20 years or so (Binks and Horozov 2006; Ngai and Bon 2015) can now be utilized, which are also highly useful for many potential oilfield applications and will be described in detail in this book.

The immense potential of nanoparticles to perform tasks deep in subsurface porous media can be heuristically compared with the critical role played by honey bees to help maintain nature's ecological balance, which is based on their two important characteristics: As a swarm, they spread out to the field to collect the honey and come back home to download their harvest. And they repeat this process during their lifetime. When properly designed

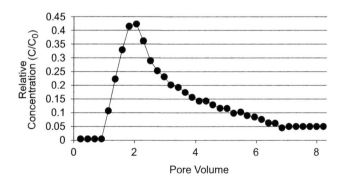

FIGURE 1.2
When an 18.7 wt% aqueous dispersion of 5-nm silica nanoparticles is injected into a 10-md limestone core, 96% of injected particles were recovered. Figure generated using data from Rodriguez et al. (2009).

and controlled, a dispersion of nanoparticles can function like honey bees for many oilfield operations, carrying out their design function either deep in the reservoir or downhole at the wellbore; returning to the surface to download the specific chemical or information that they collected; and then being gathered together to be regenerated and sent out again to repeat the process. As will be further elaborated below, such honey bee-like capability is especially true for the superparamagnetic nanoparticles. This challenging but exciting vision of making the oilfield operations as environmentally green as possible has to be the future of the upstream oil industry and the use of nanoparticles offers an important step to achieve the goal.

1.2 What Is There with Nanotechnology for Petroleum Engineers?

Even though the details of the nanotechnology applications development for the upstream oil industry will be described in Part II of this book, a brief list of what is becoming available and what to expect in near term is given below, with relevant sections of this book being shown in parenthesis:

1. Nanoparticle rheology modifiers
   a. Nanoparticle-polymer/surfactant mixture for improved formulation of drilling fluids and fracturing fluids (Section 7.2).
   b. Nanoparticle-polymer mixture for improved polymer flooding formulation (Section 11.3).
2. Nanoparticle-added metals/ceramics
   a. Improved strength and reduced weight of hardware materials for drilling and completions and for surface facilities (Section 7.5) and pipelines.
   b. Improved ball packers for multiple hydraulic fracturing (Section 7.5).
3. Nanoparticle-polymer composites
   a. Improved elastomers for surface facilities (Section 7.5).
   b. Improved near-wellbore conformance control using temperature-responsive polymer gels (Section 8.6).
4. Smart coatings
   a. Abrasion/corrosion-resistant and self-healing linings for hardware materials (Section 7.6).
   b. Subsea pipeline inner lining that prevents wax or hydrate deposition (Section 8.4).
   c. Reservoir rock wettability alteration for EOR (Section 11.2).

 5. Nanoparticle- or nanofiltration-based water treatment

  a. Removal of micron-scale oil droplets from produced water (Section 8.2).

  b. Conversion of hard brine to soft brine (Section 8.2).

  c. Removal of remnant EOR polymer from produced water (Section 8.2).

 6. Nanosensors

  a. Magnetic nanoparticles as remote-sensing contrast agents (Section 9.1).

  b. Nanoparticles and quantum dots as intelligent tracers (Section 9.2).

  c. Downhole nanosensors and monitors (Section 9.3).

 7. Nanoparticle-stabilized emulsions and foams

  a. Improved mobility control agents for gas-based EOR (Section 10.2).

  b. "Waterless" foam fracturing fluids (Section 7.3).

  c. Improved mobility control agents for heavy oil recovery (Section 12.2).

  d. Delivery medium for chemicals, catalysts, and sensors (Section 12.2).

 8. Nano catalysts

  a. In-situ upgrading of heavy oil (Section 12.3).

 9. Nano motors

  a. Self-propelling nanoparticle assembly that delivers chemicals to the reservoir target locations or carries out required functions there.

 10. Nano power sources

  a. Power source for functionalized nanoparticles, placed deep in the reservoir, to perform intended tasks or to emit back signal to detectable location.

 In developing the above exciting applications, two hurdles need to be overcome. First, the nanomaterials are still relatively expensive for certain applications such as EOR. On the other hand, those reservoir applications require a large volume of material, which potentially could lower the manufacturing cost. Another hurdle is the environmental concern when we consider injecting nanoparticles into the reservoirs. Careful life-cycle considerations should be made in developing those injectants which should be environmentally benign, such as silica and iron-oxide nanoparticles. Learning from other industries, which have carried out extensive studies in this regard, should again be carefully studied and utilized. With careful adaptation of the huge advances made by other industries, the future of nanotechnology use for the upstream oil industry is judged to be bright.

1.3 Brief Summary of Chapters of this Book

As mentioned above, because the basic building blocks of nanotechnology are nanoparticles, a tremendous variety of their synthesis, surface functionalization, and characterization methods have been developed. While it is difficult to present a concise description of these subjects due to the immensity of the available literature, in view of their importance in developing the oilfield-specific applications, an overview of the synthesis, surface functionalization, and characterization of the nanoparticles is first given in Chapter 2. The specific, step-by-step examples of the synthesis, surface coating, and characterization are also given, which should be useful to developers and evaluators of new oilfield applications.

An important capability of nanoparticle that is highly useful for reservoir application is its ability to reach the target location far and deep in the reservoir. As described in more detail in Section 2.1, the conditions in oil and gas reservoirs are generally quite harsh (*e.g.*, high temperature, high salinity), and maintaining the nanoparticle's individual integrity for a long period of time (in months, years) and making them to travel a long distance (in miles) to reach the target location require a good understanding of relevant mechanisms for nanoparticle's dispersion stability and transport in porous media. Both experimental and modeling aspects of these topics, which have been extensively studied in subsurface hydrology and environmental engineering disciplines, are described in Chapter 3.

As mentioned above, the use of nanoparticles as stabilizers for Pickering emulsions and foams has an immense potential utility for oilfield applications development. Colloid and interface chemists from many branches of chemical industry and other related industries extensively carried out research and development work to generate a wide variety of multi-scale dispersions of sophisticated internal structure. This is also comprehensively supported by theoretical modeling work. The latter aspect, relevant for oilfield applications development, will be mainly described in Chapter 4.

Superparamagnetic nanoparticles (SNPs) are normally benign but become nano-size magnets when subjected to an external magnetic field. Consequently, they can be moved in a desired direction by applying a magnetic field gradient, allowing them (and what's attached to them) to be collected in a controlled and/or selective manner. When exposed to an external magnetic field, they generate an induced field, allowing them to be detected remotely. Another important property of SNP is ability to generate intense, highly localized heat under a certain range of magnetic oscillation, which is known as "hyperthermia" in medical discipline. Therefore, SNP's (commonly) magnetite core serves a quite active role, and its unique properties above are highly useful for oilfield applications development, as can be seen in Part II of this book. The relevant basics of nanomagnetism are provided in Chapter 5.

Practical implementation of nanotechnology has always been under rigorous scrutiny due to the public's concern on its human and environmental impact, and researchers in the environmental engineering discipline have carried out comprehensive studies (*e.g.*, Wiesner and Bottero 2007). As the oilfield applications of nanotechnology are only just beginning to be developed, there is no published information available so far on the oil and gas-specific investigations on the human and environmental impacts resulting from the oilfield use of nanoparticles. Nevertheless, in view of its importance, the studies from other disciplines are reviewed in Chapter 6.

As can be noticed from the above list and will be seen later in Chapter 7, drilling and completions areas saw more of the development of practical applications of nanotechnology than other oil and gas areas. This is probably because, unlike the reservoir applications which require a long lead time and carry inherent uncertainties, the development in improvement of drilling or fracturing fluids, or of hardware, can be quickly assessed. A suite of recently developed applications is described in Chapter 7. Notable among them is the addition of nanoparticles to drilling fluids and fracturing fluids for their improved rheological behavior and to enhance the shale zone integrity during drilling. Recent progress on the property improvement for hardware materials, such as the longer-life drill bits, higher-strength and lighter-weight tubings and casings, and intelligent packer balls, is also described.

For an efficient production of oil and/or gas for the economic life span of a well (or a reservoir unit), maintaining the downhole near-wellbore zone, tubing/casing, surface facilities, and transport pipeline trouble-free is a critical requirement. Chapter 8 mainly deals with a number of novel applications of SNPs for trouble-free production operation and flow assurance. The first set uses SNPs to treat the oilfield produced water by functionalizing their surface so that micron-scale oil droplets or the EOR polymer left over from the EOR process can be attached to the particles and removed from the water. Specially functionalized SNPs can also be used to attach multi-valent cations in brine to make hard water soft. In another application, SNPs are employed to create a temperature-responsive polymer gel at the near-wellbore reservoir layers for improved conformance control. In still another application, SNPs are imbedded in the inner lining of the subsea oil (or gas) pipelines so that the deposition of wax (or hydrate from gas) can be prevented. As described above, such uses of SNPs are an important first step to make the oilfield operations environmentally "green."

In attempting to control the movement of fluids at far and deep down the reservoir, sensing the reservoir structure and the location and dynamics of the fluids in it is a challenging but absolutely necessary task for oilfield development. In Chapter 9, three different approaches to help achieve the above objective employing nanoparticles are described. The first is to inject into the reservoir a slug of an aqueous dispersion of paramagnetic or electrically conductive nanoparticles, so that their location and dynamics can be remotely detected. The second is the injection of "intelligent" tracers (or "nano-reporter") whose surface is specially functionalized so that, at the target location in the reservoir, certain fluid or formation properties are detected and recorded on them. When produced to the surface, they can be debriefed to retrieve the collected information. The third approach is the placement of nano- or micron-scale sensors at a downhole wellbore for continuous sensing/monitoring there.

Among various EOR processes, CO_2 flooding is probably the most reliable and economic EOR method when a steady CO_2 source is available near the oilfield. A key weakness of CO_2 flooding is its poor volumetric sweep efficiency, and to remedy the problem, use of CO_2 foam has been extensively studied. As nanoparticles can stabilize CO_2 foam with better tolerance to high temperature and high salinity/hardness than surfactants, this application has received considerable attention recently. In Chapter 10, in addition to the review of the recent progress on nanoparticle-stabilized CO_2 foam, the detailed steps for the optimal design of foams for specific reservoir application are also described, together with the relevant mechanisms. Recently, use of nanoparticle-stabilized oil-in-water emulsions is drawing an active interest as a way to enhance recovery of heavy oil. In particular, the use of the currently abundant natural gas liquids due to the active shale oil and gas development, as the emulsion's internal phase, is a very attractive option; and such development is described in the chapter.

"Smart coating" by deposition of functionalized nanoparticles on the surface of a variety of solids (thereby making the surface super-hydrophilic, super-hydrophobic, scratch- and corrosion-free, or self-healing) is an active research and development area in materials engineering. Picking up the concept, many researchers investigated the effectiveness of injecting aqueous dispersions of nanoparticles into a reservoir rock to alter its wettability for EOR, which is described in Chapter 11. Use of silica nanoparticles for a novel method of low-cost CO_2 sequestration is also described, which is to generate a highly porous CO_2 hydrate structure, supported by the nanoparticles, so that a large volume of CO_2 can be placed for long-term storage at subsea shallow geologic zones.

In view of the huge amounts of the heavy oil reserve available in Alberta, Canada, and elsewhere in the world, and because the oil that is usually produced by thermal means still requires upgrading either on site or at the refinery, the in-situ upgrading, if successful, will have a tremendous societal and business impact. Chapter 12 mainly addresses the exciting possibility of utilizing nanoparticle catalysts for in-situ upgrading of heavy oil, especially ultra-heavy bitumen, economically in large scale with minimal environmental disruption. While this is a truly daunting task, development of a new production technology may be an achievable possibility if we consider the recent tremendous success in producing shale oil and gas after the persistent development of the combined technology of horizontal well and multiple hydraulic fracturing.

References

Agista, M. N., Guo, K., and Yu, Z. (2018) A state-of-the-art review of nanoparticles application in petroleum with focus on enhanced oil recovery. *Appl. Sci.*, 8, 871.

Amanullah, M., and Ramasamy, J. (2014) Nanotechnology Can Overcome the Critical Issues of Extremely Challenging Drilling and Production Environments. (SPE 171693), SPE Abu Dhabi International Petroleum Exhibition and Conference, Abu Dhabi, UAE.

Bennetzen, M. V., and Mogensen, K. (2014) Novel Applications of Nanoparticles for Future Enhanced Oil Recovery. (IPTC 17857), International Petroleum Technology Conference, Dec. 10–12, Kuala Lumpur, Malaysia.

Bera, A., and Belhaj, H. (2016) Application of nanotechnology by means of nanoparticles and nanodispersions in oil recovery-A comprehensive review. *J. Nat. Gas Sci. Eng.*, 34, 1284–1309.

Binks, B. P., and Horozov, T. S. (2006) *Colloidal Particles at Liquid Interfaces.* Cambridge University Press, Cambridge, UK.

El-Diasty, A. I., and Ragab, A. M. S. (2013) Applications of Nanotechnology in the Oil and Gas Industry: Latest Trends Worldwide and Future Challenges in Egypt. (SPE 164716), SPE North Technical Conference and Exhibition, Apr. 15–17, Cairo, Egypt.

Hosokawa, M., Nogi, K., Naito, M., and Yokayama, T. (2012) *Nanoparticle Technology Handbook.* Elsevier BV, Oxford, UK.

Kapusta, S., Balzano, L., and Te Riele, P. M. (2011) Nanotechnology Applications in Oil and Gas Exploration and Production. (IPTC 15152), International Petroleum Technology Conference, Nov. 15–17, Bangkok, Thailand.

Kazemzadeh, Y., Shojaei, S., Riazi, M., and Sharifi, M. (2018) Review on application of nanoparticles for EOR purposes; a critical of the opportunities and challenges. *Chinese J. Chem. Eng.*, doi:10.1016/j.cche.2018.05.022.

Klabunde, K. J., and Richards, R. M. (2009) *Nanoscale Materials in Chemistry*, 2nd ed, John Wiley, Hoboken, NJ, USA.

Kong, X., and Ohadi, M. (2010) Applications of Micro and Nano Technologies in the Oil and Gas Industry-Overview of the Recent Progress. (SPE 138241), Abu Dhabi International Petroleum Technology Conference, Nov. 1–4, Abu Dhabi, UAE.

Krishnamoorti, R. (2006) Extracting the benefits of nanotechnology for the oil industry. *J. Petrol. Tech.*, 58(11), 24–26.

Matteo, C., Candido, P., Vera, R., and Francesca, V. (2012) Current and future nanotech applications in the oil industry. *Am. J. Appl. Sci.*, 9(6), 784–793.

Mokhatab, S., Fresky, M. A., and Islam, M. R. (2006) Applications of nanotechnology in oil and gas E&P. *J. Petrol. Tech.*, 58(4), 48–51.

Ngai, T., and Bon, S. A. (2015) *Particle-Stabilized Emulsions and Colloids.* Royal Society of Chemistry, London, UK.

Negin, C., Ali, S., and Xie, Q. (2016) Application of nanotechnology for enhancing oil recovery—A review. *Petroleum*, 2(4), 324–333.

Pourafshary, P., Azimpour, S., Motamedi, P., Samet, M., Taheri, S., Bargozin, H., and Hendi, S. (2009) Priority Assessment of Investment in Development of Nanotechnology in Upstream Petroleum Industry. (SPE 126101), SPE Saudi Arabia Section Technical Symposium, May 9–11, Al-Khobar, Saudi Arabia.

Roberts, M., Aminzadeh, B., DiCarlo, D. A., Bryant, S. L., and Huh, C. (2012) Generation of Nanoparticle-Stabilized Emulsions in Fractures. (SPE 154228), SPE Improved Oil Recovery Symposium, Apr. 14–18, Tulsa, OK, USA.

Rodriguez, E., Roberts, M., Yu, H., Huh, C., and Bryant, S. L. (2009) Enhanced Migration of Surface-Treated Nanoparticles in Sedimentary Rocks. (SPE 124418), SPE Annual Technical Conference, Oct. 4–7, New Orleans, LA, USA,

Saggaf, M. (2008) A vision for future upstream technologies. *J. Petrol. Tech.*, 60(3), 54–98.

Schaefer, H.-E. (2010) *Nanoscience: The Science of the Small in Physics, Engineering, Chemistry, Biology and Medicine*. Springer Science & Business Media, Berlin, Germany.

Schmid, G., Brune, H., Ernst, H., Grünwald, W., Grunwald, A., Hofmann, H., Janich, P., Krug, H., Mayor, M., and Rathgeber, W. (2006) *Nanotechnology: Assessment and Perspectives*. Gethmann Springer-Verlag, Berlin, Germany.

ShamsiJazeyi, H., Miller, C. A., Wong, M. S., Tour, J. M., and Verduzco, R. (2014) Polymer-coated nanoparticles for enhanced oil recovery. *J. Appl. Polymer Sci.*, 131(15), 40576.

Skandon, G., and Singhal, A. (2004) *Nanomaterials; Recent Advances in Technology and Industry*. Taylor & Francis, Boca Raton, FL, USA.

Wiesner, M. R., and Bottero, J.-Y. (2007) *Environmental Nanotechnology-Applications and Impacts of Nanomaterials*. McGraw-Hill, New York, NY, USA.

Zhang, T. C., Surampalli, R. Y., Lai, K. C. K., Hu, Z., Tyagi, R. D., and Lo, I. M. C. (2009) *Nanotechnologies for Water Environment Applications*. The American Society of Civil Engineers, Reston, VA, USA.

2

Nanoparticle Synthesis and Surface Coating

2.1 Introduction

As the basic building block of nanotechnology is the nanoparticle, a tremendous variety of their synthesis methods have been developed, producing nanoparticles of different size, shape, composition, internal structure, and so on, to meet the requirements of the specific applications. Additionally, depending on the application's needs, a wide range of surface coating methods have been developed. There is an immense literature on nanoparticle synthesis and surface coating, and a recent monograph by Bensebaa (2013) is a good reference source. A comprehensive handbook on various nanoparticle applications by Hosokawa et al. (2012) is also an excellent resource.

Nanoparticle synthesis methods can be broadly divided into "top-down" and "bottom-up" methods. Top-down methods start with a bulk material that is broken into smaller pieces, such as by ball-milling (Lee et al. 2015), as described in Section 10.3. Because they tend to produce particles of broad size distribution and varied shape, the top-down approach is less taken, as most nanoparticle applications require particles of fairly uniform size and shape. The bottom-up methods form nanoparticles from different kinds of precursors mainly via a chemical reaction. Homogeneous nucleation and particle growth from a liquid or vapor, or heterogeneous nucleation and growth on a substrate, are typical ways of synthesizing nanoparticles using the bottom-up approach. It can be further divided into wet chemical processes and high-temperature dry processes. In this brief review on synthesis, only some of the bottom-up methods will be described.

2.2 Requirements for Nanoparticles for Oilfield Use

In synthesizing nanoparticles for oilfield operations, a number of important and critical requirements need to be satisfied, distinct from the previously mentioned medical, biological, and other industrial applications. The extensive technology already developed in those disciplines cannot be directly utilized because of the unique nature of oil and gas reservoirs and fluids in them. Unlike human bodies, where temperature and body fluid salinity are uniform (see Figure 2.1), the temperature of oil reservoirs can vary from ambient to above boiling point temperature and water can be from almost fresh water to a brine with salinity as high as 22 wt%. Therefore, any nanoparticles that are to carry out their design function should be able to withstand such harsh environments, remaining as individual *nano-size* particles without aggregation and maintaining their surface functionality intact for a sufficiently long enough time during their stay inside the reservoir or downhole at the wellbore. Another important distinction from human body application is that the length scale and time scale for reservoir applications are typically in kilometers and months, respectively. Therefore, the nanoparticles should be able to survive the long-distance transport which may take months. This means that virtually no adsorption/retention of nanoparticles should occur in the rock pores. For practical oilfield applications, the third distinction is probably most important: While the biomedical application generally requires only a small amount of nanoparticles and their recovery and regeneration for re-use is a non-issue, for oilfield applications for which use of a large volume of nanoparticles would be generally required, their regeneration and re-use is important not only for their economic usage, but also for their environmental and safety benefits. The possibility of *repeatedly re-using the nanoparticles* for certain oilfield operations, instead of the chemicals that are currently used, and which may not be "green," opens up the exciting potential of making many oilfield operations environmentally safe and friendly. This is further described in more detail below.

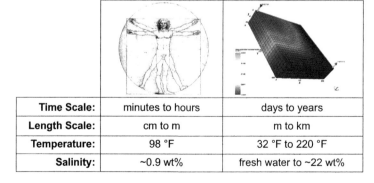

Time Scale:	minutes to hours	days to years
Length Scale:	cm to m	m to km
Temperature:	98 °F	32 °F to 220 °F
Salinity:	~0.9 wt%	fresh water to ~22 wt%

FIGURE 2.1
Differences in key properties of the human body (for which many biomedical applications using nanoparticles are developed) and typical oil reservoirs.

2.3 Synthesis in the Liquid Phase

There are a number of different processes whose synthesis reactions are carried out in the liquid phase ("wet" methods). Compared with the various methods of synthesizing nanoparticles in the gas phase ("dry" methods), "wet" methods are easier to control the synthesis reaction steps so that products with uniform particle size and shape, for example, can be obtained. A wide variety of wet chemical processes have been developed and reported, some of which are (Bensebaa 2013):

1. Salt reduction and co-reduction (precipitation method)
2. Sol-gel method
3. Microemulsion method
4. Sonochemical method
5. Photochemical reduction
6. Thermal decomposition of organometallic compounds.

2.3.1 Precipitation Method

This is probably the most widely employed technique for nanoparticle synthesis. In the precipitation reactions, metal precursors such as chloride or nitrate salts are dissolved in a liquid medium such as water, and a base such as sodium hydroxide or ammonium hydroxide is added to cause precipitation due to chemical reduction of metal cation, and subsequent formation of nanoparticles. The nucleation step should be able to generate a large number of seeds, which will subsequently grow, coarsen, and/or agglomerate. Oswald ripening and aggregation determine the size, shape, morphology, and properties of the nanoparticles generated (Gupta and Gupta 2005; Laurent et al. 2008; Tartaj et al. 2003; Wu et al. 2008, 2016).

In view of the importance of the method, a detailed example of synthesizing magnetite nanoparticles by the co-precipitation method is described in Section 2.4.

2.3.2 Sol-Gel Method

This is another wet chemical process which is frequently employed for synthesis of metal oxide nanoparticles. The sol-gel method involves conversion of monomers into a colloidal "sol" that acts as the precursor for formation of an integrated network ("gel") of either discrete particles or polymers; and it has traditionally been used widely to produce colloids of ceramics and glasses at a relatively low temperature. Precursor chemical is used, usually together with a catalyst, to induce hydrolysis and condensation in the solution. The most commonly used

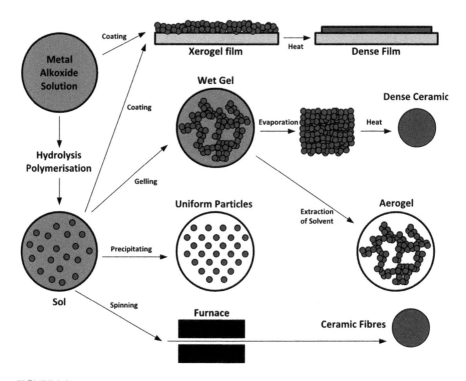

FIGURE 2.2
Use of the sol-gel method to produce various forms of colloidal and nanoscale materials, including mono-dispersed nanoparticle (lower middle center). From Wikipedia (2018).

precursors are metal alkoxides $[M(OR)_n]$ where $M = Fe$, Si, Al, Ti, Zn; and R is a lipophilic group, which forms bridging hydroxyl $[M–(OH)_n–M]$ or oxo (M–O–M) bonds. Both hydrolysis and condensation processes lead to formation of a gel, which is then dried, sometimes with a thermal treatment. With a careful control of sol preparation and gel formation, various oxide nanoparticles, such as Fe_2O_3, SnO_2, Al_2O_3, TiO_2, and ZnO, have been prepared (Dai et al. 2005; Deng et al. 2005; Duraes et al. 2005; Ismail 2005; Laurent et al. 2008). Figure 2.2 shows the use of the sol-gel method to produce various forms of colloidal and nanoscale materials, including mono-dispersed nanoparticle.

2.3.3 Microemulsion Method

This is a variation of the precipitation method where the reaction solution is confined to the inside of a microemulsion droplet, so that nanoparticle growth can be precisely controlled. While most of the synthesis methods require a quenching step to stop the further growth of the particle, which is difficult to control precisely, carrying out the precipitation reaction inside a well-defined, confined reaction space allows production of mono-dispersed nanoparticles of a prescribed size. The method utilizes the well-established technology of

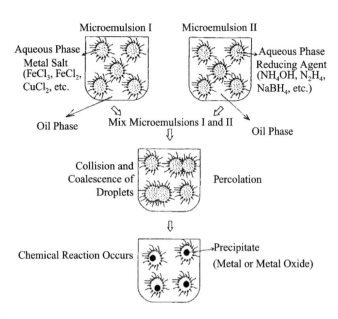

FIGURE 2.3
Schematics of metal-oxide nanoparticle synthesis by the microemulsion method. From Capek (2004).

generating microemulsion phases of desired internal structure (*e.g.*, droplet size) from oil, surfactant (and many times co-surfactant), and water with a metal salt of controlled ionic strength. From the above mixture of prescribed volume fractions and compositions, the microemulsion droplets of known size with prescribed amount of the precipitation reactants can be generated. Water-in-oil microemulsion ("inverse microemulsion") remains the most used medium. Typically, two separate microemulsion aqueous phases consisting of metal salts (*e.g.*, $FeCl_2$ and $FeCl_3$) and reducing agent (*e.g.*, NH_4OH, $NaBH_4$) are prepared. These two solutions are mixed, allowing formation of precipitate aggregation inside the microemulsion droplets (Antonietti and Landfester 2002; Capek 2004; Munoz-Espi et al. 2012). Figure 2.3 shows the schematics of metal-oxide nanoparticle synthesis by the microemulsion method.

2.3.4 Mini-Emulsion Method

This method is quite similar to the microemulsion method above, except that a dynamically stable emulsion droplet, instead of a thermodynamically stable microemulsion droplet, is employed as the confined reaction space. (The terms, microemulsion method and mini-emulsion method, seem to be frequently used without clear distinction.) Since an emulsion droplet is generally larger than a microemulsion droplet, this method has the flexibility of carrying out either the precipitation process or the sol-gel process (Antonietti and Landfester 2002; Munoz-Espi et al. 2012).

2.3.5 Sonochemical Method

In this technique, the precursor molecules undergo the synthesis reaction due to the powerful ultrasound energy input, which causes microscopic cavitations in the solution. The extremely high temperature and pressure consequently generated locally promote the formation and growth of nanoparticles (Dhas et al. 1998; Laurent et al. 2008; Wu et al. 2007, 2008).

2.4 Synthesis in the Gas Phase

While the "wet" methods are generally employed to produce nanoparticles in small quantities for research purposes, the "dry" methods, i.e., synthesis in the gas phase, are usually employed for mass production of nanoparticles. While the dry nanoparticle synthesis process is often easier to scale up cost-effectively, producing mono-disperse nanoparticles, and applying the surface coating during the particle synthesis step, are known to be generally more difficult than the wet processes (Wooldridge 1998). As described well in a recent overview on vapor-phase synthesis of nanoparticles by Swihart (2003), the first key requirement for the "dry" methods is the creation of the conditions in which the vapor-phase mixture is thermodynamically unstable so that a solid phase is created in nano-particulate form. This includes not only the usual "supersaturated" vapor state but also the "chemical supersaturation" state in which it is thermodynamically favorable for the vapor-phase molecules to react chemically to create a solid phase in nano-particulate form. With sufficient supersaturation (and if the reaction/condensation kinetics permit), particles will nucleate homogeneously. Once nucleation occurs, condensation or reaction of the vapor-phase molecules onto the nucleated particle, rather than further nucleate generation, will make the particles grow. To make nano-size particles, therefore, a high degree of supersaturation is created to induce a high nucleation density; and then the system is quickly quenched so that the nanoparticles do not grow further. In most of the vapor-phase methods, the nucleation and quenching happens in milliseconds to seconds, allowing the method to be a continuous process.

At a sufficiently high temperature (500~1000°C), the particles sinter faster than they aggregate, with the resulting formation of spherical nanoparticles. At low temperatures, loose aggregates of small particles are formed. Therefore, to produce spherical nanoparticles with narrow size distribution, the precise control of the supersaturation, quenching, and temperature conditions, and the associated flow conditions for the continuous synthesis operation, is critical. In contrast to the "wet" methods in which the dispersion stability of the newly synthesized nanoparticles can be fairly easily maintained with the simultaneous or subsequent application of surface coating, during the fast, continuous generation of nanoparticles in the vapor phase, the nanoparticle aggregation almost always occurs; and the usual technique to produce a relatively narrow size distribution is how to make the particles only loosely aggregated so that they can be re-dispersed without too costly an effort.

Broadly, there are two general classes of "dry" methods of synthesis. In the first class, to achieve the supersaturation necessary to induce homogeneous nucleation, the material to be made into nanoparticles is vaporized into a background gas, which is then cooled. In the second class, the supersaturation required to induce homogenous nucleation is brought by chemical reaction. That is, the chemical precursors are mixed and heated. Figure 2.4 shows the process flow chart for the vapor-phase synthesis methods of particle formation in multi-component system: (a) vapor condensation and (b) gas-phase reaction. Various gas-phase synthesis techniques in each class will be briefly described below.

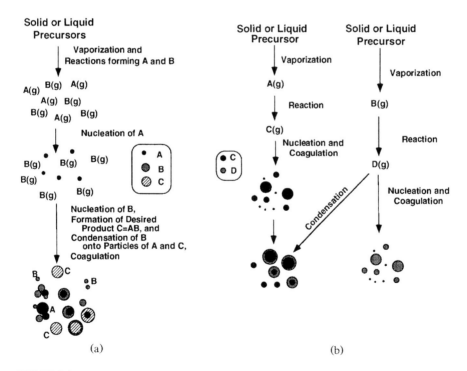

FIGURE 2.4
Process flow chart for the vapor-phase synthesis methods: Particle formation in multi-component system leading to segregation of species: (a) vapor condensation and (b) gas-phase reaction. From Gurav et al. (1993).

2.4.1 Methods Using the Nanoparticle Material Directly as Precursors

2.4.1.1 Inert Gas Condensation

In this widely employed method to make metal or metal-oxide nanoparticles, the supersaturation is achieved by directly heating a solid to evaporate it and then the vapor is mixed with a cold gas to reduce the temperature. Many metals evaporate at reasonable rates at economically attainable temperature. By including oxygen in the cold gas stream, oxides of the evaporated solid material can be formed as nanoparticles. A good example of the method is the synthesis of MnO nanoparticles from the $MnCl_2$ powders (Mohanty et al. 2007).

2.4.1.2 Pulsed Laser Ablation

In this technique, a pulsed laser vaporizes a plume of material that is tightly confined. This method is employed to vaporize materials that cannot be readily evaporated.

2.4.1.3 Spark Discharge Generation

In this technique, the metal to be vaporized is made as electrodes; and in the presence of background inert gas, a high voltage is applied to generate sparks across the electrodes, resulting in the vaporization of the metal.

2.4.2 Methods Using Chemical Precursors Which React to Produce Nanoparticle

2.4.2.1 Chemical Vapor Condensation

In this technique, vapor-phase precursors are brought into a hot-wall reactor under conditions that favor homogeneous nucleation in the vapor phase rather than deposition of nucleates on the wall. It is called chemical vapor condensation process in analogy to the chemical vapor deposition (CVD) processes that are widely used to deposit thin solid films on surfaces. This method takes advantage of the huge database of precursor chemistries that have been developed for CVD processes. The precursors can be solid, liquid, or gas at ambient conditions, but are delivered to the reactor as a vapor (from a bubbler or sublimation source, as necessary). Another key feature of synthesis by chemical vapor condensation is that it allows formation of doped or multi-component nanoparticles by use of multiple precursors.

2.4.2.2 Spray Pyrolysis

In this technique, instead of delivering the nanoparticle precursors into a hot reactor as a vapor, a nebulizer is used to directly inject very small droplets of precursor solution. This has been called spray pyrolysis, aerosol decomposition synthesis, or droplet-to-particle conversion (Okuyama and Lenggoro 2003). Reaction often takes place in solution in the droplets, followed by solvent evaporation.

2.4.2.3 Flame Synthesis

This process is by far the most commercially successful approach to nanoparticle synthesis—producing millions of metric tons per year of carbon black and metal oxides. Unlike other techniques with which the thermal energy is externally supplied to induce reaction and particle nucleation, this method carries out the particle synthesis within a flame, so that the heat needed is produced in-situ by the combustion reactions. However, the coupling of the particle production to the flame chemistry makes this a complex process which is difficult to control. It is primarily used to make oxides, since the flame environment is quite oxidizing.

2.4.2.4 Flame Spray Pyrolysis

In this technique, instead of injecting vapor precursors into the flame as is done with the above "Flame synthesis" technique, the liquid precursor is directly sprayed into the flame. Thus, it is a combination of the "Spray pyrolysis" and "Flame synthesis" techniques. This method allows use of precursors that do not have sufficiently high vapor pressure to be delivered as a vapor. This technique is commonly used for large-scale commercial production of nanoparticles of, for example, titania and silica, because the oxides of such metals can be easily generated from their water-soluble precursors at low cost. Typically, $TiCl_4$ is used for TiO_2 and $SiCl_4$ is used for SiO_2. The precursor materials are injected into the burner as a gas, droplets, or solid particles, which rapidly evaporate as they are exposed to the high-temperature flame. The oxidized metal vapor then nucleates to form nanoparticles. After the high-temperature step, the aerosol stream slowly cools to a lower temperature, causing the aggregation of the particles by collision and coalescence mechanism, to form the desired-size nanoparticles (Ahonen et al. 2001; Gurav et al. 1993; Iskandar 2009; Kim et al. 2002; Kruis et al. 1998; Okuyama and Lenggoro 2003; Tartaj et al. 2003).

2.5 Nanoparticle Surface Modification and Functionalization

Probably the most important feature of the nanoparticle is its extremely high surface-to-volume ratio. Consequently, nanoparticles carry very large excess surface free energy and they invariably try to aggregate to form a bulk state to lower the free energy. Therefore, for any nanoparticle production and its use, it is paramount to maintain its dispersion stability, which almost always requires the surface "modification." This is carried out either during the nanoparticle synthesis stage or right after the synthesis. For many nanoparticle applications, certain chemicals are attached ("grafted") to the nanoparticle surface so that, when the nanoparticles are delivered to the target location, those "grafts" carry out their intended task(s) or "functions." The above two reasons are why the surface modification and functionalization (which is often simply called "surface coating") is so important for any nanoparticle application development. There is an immense literature on the nanoparticle surface coating; and excellent recent reviews by Kango et al. (2013), Mallakpour and Madani (2015), Neouze and Schubert (2008), and Sperling and Parak (2010) are available.

2.5.1 Surface Modification of Nanoparticles

The discussion here will be limited to surface modification of metal oxide nanoparticles, not only because the metal oxides, for example, iron oxide, are main nanoparticles that are being employed for the oil and gas applications development, but also because it is difficult to cover the immense range of developments concisely; and for those who are interested, the above review articles are referred to. Surface modification of nanoparticles can be made by physical or chemical routes. Physical modification is accomplished with surfactants or macromolecules adsorbed on the surface of metal oxide nanoparticles. The polar groups of surfactant/polymer can adsorb on the particle surface by electrostatic attraction. The adsorbed ligands then decrease the particle–particle attraction thus diminishing the agglomerate formation. A drawback of physical modification is that they are not permanent, as the adsorbed molecule could be desorbed depending on the thermodynamic equilibrium with the surfactant/polymer in the solvent in which the nanoparticles are dispersed. This is especially true when the nanoparticles are injected into the oil reservoir in which there is no surfactant/polymer that can provide its thermodynamic equilibrium. Chemical modification involves covalent grafting of organic compounds of low molecular weight to the nanoparticle surface. The covalent grafting is preferred because the attached ligands will not be easily detached (even in oil reservoirs), ensuring that the ligands provide the nanoparticle's dispersion stability. The modification of the nanoparticle surface can be carried out either during the particle synthesis step or immediately after the synthesis, as mentioned above. The latter method is usually more versatile than the in-situ, or "one-pot," method. Another novel scheme of surface modification, as developed by Yoon et al. (2011), is the formation of a crosslinked polymer net "wrapping around" one or a small aggregate of nanoparticles. Because the polymer network completely encloses the nanoparticle core, it does not need to be attached to the particle either physically or chemically.

While there are a wide variety of surfactants and polymers (and other chemicals) used as surface modifiers, the following three groups of chemicals are most commonly employed.

2.5.1.1 Carboxylic Acids

Carboxylate ligands, mainly fatty acids, are often used to modify metal-oxide nanoparticle surfaces. For the biomedical application use of magnetite nanoparticles, oleic acid is commonly employed as the dispersion stabilizer (Sperling and Parak 2010). For the oilfield application, oleic acid has also been employed (Ingram et al. 2010), and citric acid is also shown to be a good surface modifier (Kotsmar et al. 2010). Use of these and other coating materials for an example oilfield application is described in detail in Section 2.7.2.

2.5.1.2 Silanes

Silanes are the most often used modifiers for metal oxide surfaces. Silanes have two functional groups, one that can be attached to the nanoparticle surface, and another that can either serve as a dispersion stabilizer or attach a molecule or ligand that carries out specific function(s). The silanes include alkoxysilane ($\equiv$Si–OR, *where R*=alkyl), hydrogenosilane ($\equiv$Si–H), or chlorosilane ($\equiv$Si–Cl), and others are readily synthesized or modified. The main advantage of silanes is that they can carry a variety of functionalities, such as amino, cyano, carboxylic, and epoxy groups. The post-synthesis grafting of silyl groups on a metal oxide surface is generally straightforward. Silanes can also be used for "one-pot" synthesis, where the silane is introduced at the same time as the metal oxide precursors. Silanes react with –OH groups on the metal oxide surface through a condensation. Figure 2.5 shows schematically the chemical grafting of organosilanes onto the TiO_2 nanoparticle surface.

2.5.1.3 Phosphonates

Modification of a metal oxide surface by phosphonate groups create M–O–P bonds. Phosphonate groups can be introduced by P–OH, P–OR (*R*=alkyl), or P–O$^-$ containing precursors.

2.5.2 Surface Functionalization of Nanoparticles

Because an individualized strategy for surface functionalization to meet the specific requirements for a particular nanoparticle application is needed, and also because an immense variety of nanoparticle applications are available in literature, a general but concise description of the surface functionalization

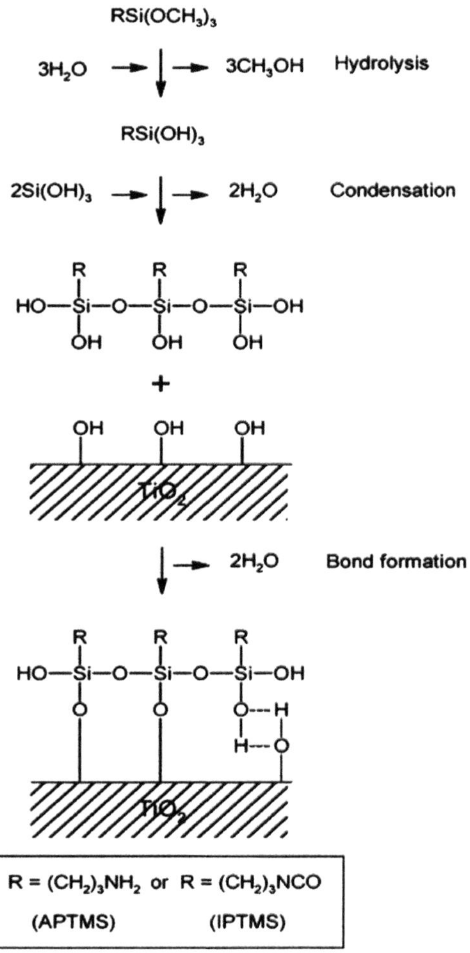

FIGURE 2.5
Chemical grafting of organosilanes onto the TiO_2 nanoparticle surface. From Mallakpour and Madani (2015).

is difficult. Broadly, there are two strategies for the introduction of functional groups, as schematically shown in Figure 2.6. In the first method, a whole functional ligand is introduced in a single step. This requires bifunctional organic compounds, where one functionality (X) is used to attach to the nanoparticle surface and the second (Z) is used for the nanoparticle functionalization. In the second method, a bifunctional compound X–Y is reacted first, where the group Y acts as a coupling site for the final functionality Z in the second step. While the first method is preferred for its simplicity, the reason for the second method is incompatibility of the functional group Z with the synthesis process, for example, when the Z can also react with the particle surface.

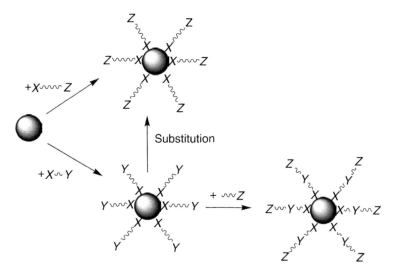

FIGURE 2.6
Two ways to functionalize nanoparticles: In Method 1 (*top*), ligand with the Z functionality reacts directly with nanoparticle; and in Method 2 (*bottom*): a ligand with a Y functionality reacts first with the nanoparticle which then attaches the functionality Z. From Neouze and Schubert (2008).

In Section 2.7.2, the surface functionalization is further described with a detailed example of how to find a coating material that can be used for a particular oil reservoir application.

2.6 Other Forms of Nanomaterials

In Sections 2.3 and 2.5, the synthesis and the surface modification and functionalization of the most commonly and widely used nanomaterial, i.e., spherical nanoparticles and their small aggregates, are described. This is because those (more or less) spherical nanoparticles have probably most practical applicability for oil and gas industry, as was the case with other industries. There is, however, a tremendous variety of other forms of nanomaterials, which are only briefly described in this section, with no attempt even in providing an overview in view of the vast scope of the topic.

2.6.1 Carbon Nanotubes and Fullerenes

Carbon nanotubes (CNTs) are a monolayer of carbon molecules with a cylindrical structure, with diameter as small as 0.4 nm. These cylindrical arrangements of carbon molecules exhibit highly unusual mechanical, thermal, electrical, and other properties. CNTs are the strongest and stiffest materials yet discovered in terms of tensile strength and elastic modulus, respectively. For this reason, CNTs are used as additives to materials to modify their mechanical properties. CNTs also have very high electrical and thermal conductivities along their axis. There are single-walled CNTs (SWNTs) and multi-walled CNTs (MWNTs), the latter of which is easier to synthesize and accordingly of lower cost, and more widely used for commercial applications.

A fullerene is any of carbon molecular structures in the form of a hollow sphere, ellipsoid, tube, and other shapes. CNT is thus a kind of fullerene, even though hollow sphere is usually known as fullerene. Spherical fullerene, which is also known as a buckyball or Buckminsterfullerene (C_{60}), was the first fullerene to be discovered, and was manufactured in 1985 by Richard Smalley and his group at Rice University. In a way, the fullerene's discovery started the general interest on the tremendous potential of nanotechnology, and the big research and development efforts at many academic and government institutions.

2.6.2 Quantum Dots

Quantum dots (QDs) are nanometer-scale semiconductor particles whose optical and electronic properties differ from those of larger particles. If light or electricity is applied to them, due to their small size, they will emit light of specific frequencies which can be precisely tuned by changing the QD's size, shape, and material, allowing them to be useful for a wide variety of applications. For example, for the oil reservoir management, QDs can be employed as highly sensitive multiple tracers that can be detected at extremely low concentrations (Murugesan et al. 2016; Turkenburg et al. 2012).

2.6.3 Core-Shell Nanoparticles

As described in Chapter 5, the superparamagnetic iron-oxide nanoparticles are widely employed in many different scientific and industrial disciplines because of their unique properties. During their application use, if the iron-oxide comes in contact with certain chemicals, it may lose its unique properties, and thus its effectiveness, due to its chemical degradation. In order to protect the iron-oxide material from contacting the chemical(s), a silica shell can be built around the iron-oxide core, because the superparamagnetic iron-oxide core can still carry out its design function without being hampered by the presence of the shell. An example of the synthesis of such core-shell nanoparticles is given in Section 2.7.1. While the Fe_3O_4-SiO_2 core-shell nanoparticles are most commonly synthesized and employed (*e.g.*, Cendrowski et al. 2017; Stjerndahl et al. 2008), there are nanoparticles with many different combinations of core and shell materials for their specific application's needs.

2.6.4 Janus Nanoparticles

As described in Chapters 4 and 10, one important property of nanoparticle is its ability to serve as a stabilizer for emulsions and foams, just like surfactants (Binks 2002). To serve as an effective stabilizer, nanoparticle's surface wettability needs to be tuned properly so that when a nanoparticle is adsorbed at an oil (or CO_2)/water interface, different relative fractions of the nanoparticle's surface area are immersed in the lipophilic and hydrophilic fluids, respectively. This is different from surfactants which have two distinctly different parts, lipophilic and hydrophilic, on the same molecule. To make a nanoparticle more like a surfactant molecule, nanoparticles with two faces ("Janus"), i.e., a part of its surface is lipophilic and another part hydrophilic, have been synthesized and employed (Perro et al. 2009; Pradhan et al. 2007).

2.6.5 Graphene and Graphene Oxide

Graphene is a sheet of carbon molecular monolayer, and graphene oxide is its oxidized state. Because of the huge surface area that is available, in macroscopic scale, per mass of carbon which can be readily obtained from, *e.g.*, low-cost graphite, graphene and graphene oxide are drawing tremendous interest for various industrial applications. While the major application area is to form composites (Jang and Zhamu 2008; Ramanathan et al. 2008; Stankovich et al. 2006), graphene and graphene oxide are also extensively employed as substrates to support other functional materials such as nanoparticles. Because graphene and graphene oxide have two sides, they have been functionalized as "Janus" materials (Luo et al. 2016; Perro et al. 2009). Graphene and graphene oxide in the form of micron-size platelets, dispersed either in water or in organic solvent, have been used as an additive to make a composite material, and as a stabilizer for emulsions (Yoon et al. 2013). For the latter application, the fact that the adsorption at oil/water interface of a very small mass of graphene oxide platelets is needed to effectively cover the surface of emulsion droplets to stabilize them, makes graphene oxide platelets a promising emulsifying agent.

2.6.6 Nanomilled Fly Ash

As described at the beginning of this chapter, nanoparticles can be produced by the "top-down" method, that is, crushing a bulk solid to nano size, *e.g.*, by way of nanomilling. While a variety of materials has been employed, one interesting example is the generation of nanomilled fly ash which is a waste product from coal-burning power generation plants (Lee et al. 2015; Paul et al. 2007). In Section 10.3, how the fly ash nanoparticles are generated and used to stabilize CO2-in-water foam for EOR purpose will be described in detail.

2.7 Characterization of Produced Nanoparticles

Once a nanoparticle is synthesized and its surface functionalized, we have to ensure that it is indeed the intended product. Therefore, the characterization of the produced nanoparticles is an important step in applications development. While some of the conventional tools of colloidal particle characterization can be still utilized, due to their nano size, other specialized tools, such as the transmission electron microscopy (TEM) and atomic force micrography (AFM), need to be also used. Some key characterization parameters are:

1. Nanoparticle size and its distribution, and shape;
2. Nanoparticle surface charge, and amount of coating material on particle surface;
3. Nanoparticle composition; and
4. Magnetic and other properties.

As the best way to show how the characterization tasks are carried out is with an example, the effort made by the University of Texas at Austin team is described in some detail below. The objective for the team was to develop a superparamagnetic nanoparticle (MNP) where divalent cations such as Ca are selectively adsorbed on its surface. For the purpose, after the MNPs are synthesized, the NH_2 group is first attached to the surface to make it positively charged; and then polyacrylic acid (PAA) of different molecular weight (8 k, 100 k, 450 k dalton) was introduced to the MNP surface, so that the anionic carboxyl groups of PAA are not only attached to the MNP surface's NH_2 group but also sufficiently and freely available for the Ca's selective adsorption (Wang et al. 2017).

2.7.1 Nanoparticle Size and Its Distribution

Since the key property of nanoparticles for oilfield applications is its nano size, allowing them to flow in porous reservoir rock without being trapped, the determination of the average size and its distribution is usually the first characterization task. Typically, the hydrodynamic diameter of the particles is obtained from the dynamic light scattering (DLS) and frequently the aggregation state of the nanoparticles is examined by the transmission electron microscopy (TEM). Figures 2.7 and 2.8 from Wang et al. (2017) show, respectively, examples of the hydrodynamic diameter (D_H) distribution and the TEM images of the magnetite nanoparticles which have no surface coating and four different surface coatings, as described above. We note that the average D_H of bare MNP nanocluster is ~89.5 nm, and as the amine group is added to the MNP surface, D_H increases to 131.0 nm. Coating nanoparticles with PAA of higher molecular weight further increases D_H. Not only

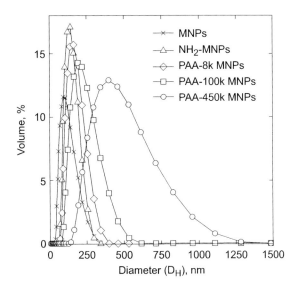

FIGURE 2.7
Hydrodynamic diameter (D_H) distribution of MNPs, NH$_2$-MNPs, PAA-8k-MNPs, PAA-100k-MNPs, and PAA-450k-MNPs in DI water. From Wang et al. (2017).

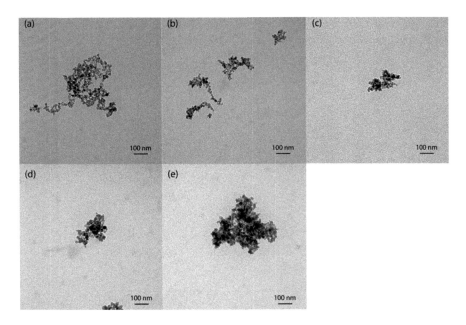

FIGURE 2.8
TEM images of (a) MNPs; (b) NH$_2$-MNPs; (c) PAA-8k-MNPs; (d) PAA-100k-MNPs; and (e) PAA-450k-MNPs. From Wang et al. (2017).

much larger average D_H but also wider D_H distribution were observed for PAA-450k-MNP nanoclusters since, with the longer chain length of the poly-acrylic acid, the nanoclusters might tangle with each other, contributing to the wider D_H distribution. We also note from the TEM images that, while the MNP clusters are larger than the maximum size (~25 nm) that brings the superparamagnetism (Guimaraes 2009; Laurent et al. 2008), they consist of smaller "primary" Fe_3O_4 crystals that are wrapped around with the coating material (Yoon et al. 2016).

2.7.2 Nanoparticle Surface Charge, Amount of Surface Coating, and Surface Area

Because the electrostatic repulsion between the particles is an important mechanism for the nanoparticle dispersion stability, the surface charge state is invariably measured with a zetameter, and is typically expressed as a zeta potential. Often, in order to determine how much coating material is attached to a solid nanoparticle, thermogravimetric analysis is made which measures the weight of the organic material (i.e., coating) that is lost due to burning. Figure 2.9 from Wang et al. (2017) shows the percent weight loss of MNP versus temperature. The weight loss of bare MNP (5.3%) is mainly due to the citric acid used to stabilize it. The overall weight loss of the PAA-MNPs is 13.7 wt% and subtracting the weight loss due to the decomposition of the amine group and citric acid from the overall weight loss of the PAA-MNPs, the amount of PAA is 5.2 wt%. The surface area of the nanoparticles is also sometimes measured employing the nitrogen adsorption and BET isotherm.

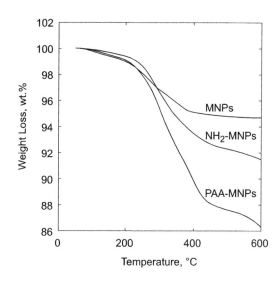

FIGURE 2.9
TGA curves of MNPs, NH_2-MNPs, and PAA-MNPs. From Wang et al. (2017).

2.7.3 Nanoparticle Composition

In order to ensure that the nanoparticle core indeed has the desired composition, *e.g.*, Fe_3O_4 crystalline structure, X-ray diffraction analysis is made employing an X-ray diffractometer (XRD). Figure 2.10 from Wang et al. (2017) shows an example of the XRD patterns of the above magnetite nanoparticles. The characteristic peak locations for magnetite remain unchanged with different polymer coatings, showing that the coating does not affect the superparamagnetic character of the MNPs. The sharpness of the peaks also indicates the high crystallinity of the PAA-MNPs.

2.7.4 Magnetic and Other Properties

An important requirement for a magnetite nanoparticle is whether it indeed carries the superparamagnetic property (see Chapter 5). For the confirmation, the magnetization curve, known as the Langevin curve (Guimaraes 2009, Laurent et al. 2008, Lu et al. 2007), needs to be produced employing either SQUID (superconducting quantum interference device) magnetometer or VSM (vibrating sample magnetometer). Figure 2.11 again from Wang et al. (2017) shows an example of Langevin curves for the above magnetite nanoparticles. For the synthesized nanoparticles to be superparamagnetic, the curves should pass through the coordinate origin, i.e., the magnetization should not show any hysteresis. Depending on the requirements of the specific applications being developed, other properties of the newly produced MNPs need to be obtained; and some of those are discussed in the coating optimization example given in Section 2.7.2.

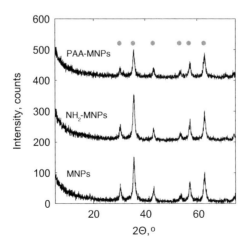

FIGURE 2.10
XRD patterns of MNPs, NH_2-MNPs, and PAA-MNPs with the location of the peaks of the reference magnetite (grey dots). From Wang et al. (2017).

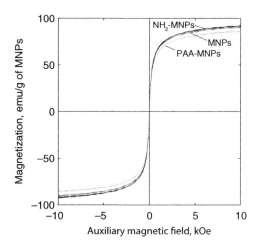

FIGURE 2.11

Langevin curves of MNPs, NH$_2$-MNPs, and PAA-MNPs at room temperature. From Wang et al. (2017).

2.8 Examples of Nanoparticle Synthesis and Surface Coating

Extensive research efforts have been and are still being made on the use of superparamagnetic nanoparticles (SPM-NPs) for various biomedical, electronic, and other applications. As will be described in Chapter 5, the superparamagnetic properties of the SPM-NPs can also be utilized for a number of petroleum engineering applications. The most commonly used SPM-NP is a magnetite (Fe$_3$O$_4$) sphere of 10–20 nm diameter, where its surface is coated with a thin layer of polymer or surfactant, or many times, their small aggregate of which the average diameter is generally smaller than 100 nm, as shown above with Figures 2.7 and 2.8. It is important that the "primary" particles in the aggregate are separated by the coating so that the magnetite cores are not in direct contact with each other. The individual "primary" particles then maintain a magnetic single domain (Guimaraes 2009) as will be described below in Chapter 5. While magnetite is a most common core material, it could be a variety of alloy metal oxides, *e.g.*, Fe/Pt, Fe/Pd, MFe$_2$O$_4$ (M = Ni, Co, Zn, Mn) (Hyeon 2003; Sabale et al. 2015).

2.8.1 Synthesis of Magnetite Nanoclusters and Surface Coating

The SPM-NP's synthesis methods, as carried out recently at the University of Texas at Austin, are described here in some detail. SPM-NPs are prepared via a co-precipitation method (Bagaria et al. 2013a; 2013b; Massart 1981; Wang et al. 2014; Xue et al. 2014). Ferrous chloride (FeCl$_2$·4H$_2$O) and ferric chloride (FeCl$_3$·6H$_2$O) at a molar ratio of 1:2 are added to an aqueous solution under nitrogen purging. Citric acid monohydrate is added to the above mixture as a stabilizer. After heating up to 90°C under vigorous magnetic stirring and nitrogen purging, ammonium hydroxide is quickly injected into the hot mixture to induce the nucleation of SPM-NPs. The reaction is continued for 2 hours at 90°C under a nitrogen gas environment, as shown step-by-step in Figure 2.12.

As described in Section 2.5, a nanoparticle with core-shell structure is synthesized for many applications. For example, for magnetite nanoparticles, it is not good for the chemical or ion (that is selectively attached to the nanoparticle) to be in direct contact with the iron-oxide core. Therefore, a thin shell of silica is inserted between the iron-oxide core and the surface coating (Si-SPM-NP), as shown in Figure 2.13. SPM-NP coating with silica is carried out in basic ethanol/water mixture at room temperature, following the Stöber method (Stöber et al. 1968) with some modification to protect the SPM-NP from unwanted redox reactions (Deng et al. 2005; Ko et al. 2017a, 2017b; Kobayashi et al. 2011; Liu et al. 1998; Lu et al. 2002). Amine-functionalization of Si-SPM-NPs and SPM-NPs is accomplished by 3-amino propyltriethoxysilane (3-APTES) coating process (Bagaria et al. 2013a, 2013b; Wang et al. 2014; Xue et al. 2014). After 1 hour of hydrolysis of APTES at acidic condition (pH≈4), pH of the solution is adjusted to ~8. Si-SPM-NP or SPM-NP is slowly added to APTES solution drop-wise. The reaction is continued at 65°C for 24 hours and then the reaction medium is cooled down to

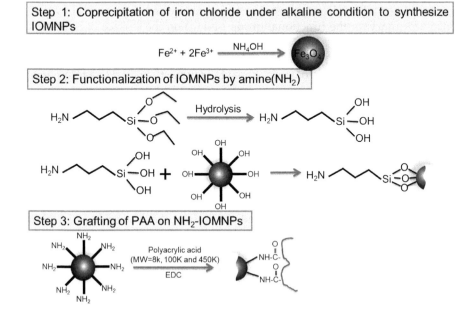

FIGURE 2.12
Magnetite nanoparticle synthesis by co-precipitation method (Step 1); surface modification with attachment of NH$_2$ ligand (Step 2); and surface functionalization with grafting of PAA (Step 3).

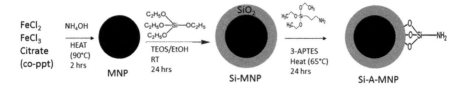

FIGURE 2.13
Synthesis steps for MNP with a magnetite core and silica shell to protect the superparamagnetic core from exposure to chemicals. From Ko et al. (2017b). Copyright 2017, Society of Petroleum Engineers. Reproduced with permission of SPE. Further reproduction prohibited without permission.

room temperature while stirring. Polyacrylic acid (PAA) or other polymers having a terminal carboxylic acid group can be grafted onto the surface of amine-functionalized SPM-NPs at pH 4.7 where carboxyl groups are activated with 1-ethyl-3-(3-dimethylaminopropyl) carbodiimide (EDC) at room temperature (Wang et al. 2014). To screen the charge of polymers and to improve the incorporation of polymer on the surface, NaCl solution is added into the reaction medium slowly to reach the desired salinity. At the end, pH is adjusted back to 4.7, and the reaction is continued for 24 hours under the magnetic stirring at room temperature.

Zeta potential of silica coated SPM-NP is negative due to a dissociation of a proton from a terminal hydroxyl group of silicate (SiO_4^{4-}) on the surface. Amine-functionalized SPM-NPs has a positive surface charge, mainly caused by a protonated primary amine functional group ($-NH_3^+$) on the surface of SPM-NPs, with pKa value of 38. SPM-NP, NH_2-SPM-NP, Si-SPM-NP, and Si-NH_2-SPM-NP are small aggregates of primary particles, whose diameter is about 10 nm. As SPM-NPs have a large ratio of surface area to volume, they tend to agglomerate to minimize high surface energy by strong magnetic dipole-dipole attractions along with van der Waals force (Gupta and Gupta 2005).

2.8.2 Example of Magnetite Nanoparticle Surface Coating Optimization

As described in Section 2.1 above, the key objective of the nanoparticle surface coating for the upstream oil and gas applications is to make it survive the long-period, long-distance travel through the reservoir rock pores, so that it reaches the target location and carries out the design function. For emphasis, the three key requirements are repeated here: (i) Long-term dispersion stability; (ii) minimum retention in reservoir rock; and (iii) functionalization to carry out the design function at the target location. Again, as the best way to show how such surface coating task is carried out is with an example, the optimization effort made by the University of Texas at Austin team is described in some detail below. The task for the team was to develop a superparamagnetic nanoparticle (i) that can stay stably dispersed at least for a month in a water whose salinity is 8 wt% NaCl and 2 wt% $CaCl_2$; (ii) that shows a negligible retention (<5% of injected concentration per injection pore volume) when flown in a sandstone core with ~200 md permeability and 1 ft. length; and (iii) that readily adsorbs at the oil/water interface of the resident oil so that the presence of oil in the reservoir could be remotely detected with application of external magnetic oscillation (Ryoo et al. 2012).

The surface coating optimization effort proceeded with the following steps:

1. After the nanoparticle is synthesized, a candidate surfactant or polymer is either covalently attached or adsorbed on the nanoparticle surface. Sometimes, as briefly described above, the synthesis and coating steps are combined as an "one-pot" process.

2. The prepared nanoparticle is dispersed in the test brine and the progression of aggregate formation is observed with time (Kim et al. 2015).

3. If the dispersion is proven to be stable for a prescribed time (*e.g.*, one month), a series of concentrations of nanoparticles are mixed with equal volumes of the test brine and test oil, and sonicated to form emulsion. The progression of emulsion flocculation and coalescence is observed with time.

4. If a stable emulsion is shown to form suggesting that the nanoparticles properly adsorbed at oil/water interface, an aqueous dispersion of prescribed concentration and volume is injected into a reservoir rock sample which is initially filled with the test brine; the nanoparticle concentration in the effluents from the coreflood is measured as a function of time. By integrating the normalized effluent concentration profile, the number of nanoparticles retained in the core is determined.

5. If the nanoparticle retention in the core is shown to be below the target value, the aqueous dispersion of prescribed concentration and volume is injected into a rock core which now contains some oil in addition to the brine. The nanoparticle concentration in the effluents is again measured and the amount of nanoparticle retention is determined, which is compared with the oil-free value, to estimate the number of nanoparticles adsorbed at the oil/water interfaces of the oil in the rock core (Yu et al. 2014).

6. Concurrently with the above tests, the magnetic susceptibility of the nanoparticles is measured as a function of nanoparticle concentration, employing a magnetometer. This is to estimate the effective magnetic susceptibility of the nanoparticles in the rock so that the magnetic detectability of those particles from a remote location can be evaluated.

To carry out the above steps, the team needed not only an expert on nanoparticle synthesis, coatings, and characterization, but also an organic chemist who could synthesize specific polymer molecules to be tried out as a coating material, as well as a petroleum engineer who could run reservoir rock coreflood tests to see if an aqueous dispersion of the newly synthesized nanoparticles flows freely without being trapped at rock pores, and the particles are indeed attached at the oil/water interfaces of the oil distributed in the rock pores. For the testing of the last step, an expert on the electromagnetic wave propagation was needed. When any one of the above steps turns out to be unsatisfactory, the testing with the nanoparticle with the particular surface coating is stopped; and after review of the available data, the team comes up with a more promising coating material which the organic chemist either purchases or synthesizes. For an efficient optimization, a rapid feedback was an essential element.

Since the water not only has a high salinity but also contains a high concentration of Ca cations, even satisfying the first requirement was difficult. This is because the electrostatic repulsion between the nanoparticles, which arises from the surface charges on particle surface, becomes ineffective due to the buffering effect of the high salinity environment. As reported by Yu et al. (2010), the early efforts (Ingram et al. 2010; Kotsmar et al. 2010) to utilize the coating materials that are widely utilized in biomedical industry, such as citric acid and oleic acid, turned out to be ineffective. Therefore, the nanoparticle's dispersion stability cannot be maintained with the conventional strategy of charging the nanoparticle surface. Instead, an alternative repulsion scheme between the particles, such as the entropic repulsion rendered by the polymer chain attached to the nanoparticle surface, should be sought.

Four different polymers were employed for the second-stage coating effort: poly(acrylic acid-r-butylacrylate) [pAA-r-pBA] with 3:1 molar ratio; poly(acrylic acid-b-styrenesulfonate) [pAA-b-pSS] with 3:1 molar ratio; poly(acrylic acid-r-butylacrylate-b-styrenesulfonate) [pAA-r-pBA-b-pSS] with 1:1:2 molar ratio; and poly(4-styrenesulfonic acid-alt-maleic acid) [pSS-alt-pMA]. Figure 2.14 shows the chemical structures of the polymers. The MNPs coated with these polymers all showed good dispersion stability and produced stable emulsions, i.e., adequately adsorbed at oil/water interfaces [see Steps (2) and (3) above] (Yoon et al. 2011, 2012). Table 2.1 lists the hydrodynamic diameter of the MNPs coated with the polymers and their zeta potential, and the pH and salinity of the resident water and the injection water in which the MNP of the listed concentration was added, for the coreflood testing of the newly produced MNPs [Steps (4) and (5)].

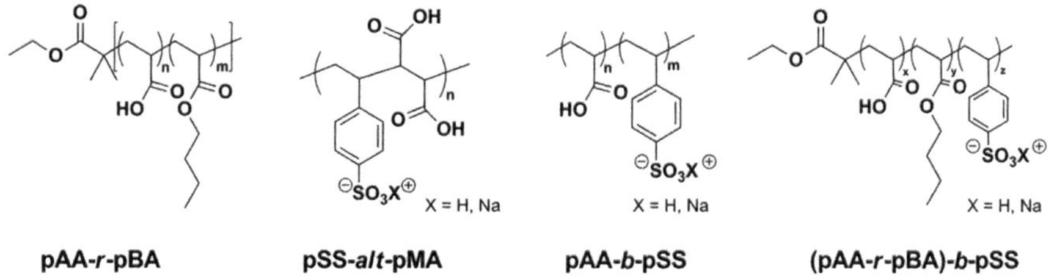

pAA-r-pBA   pSS-*alt*-pMA   pAA-*b*-pSS   (pAA-*r*-pBA)-*b*-pSS

FIGURE 2.14
Chemical structure of polymers employed for coating. From Yu et al. (2014).

TABLE 2.1

Adsorption of Nanoparticles with Different Surface Coatings During Corefloods with Water Only and with Oil in the Core

Coating Polymer	MNP Diameter (nm)	Zeta Potential (mV)	Injection MNP Concentration (wt%)	pH	Water-only Coreflood Specific Adsorption (mg/m²)	With-oil Coreflood Specific Adsorption (mg/m²)
pAA-r-pBA	130	−45.1	0.37	8	0.92	4.50
pAA-b-pSS	133	−24.0	0.2	8	0.32	–
pAA-r-pBA-b-pSS	113	−50.4	0.2	8	0.12	0.39
pSS-alt-pMA	80	−58.6	0.2	9	0.02	1.20

In Table 2.1, the specific adsorption of MNPs from the corefloods with only water in the core [Step (4)] and with water and the oil in its waterflood residual saturation [Step (5)] is shown. When the MNP with [pAA-r-pBA] coating was injected into the core, the specific adsorption was quite large for both "water-only" and "with-oil" corefloods (0.92 and 4.50 mg/m², respectively). Because the rock surface of the sandstone core employed is known to be negatively charged, in order to reduce the MNP adsorption on rock surface, it was decided to replace the non-ionic butylacrylate with the strongly anionic styrenesulfonic acid, i.e., [pAA-r-pBA] to [pAA-b-pSS]. The oil-free coreflood with the [pAA-b-pSS]-coated MNPs indeed showed much lower adsorption (0.32 mg/m²). The [pAA-b-pSS] coating, however, made the MNP too hydrophilic, as the oil/water emulsions made with the MNP were not as stable as those made with the MNPs coated with the other three polymers. It was, therefore, decided to combine the benefits of the two polymers; that is, the MNP is now coated with [pAA-r-pBA-b-pSS]. As shown in Table 2.1, the specific adsorption was now quite low for both oil-free and with-oil corefloods (0.12 and 0.39 mg/m², respectively). While the low adsorption for the oil-free coreflood was quite satisfactory, the adsorption for the with-oil coreflood could have been somewhat larger, so that the amount of MNPs adsorbed at the oil/water interfaces in rock pores could be more clearly estimated, allowing us to quantify the total oil/water interfacial area per mass of rock (Yu et al. 2014).

It was decided that the ability of pSS to bring the adsorption low should be kept, while the [pAA-r-pBA] portion somehow needs to be modified. Based on the knowledge gained from the above trials, the [pSS-alt-pMA] coating was developed which utilizes the pSS's ability to maintain a good dispersion stability and minimal adsorption in rock pores and which, with pMA replacing pAA and/or pBA, brings more hydrophobicity to the nanoparticle surface wettability. As shown in Table 2.1, the [pSS-alt-pMA] coating not only resulted in the minimal adsorption in rock pores but also a sufficient adsorption at the oil/water interfaces of oil ganglia in rock pores. Figure 2.15 shows the normalized, effluent concentrations of MNPs coated with [pSS-alt-pMA]

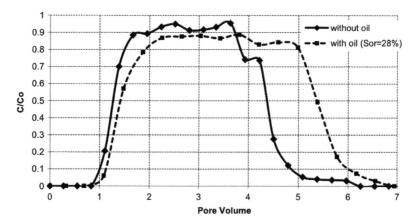

FIGURE 2.15

Normalized effluent concentration of MNPs coated with [pSS-alt-pMA] polymer versus pore volume for Boise sandstone, for "water-only" and "with-oil" corefloods of Table 2.1. Three pore volumes of nanoparticle dispersion (0.2 wt%, pH=8) were injected, followed by brine (1 wt% NaCl) with the same pH. From Yu et al. (2014).

polymer versus pore volume for Boise sandstone, for "water-only" and "with-oil" corefloods of Table 2.1. Three pore volumes of nanoparticle dispersion (0.2 wt%, pH=8) were injected, followed by brine (1 wt% NaCl) with the same pH.

2.9 Concluding Remarks

As pointed out at the beginning of this chapter, the nanoparticle is the basic building block for virtually all nanotechnology applications, and accordingly, a good understanding of various techniques for nanoparticle synthesis and surface modification and functionalization is an important first step in developing new applications or evaluating their technical and business benefits. As shown with the above examples of magnetite nanoparticle synthesis and surface coating, depending on the specific application, the synthesis may require additional processing; for example, the addition of the silica shell to protect the superparamagnetic core from exposure to chemicals that degrade it. The example of the surface functionalization to develop the desired nanoparticle functionality demonstrates that achieving the goal requires participation of different experts, as the nanoparticle application development for the upstream oil and gas industry is "getting into an unexplored territory," with potential for huge technical and business benefits.

References

Ahonen, P. P., Joutsensaari, J., Richard, O., Tapper, U., Brown, D. P., Jokiniemi, J. K., and Kauppinen, E. I. (2001) Mobility size development and the crystallization path during aerosol decomposition synthesis of TiO2 particles. *J. Aerosol Sci.*, 32(5), 615–630.

Antonietti, M., and Landfester, K. (2002) Polyreactions in miniemulsions. *Prog. Polym. Sci.*, 27, 689–757.

Bagaria, H. G., Xue, Z., Neilson, B. M., Worthen, A. J., Yoon, K. Y., Nayak, S., Cheng, V., Lee, J. H., Bielawski, C. W., and Johnston, K. P. (2013a) Iron oxide nanoparticles grafted with sulfonated copolymers are stable in concentrated brine at elevated temperatures and weakly adsorb on silica. *ACS Appl. Mater. Interfaces*, 5(8), 3329–3339.

Bagaria, H. G., Yoon, K. Y., Neilson, B. M., Cheng, V., Lee, J. H., Worthen, A. J., Xue, Z., Huh, C., Bryant, S. L., Bielawski, C. W., and Johnston, K. P. (2013b) Stabilization of iron oxide nanoparticles in high sodium and calcium brine at high temperatures with adsorbed sulfonated copolymers. *Langmuir*, 29, 3195–3206.

Bensebaa, F. (2013) *Nanoparticle Technologies: From Lab to Market*. Academic Press, Amsterdam, the Netherlands.

Binks, B. P. (2002) Particles as surfactants—similarities and differences. *Curr. Opin. Colloid Interface Sci.*, 7, 21–41.

Capek, I. (2004) Preparation of metal nanoparticles in water-in-oil (w/o) microemulsions. *Adv. Colloid Interface Sci.*, 110, 49–74.

Cendrowski, K., Sikora, P., Zielinska, B., and Horszczaruk, E. (2017) Chemical and thermal stability of core-shelled magnetite nanoparticles and solid silica. *Appl. Surf. Sci.*, 407, 391–397.

Dai, Z., Meiser, F., and Mohwald, H. (2005) Nanoengineering of iron oxide and iron oxide/silica hollow spheres by sequential layering combined with a sol–gel process. *J. Colloid Interface Sci.*, 28(1), 298–300.

Deng, Y.-H., Wang, C.-C., Hu, J.-H., Yang, W.-L., and Fu, S.-K. (2005) Investigation of formation of silica-coated magnetite nanoparticles via sol–gel approach. *Colloids Surf. A: Physicochem. Eng. Asp.*, 262(1), 87–93.

Dhas, N. A., Raj, C. P., and Gedanken, A. (1998) Synthesis, characterization, and properties of metallic copper nanoparticles. *Chem. Mater.*, 10(5), 1446–1452.

Duraes, L., Costa, B., Vasques, J., Campos, J., and Portugal, A. (2005) Phase investigation of as-prepared iron oxide/hydroxide produced by sol–gel synthesis. *Mater. Lett.*, 59(7), 859–863.

Guimaraes, A. P. (2009) *Principles of Nanomagnetism*. Springer Science & Business Media, Cham.

Gupta, A. K., and Gupta, M. (2005) Synthesis and surface engineering of iron oxide nanoparticles for biomedical applications. *Biomaterials*, 26(18), 3995–4021.

Gurav, A., Kodas, T., Pluym, T., and Xiong, Y. (1993) Aerosol processing of materials. *Aerosol Sci. Tech.*, 19, 411–452.

Hosokawa, M., Nogi, K., Naito, M., and Yokayama, T. (2012) *Nanoparticle Technology Handbook*. Elsevier BV, Oxford, UK.

Hyeon, T. (2003) Chemical synthesis of magnetic nanoparticles. *Chem. Commun.*, (8), 927–934.

Ingram, D. R., Kotsmar, C., Yoon, K. Y., Shao, S., Huh, C., Bryant, S. L., Milner, T. E., and Johnston, K. P. (2010) Superparamagnetic nanoclusters coated with oleic acid bilayers for stabilization of emulsions of water and oil at low concentration. *J. Colloid Interface Sci.*, 351, 225–232.

Iskandar, F. (2009) Nanoparticle processing for optical applications—A review. *Adv. Powder Technol.*, 20(4), 283–292.

Ismail, A. A. (2005) Synthesis and characterization of Y2O3/Fe2O3/TiO2 nanoparticles by sol–gel method. *Appl. Catal. B: Environ.*, 58(1), 115–121.

Jang, B. Z., and Zhamu, A. (2008) Processing of nanographene platelets (NGPs) and NGP nanocomposites: A review. *J. Mater. Sci.*, 43, 5092–5101.

Kango, S., Kalia, S., Celli, A., Njuguna, J., Habibi, Y., and Kumar, R. (2013) Surface modification of inorganic nanoparticles for development of organic-inorganic nanocomposites-A review. *Prog. Polym. Sci.*, 38, 1232–1261.

Kim, I., Taghavy, A., DiCarlo, D., and Huh, C. (2015) Aggregation of silica nanoparticles and its impact on particle mobility under high-salinity conditions. *J. Petrol. Sci. Eng.*, 133, 376–383.

Kim, J. H., Germer, T. A., Mulholland, G. W., and Ehrman, S. H. (2002) Size-monodisperse metal nanoparticles via hydrogen-free spray pyrolysis. *Adv. Mater.*, 14(7), 518–521.

Ko, S., Kim, E., Park, S., Daigle, H., Milner, T., Huh, C., Bennetzen, M. V., and Geremia, G. (2017a) Amine functionalized magnetic nanoparticles for removal of oil droplet from produced water and accelerated magnetic separation. *J. Nanopart. Res.*, 19(4), 132.

Ko, S., Lee, H., and Huh, C. (2017b) Efficient removal of EOR polymer from produced water using magnetic nanoparticles and regeneration/re-use of spent particles. *SPE Prod. Oper.*, 32(03), 374–381.

Kobayashi, Y., Inose, H., Nakagawa, T., Gonda, K., Takeda, M., Ohuchi, N., and Kasuya, A. (2011) Control of shell thickness in silica-coating of Au nanoparticles and their X-ray imaging properties. *J. Colloid Interface Sci.*, 358(2), 329–333.

Kotsmar, C., Yoon, K. Y., Yu, H., Ryoo, S., Barth, J., Shao, S., Milner, T. E., Bryant, S. L., Huh, C., and Johnston, K. P. (2010) Stable citrate coated iron oxide superparamagnetic nanoclusters at high salinity. *Ind. Eng. Chem. Res.*, 49, 12435–12443.

Kruis, F. E., Fissan, H., and Peled, A. (1998) Synthesis of nanoparticles in the gas phase for electronic, optical and magnetic applications—A review. *J. Aerosol Sci.*, 29(5–6), 511–535.

Laurent, S., Forge, D., Port, M., Roch, A., Robic, C., Vander Elst, L., and Muller, R. N. (2008) Magnetic iron oxide nanoparticles: Synthesis, stabilization, vectorization, physicochemical characterizations, and biological applications. *Chem. Rev.*, 108(6), 2064–2110.

Lee, D., Cho, H., Lee, J., Huh, C., and Mohanty, K. (2015) Fly ash nanoparticles as a CO2 foam stabilizer. *Powder Technol.*, 283, 77–84.

Liu, Q., Xu, Z., Finch, J., and Egerton, R. (1998) A novel two-step silica-coating process for engineering magnetic nanocomposites. *Chem. Mater.*, 10(12), 3936–3940.

Lu, A. H., Salabas, E. L., and Schüth, F. (2007) Magnetic nanoparticles: Synthesis, protection, functionalization, and application. *Angew. Chem. Intern. Ed.*, 46(8), 1222–1244.

Lu, Y., Yin, Y., Mayers, B. T., and Xia, Y. (2002) Modifying the surface properties of superparamagnetic iron oxide nanoparticles through a sol-gel approach. *Nano Lett.*, 2(3), 183–186.

Luo, D., Wang, F., Zhu, J., Cao, F., Liu, Y., Li, X., Wilson, R. C., Yang, Z., Chu, C.-W., and Ren, Z. (2016) Nanofluid of graphene-based amphiphilic Janus nanosheets for tertiary or enhanced oil recovery: High performance at low concentration. *PNAS*, 113, 7711–7716.

Mallakpour, S., and Madani, M. (2015) A review of current coupling agents for modification of metal oxide nanoparticles. *Prog. Org. Coat.*, 86, 194–207.

Massart, R. (1981) Preparation of aqueous magnetic liquids in alkaline and acidic media. *IEEE Trans. Magn.*, 17(2), 1247–1248.

Mohanty, P., Yoon, I., Kang, T., Seo, K., Varadwaj, K. S. K., Choi, W., Park, Q. H., Ahn, J. P., Suh, Y. D., Ihee, H., and Kim, B. (2007) Simple vapor-phase synthesis of single-crystalline Ag nanowires and single-nanowire surface-enhanced raman scattering. *J. Am. Chem. Soc.*, 129(31), 9576–9577.

Munoz-Espi, R., Weiss, C. K., and Landfester, K. (2012) Inorganic nanoparticles prepared in miniemulsion. *Curr. Opin. Colloid Interface Sci.*, 17, 212–224.

Murugesan, S., Kuznetsov, O., Suresh, R., Agrawal, D., Monteiro, O., and Khabashesku, V. N. (2016) Carbon Quantum Dots fluorescent Tracers for Production and Well Monitoring. (SPE-181503), SPE Annual Technical Conference, September 26–28, Dubai, UAE.

Neouze, M., and Schubert, U. (2008) Surface modification and functionalization of metal and metal oxide nanoparticles by organic ligands. *Monatsh. Chem.*, 139, 183–195.

Okuyama, K., and Lenggoro, I. W. (2003) Preparation of nanoparticles via spray route. *Chem. Eng. Sci.*, 58(3–6), 537–547.

Paul, K. T., Satpathy, S. K., Manna, I., Chakraborty, K. K., and Nando, G. B. (2007) Preparation and characterization of nano structured materials from fly ash: A waste from thermal power stations, by high energy ball milling. *Nanoscale Res. Lett.*, 2, 397–404.

Perro, A., Meunier, F., Schmitt, V., and Ravaine, S. (2009) Production of large quantites of "Janus" nanoparticles using wax-in-water emulsions. *Colloids Surf. A: Physicochem. Eng. Asp.*, 332, 57–62.

Pradhan, S., Xu, L.-P., and Chen, S. (2007) Janus nanoparticles by interfacial engineering. *Adv. Funct. Mater.*, 17, 2385–2392.

Ramanathan, T., Abdala, A. A., Stankovich, S., Dikin, D. A., Herrera-Alonso, M., Piner, R. D., Adamson, D. H., Schniepp, H. C., Chen, X., Ruoff, R. S., Nguyen, S. T., Aksay, I. A., Prud'Homme, R. K., and Brinson, L. C. (2008) Functionalized graphene sheets for polymer nanocomposites. *Nat. Nanotechnol.*, 3, 327–331.

Ryoo, S., Ramani, A. R., Yoon, K. Y., Prodanovic, M., Kotsmar, C., Milner, T. E., Johnston, K. P., Bryant, S. L., and Huh, C. (2012) Theoretical and experimental investigation of the motion of multiphase fluids containing paramagnetic nanoparticles in porous media. *J. Petrol. Sci. Eng.*, 81, 129–144.

Sabale, S., Jadhav, V., Khot, V., Zhu, X., Xin, M., and Chen, H. (2015) Superparamagnetic MFe_2O_4 (M = Ni, Co, Zn, Mn) nanoparticles: Synthesis, characterization, induction heating and cell viability studies for cancer hyperthermia applications. *J. Mater. Sci. Mater. Med.*, 26(3), 127.

Sperling, R. A., and Parak, W. J. (2010) Surface modification, functionalization and bioconjugation of colloidal inorganic nanoparticles. *Philos. Trans. Royal Soc. A,* 368, 1333–1383.

Stankovich, S., Dikin, D. A., Dommett, G. H. B., Kohlhass, K. M., Zimney, E. J., Stach, E. A., Piner, R. D., Nguyen, S. T., and Ruoff, R. S. (2006) Graphene-based composite materials. *Nat. Lett.,* 442(20), 282–286.

Stjerndahl, M., Andersson, M., Hall, H. E., Pajerowski, D. M., Meisel, M. W., and Duran, R. S. (2008) Superparamagnetic Fe_3O_4/SiO_2 nanocomposites: Enabling the tuning of both the iron oxide load and the size of the nanoparticles. *Langmuir,* 24, 3532–3536.

Stöber, W., Fink, A., and Bohn, E. (1968) Controlled growth of monodisperse silica spheres in the micron size range. *J. Colloid Interface Sci.,* 26(1), 62–69.

Swihart, M. (2003) Vapor-phase synthesis of nanoparticles. *Curr. Opin. Colloid Interface Sci.,* 8, 127–133.

Tartaj, P., del Puerto Morales, M., Veintemillas-Verdaguer, S., González-Carreño, T., and Serna, C. J. (2003) The preparation of magnetic nanoparticles for applications in biomedicine. *J. Phys. D: Appl. Phys.,* 36(13), R182.

Turkenburg, D. H., Chin, P. T. K., and Fischer, H. R. (2012) Use of Modified Nanoparticles in Oil and Gas Reservoir Management. (SPE 157120), SPE International Oilfield Nanotechnology Conference, June 12–14, Noordwijk, the Netherlands.

Wang, Q., Prigiobbe, V., Huh, C., Bryant, S. L., Mogensen, K., and Bennetzen, M. V. (2014) Removal of Divalent Cations from Brine Using Selective Adsorption onto Magnetic Nanoparticles. (IPTC 17901), International Petroleum Technology Conference, Dec. 10–12, Kuala Lumpur, Malaysia.

Wang, Q., Prigiobbe, V., Huh, C., and Bryant, S. L. (2017) Alkaline earth element adsorption onto PAA-coated magnetic nanoparticles. *Energies,* 10(2), 0223.

Wikipedia. 2018. Sol–gel Process. https://en.wikipedia.org/wiki/Sol–gel_process (last modified January 10, 2019).

Wooldridge, M. S. (1998) Gas-phase combustion synthesis of particles. *Prog. Energy Combust. Sci.,* 24, 63–87.

Wu, W., He, Q., Chen, H., Tang, J., and Nie, L. (2007) Sonochemical synthesis, structure and magnetic properties of air-stable Fe3O4/Au nanoparticles. *Nanotechnology,* 18(14), 145609.

Wu, W., He, Q., and Jiang, C. (2008) Magnetic iron oxide nanoparticles: Synthesis and surface functionalization strategies. *Nanoscale Res. Lett.,* 3(11), 397–415.

Wu, W., Jiang, C., and Roy, V. A. (2016) Designed synthesis and surface engineering strategies of magnetic iron oxide nanoparticles for biomedical application. *Nanoscale,* 8, 19421–19474.

Xue, Z., Foster, E., Wang, Y., Nayak, S., Cheng, V., Ngo, V. W., Pennell, K. D., Bielawski, C. W., and Johnston, K. P. (2014) Effect of grafted copolymer composition on iron oxide nanoparticle stability and transport in porous media at high salinity. *Energy Fuels,* 28(6), 3655–3665.

Yoon, K. Y., Xue, Z., Fei, Y., Lee, J. H., Cheng, V., Bagaria, H. G., Huh, C., Bryant, S. L., Kong, S. D., Ngo, V. W., Rahmani, A.-R., Ahmadian, M., Ellison, C. J., and Johnston, K. P. (2016) Control of magnetite primary particle size in aqueous dispersions of nanoclusters for high magnetic susceptibilities. *J. Colloid Interface Sci.,* 462, 359–367.

Yoon, K. Y., An, S. J., Chen, Y., Lee, J. H., Bryant, S. L., Ruoff, R., Huh, C., and Johnston, K. P. (2013) Development of graphene-oxide-stabilized oil-in-water emulsions for high-salinity conditions. *J. Colloid Interface Sci.*, 403, 1–6.

Yoon, K. Y., Li, Z., Neilson, B. M., Lee, W., Huh, C., Bryant, S. L., Bielawski, C. W., and Johnston, K. P. (2012) Effect of adsorbed amphiphilic copolymers on the interfacial activity of superparamagnetic nanoclusters and emulsification of oil in water. *Macromolecules*, 45(12), 5157–5166.

Yoon, K. Y., Kotsmar, C., Ingram, D. R., Huh, C., Bryant, S. L., Milner, T. E., and Johnston, K. P. (2011) Stabilization of superparamagnetic iron oxide nanoclusters in concentrated brine with cross-linked polymer shells. *Langmuir*, 27, 10962–10969.

Yu, H., Kotsmar, C., Yoon, K. Y., Ingram, D., Johnston, K. P., Bryant, S. L., and Huh, C. (2010) Mobility and Retention of Aqueous Dispersions of Paramagnetic Nanoparticles in Reservoir Rocks. (SPE 129887), SPE/DOE Improved Oil Recovery Symposium, Improved Oil Recovery, April 26–28, Tulsa, OK, USA.

Yu, H., Yoon, K. Y., Neilson, B. M., Bagaria, H. G., Worthen, A. J., Lee, J. H., Cheng, V., Bielawski, C. W., Johnston, K. P., Bryant, S. L., and Huh, C. (2014) Transport and retention of aqueous dispersion of superparamagnetic nanoparticles in sandstone. *J. Petrol. Sci. Eng.*, 116, 115–123.

3

Nanoparticles in Fluids

3.1 Introduction

In this chapter, the description of nanoparticle behavior in fluids is presented and a series of examples are given to help the reader to directly test the concepts. To this aim, codes written in MATLAB® are provided to calculate the forces of interaction between nanoparticles (NPs), the kinetics of NP aggregation, and the transport of NPs in porous media.

3.2 Dispersion Stability of Nanoparticles in Fluids

Nanoparticles in engineered or environmental systems are considered dispersions of primary particles. However, primary NPs tend to aggregate into clusters up to several microns in size. The thermodynamic conditions of the liquid phase and physico-chemical properties of the NPs control the kinetics of the formation of cluster and their stability (Hotze et al. 2010). In Section 3.2.1, the classical DLVO (Derjaguin, Landau, Verwey, and Overbeek) theory is explained together with its extension (XDLVO) to include the description of functionalized NPs with a magnetic core and functionalized surface. Section 3.2.2 follows where the kinetics of NP aggregation is described. Sections 3.2.3 and 3.2.4 focus on the analysis of the effect of the reservoir conditions on the stability of the NPs. Finally, Section 3.2.5 introduces the concept of NP functionalization to mitigate NP aggregation and deposition onto the porous medium walls.

3.2.1 Introduction to DLVO Theory and its Extension

In a suspension of nanoparticles, Brownian diffusion controls the long-range forces between individual nanoparticles, causing collisions that might lead to repulsion or aggregation of the particles. Brownian energy imparted to a particle originates principally from the collision of that particle with water molecules. Therefore, the smaller the size of the particle, the more significant is the effect of the Brownian energy on the overall particle energy (Wiesner and Bottero 2007). Generally, the interaction between two particles involves both the electrostatic force and the Lifshitz-van der Waals force. The former is strong at large distances between the particle surfaces and it is sensitive to the concentration of electrolytes. Contrarily, the latter is dominant when the separation between the two surfaces is small and it is not sensitive to the concentration of the electrolytes in solution. The framework to describe the change of these forces of interaction with the distance between particle surfaces and the electrolyte concentration is given by the DLVO theory (named after the Russian scientists B. Derjaguin and L. Landau, and Dutch scientists E. Verwey and J. Overbeek), which was developed in the 1940s (Derjaguin and Landau 1993; Verwey 1948). Therefore, DLVO theory includes the energies of interaction between two surfaces accounting for the electrostatic interaction energy (U_{el}) and the Lifshitz-van der Waals attraction energy (U_{lw}) (Hotze et al. 2010). At the nanoscale, DLVO theory is no longer appropriate and non-DLVO energies are considered as they have been found to be significant for nanoparticles in suspension within a fluid. This new theory to calculate the total energy of interaction between two surfaces of which at least one is of a nanoparticle takes the name of extended DLVO (XDLVO) theory which in addition to U_{lw} and U_{el} includes (Hotze et al. 2010; Israelachvili 1992; van Oss 1993), namely

- the magnetic attraction energy (U_m) if the NP has a magnetic property;
- the acid-base interaction energy (U_{ab});
- the repulsion energy due to Born repulsion (U_{bo}); and
- the osmotic repulsion energy (U_{osm}) and the elastic-steric repulsion resulting from the polymer coating (U_{elas}) if the particle surface is functionalized.

Therefore, the overall energy of interaction writes,

$$U_T = U_{lw} + U_{el} + U_{ab} + U_{bo} + U_m + U_{osm} + U_{elas} \tag{3.1}$$

In the next part of this section, each term of Equation (3.1) is presented in detail and the corresponding expressions considering the interaction both between a particle and a flat surface and between two particles are given. The Lifshitz-van der Waals interaction energy between a flat surface and a spherical particle, U_{lw}^{s-p}, can be expressed using the Derjaguin's approximation as (Gregory 1981),

$$U_{lw}^{s-p} = \frac{H_a r}{6h} \left(1 + \frac{14\,h}{\lambda} \right)^{-1} \tag{3.2}$$

with H_a the Hamaker constant, r the effective radius of the interactive particle, h the distance between the two planar semi-infinite walls representing the flat surface and the particle surface, and λ is the characteristic London wavelength, generally assumed equal to 100 nm. The van der Waals interaction energy between two spherical particles, U_{lw}^{p-p}, is given by (Gregory 1977),

$$U_{lw}^{p-p} = -\frac{H_a}{6}\left(\frac{2r^2}{\eta(4r+\eta)} + \frac{2r^2}{(4r+\eta)^2} + \ln\eta\frac{2r+\eta}{(4r+\eta)^2}\right) \quad (3.3)$$

where η is the distance between the surfaces of two interacting NPs. The electrostatic interaction energy between a flat surface and a spherical particle, U_{el}^{s-p}, is given by (Hogg et al. 1966, Sader et al. 1995),

$$U_{el}^{s-p} = \pi\varepsilon_0\varepsilon_r r\left[2\Psi_1\Psi_2\ln\frac{1+e^{-\kappa h}}{1-e^{-\kappa h}} + \left(\Psi_1^2+\Psi_2^2\right)\ln\left(1-e^{-2\kappa h}\right)\right] \quad (3.4)$$

where ε_0 is the permittivity of free space; ε_r is the relative dielectric constant of the liquid; Ψ_1 and Ψ_2 are the surface potentials of the interacting surfaces; κ is the inverse of the thickness of the electrostatic double layer, i.e., $\kappa^{-1} = \sqrt{\dfrac{\varepsilon_0\varepsilon_r k_B T}{2N_A e^2 I}}$, with N_A the Avogadro number, e the elementary charge of an electron, and I the ionic strength (Equation 3.16). In the case of the interaction between two nanoparticles, the electrostatic interaction energy, U_{el}^{p-p}, can be simplified as (de Vicente et al. 2000)

$$U_{el}^{p-p} = 2\pi\varepsilon_0\varepsilon_r r\Psi^2\ln\left(1+e^{-\kappa h}\right) \quad (3.5)$$

where the surface potential Ψ is considered to be equal on the two interacting particles, ε_0 and ε_r are the permittivity of the vacuum and the relative dielectric constant of the fluid, respectively; κ is the inverse Debye length. The magnetic attractive energy is considered when two interacting surfaces are magnetic, e.g., NPs made of magnetite (Fe_3O_4). In this case, the energy of interaction writes as (de Vicente et al. 2000),

$$U_m^{p-p} = -\frac{8\phi\mu_0 M_s^2 r^3}{9\left(\dfrac{h}{r}+2\right)^3} \quad (3.6)$$

where μ_0 is the magnetic permeability of the vacuum and M_s is the saturation magnetization. Figure 3.1 reports the change of U_{lw}^{p-p}, U_{el}^{p-p}, and U_m^{p-p}, as well as the resulting total energy of these components for two magnetic NPs. The model parameter values used later for the calculation of the total energy of interaction are given in Table 3.1.

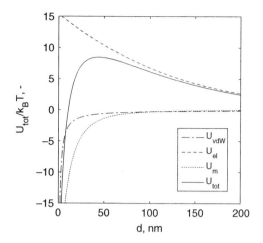

FIGURE 3.1

Total energy of interaction for a pair of magnetic NPs according to extended DLVO theory. The values of the parameters are reported in Table 3.1.

TABLE 3.1

Model Parameter Values Used for the Calculation of the Total Energy of Interaction According to Extended DLVO Theory Shown in Figure 3.1.

H_a	10^{19}
r (nm)	14
Electrolyte (mole/kg)	1/50
μ_0 (mkg/(m^{2}A^2))	1.23×10^6
M_s (A/m)	3.3×10^3

The acid-base interaction energy between a spherical particle and a flat surface, U_{ab}^{s-p}, is given by (van Oss 1993, Hoek and Agarwal 2006),

$$U_{ab}^{s-p} = 2\pi r \lambda \Delta G_{ab} e^{(y_0 - h)/\lambda} \tag{3.7}$$

where ΔG_{ab} is the acid-base free energy of interaction between the two interacting surfaces (Hoek and Agarwal 2006; van Oss 1993), λ is the decay length for acid–base interactions in water, which is generally taken equal to 0.6 nm.

The Born repulsion energy between a flat surface and a spherical particle, U_{bo}^{s-p}, was derived by Prieve and Ruckenstein (1976) and writes as,

$$U_{bo}^{s-p} = \frac{H_a \sigma_B^6}{7500} \left[\frac{2r + h}{(2r + h)^7} + \frac{6r - h}{h^7} \right] \tag{3.8}$$

where σ_B is the Born collision diameter, generally assumed equal to 0.5 nm (Hahn et al. 2004). In the case of two interacting particles, the Born repulsion energy, U_{bo}^{p-p} becomes (Sato and Ruch 1980),

$$U_{bo}^{p-p} = \frac{C_B}{R}\left(\frac{R^2 - 14R + 54}{(R-2)^7} + \frac{-2R^2 + 60}{R^7} + \frac{R^2 + 14R + 54}{(R-2)^7} \right) \tag{3.9}$$

where C_B is equal to 1.0×10^{-22} and $R = R*/2r$, with $R*$ the distance between the centers of the interacting nanoparticles. The osmotic energy of interaction between two functionalized NPs, U_{osm}^{p-p}, which is due to the presence of a polymeric coating, is given by (Phenrat et al. 2008)

$$U_{osm}^{p-p} = \begin{cases} 0 & \text{if} \quad 2d \leq h \\ \dfrac{\kappa T a 4\pi}{v} \phi^2 \left(\dfrac{1}{2} - \chi\right)\left(d - \dfrac{h}{2}\right)^2 & \text{if} \quad d \leq h < 2d \\ \dfrac{\kappa T a 4\pi}{v} \phi^2 \left(\dfrac{1}{2} - \chi\right)\left(\dfrac{h}{2d} - \dfrac{1}{4} - \ln\dfrac{h}{d}\right) & \text{if} \quad d > h \end{cases} \tag{3.10}$$

where d is the thickness of the unperturbed adsorbed polyelectrolyte layer, ϕ is the volume fraction of polymer in brush, v is the volume of solvent molecule, and χ is the Flory-Huggins solvency parameter. Finally, the elastic-steric repulsion energy between two functionalized NPs, U_{elas}^{p-p}, is given by (Phenrat et al. 2008),

$$U_{elas}^{p-p} = \begin{cases} 0 & \text{if} \quad d \leq h \\ kT\left(\dfrac{2a\pi}{M_w}\right)\phi d^2 \rho_p \dfrac{h}{d}\left\{\ln\left[\dfrac{h}{d}\left(\dfrac{3 - \dfrac{h}{d}}{2}\right)^2\right] - 6\ln\left(\dfrac{3 - \dfrac{h}{d}}{2}\right) + 3\left(1 + \dfrac{h}{d}\right)^2\right\} & \text{if} \quad d > h \end{cases}$$
$$\tag{3.11}$$

where M_w is the molecular weight of polyelectrolyte and ρ_p is its density. Overall, the magnitude of the interaction energies depends on the particle size and surface properties and the composition of the fluid. In the case of functionalized NPs with a magnetic core, the magnitude of the overall energy on interaction is controlled by the magnetic moment in conjunction with the adsorbed mass of polymers per unit surface area of NP, the polymer molecular weight, the polymer density, the adsorbed layer thickness, and the solution composition (Hotze et al. 2010).

3.2.1.1 Code for XDLVO Theory Calculations

In this section, a MATLAB® code (The MathWorks 2015) for the calculations of the total energy of interaction between two nanoparticles of magnetite is reported as an example (Figure 3.2).

```
% XDLVO Theory for a magnetic nanoparticle

% =========================================

clear all
close all
clc

kBT=1.38*10^(-23)*293.15; % m2 kg s-2 K-1 * K
etavec=linspace(1e-15,200,1000); % nm, distance between the surface of two interacting particles

% van der Waals energy
% ------------------
Ha=1e-19; % N*m for iron nanoparticles
r=13.75; % nm, effective radius of the nanoparticle
UvdW=-Ha/6.*(2*r^2./(etavec.*(4*r+etavec))+2*r^2./(4*r+etavec).^2+log(etavec).*(4*r+etavec)./(4*r+etavec).^2); % J

% Electrostatic energy
% ------------------
epsr=80.4; % (-) at 20oC, relative permittivity of water,
eps0=8.85*1e-12; % F/m, vacuum permittivity
NA=6.022140857*10^(23); % mol-1
elec=1.602176634*10^(-19); % C
z=1;
conc=1/50; % % mole/kg
I = 0.5*(conc*z^2); % mole/kg
DebyeLength =sqrt(epsr*eps0*kBT/(2*NA*elec^2*I)); %m
kappa =1/(DebyeLength*1e+9); % 1/nm
psi=38.2*1e-3; % V, surface potential
Uel=2*pi*epsr*eps0*r*1e-9*psi^2.*log(1+exp(-kappa.*etavec)); % J

% Magnetic energy
% ------------
mu0=1.25663706*1e-6; % m kg s-2 A-2, vacuum permeability
Ms=330*1e+3; % A/m, saturation magnetization
Um=-8*pi*mu0*Ms^2*(r*1e-9)^3./(9*(etavec./r+2).^3);

% Plotting
% -------
figure(1)
plot(etavec,UvdW./kBT,'-','linewidth',1.5); hold on
plot(etavec,Uel./kBT,'-','linewidth',1.5);hold on
plot(etavec,Um./kBT,'-','linewidth',1.5);hold on
plot(etavec,(UvdW+Uel+Um)./kBT,'-','linewidth',1.5);hold on
ylim([-15 15]);
axis square
ylabel('U_{tot}/k_BT, -','FontSize',14);
xlabel('d, nm','FontSize',14);
legend('U_{vdW}','U_{el}','U_{m}','U_{tot}');
```

FIGURE 3.2
Code for the calculation of the total energy of interaction for a pair of magnetic NPs according to the extended DLVO theory.

3.2.2 Nanoparticle Aggregation Kinetics

Aggregation of nanoparticles results in the formation and growth of clusters. The thermodynamics and kinetics of these clusters are regulated by the composition both of the fluid surrounding the particles and that of the surface (type of functionalization). During the transport of NPs through a porous medium, aggregation may increase retention within the pores, thus negatively affecting the operation. Similarly, during wastewater treatment, the formation of clusters reduces the specific surface area available for contaminant removal, making the treatment less effective. However, if clusters form upon removal of contaminants from wastewater, their larger size facilitate separation, and, therefore, their formation is beneficial to the process. In any case, the ability to predict and control aggregation in order to increase the performance of NP application is generally envisaged. The aggregation behavior of NPs is strongly affected by particle size (Kobayashi et al. 2005). Particles in the large size fraction (i.e., $r > 100$ nm) follow the classical DLVO theory. Whereas, particles in small size fraction (i.e., $r < 50$ nm) aggregate more slowly following the extended DLVO theory that includes acid-base and osmotic energies of interaction in addition to magnetic and elastic-steric repulsion energies if they have a magnetic core and a functionalized surface. In addition, solution composition and temperature have a strong effect on the kinetics of aggregation. The combination of XDLVO with the population balance equation (PBE) can help to account for all these influencing factors allowing, therefore, to follow the evolution of NP aggregation in space and time. PBE is the most common modeling tool used to describe and control a wide range of particulate processes, for example, precipitation, crystallization, and flocculation. In a perfectly mixed particle suspension with a constant volume, assuming neither nucleation, growth, nor breakage, the PBE accounting only for agglomeration is given by (Mersmann 2001; Randolph and Larson 1988)

$$\frac{\partial n_p(t,x)}{\partial t} = \frac{1}{2} \int_{\phi}^{\infty} \beta(t, x-y, y) n_p(t, r-y) n_p(t, y) dy$$

$$-n_p(t,r) \int_{0}^{\infty} \beta(t, r, y) n_p(t, y) dy \qquad (3.12)$$

with initial condition $n_p(x,0) = n_{p,0}(x)$, where $n_p(t,x)$ is the number concentration distribution function representing the number of nanoparticles per system volume (#/m³) and it is described by a particle size distribution (PSD) function; $n_{p,0}$ is the initial PSD; and t is the time. The first term on the right-hand side represents the birth of the nanoparticles of size x as a result of the agglomeration of nanoparticles of sizes $(x-y)$ and y. The factor 1/2 prevents the double counting of collisions of the nanoparticles. The second term describes the merging of nanoparticles of size x with any other nanoparticle. The second term is called the death term due to agglomeration. The fraction β is known to be the agglomeration kernel, which describes the kinetics of agglomeration by measuring the frequency with which a nanoparticle of size x aggregates with one of size y. It is a product of two factors, a size-dependent collision kernel, β_0, accounting for the collisions of nanoparticles and an efficiency factor ϕ, which takes into account only of the collisions that lead to a stable agglomerate, i.e.,

$$\beta(t,x,y) = \beta_0(t)\phi(x,y) \tag{3.13}$$

The challenge in agglomeration modeling is to find a suitable expression for β_0. The collision efficiency, ϕ, can be described by the ratio of the number of collisions between particles of radii x and y to the number of collisions that could result in agglomeration between them (Hunter and White 1987)

$$\phi = 2\int_2^\infty H_d \frac{e^{U_T/(k_B T)}}{s^2} ds \tag{3.14}$$

where H_d is the expression to incorporate the hydrodynamic drag forces between particles of radii x and y.

$$H_d = \frac{6(s-2)^2 + 13(s-2) + 2}{6(s-2)^2 + 4(s-2)}, \tag{3.15}$$

with $s = 2R/(x+y)$, where R is the center-to-center distance between the interacting nanoparticles. In Equation (3.14) the factor U_T is the total interaction energy, which results from the repulsive and attractive forces described within the framework of the XDLVO theory (Equation 3.1). The population balance equation (Equation 3.12) is a partial differential equation (PDE) that can be solved by various numerical methods, such as the method of moments (Vollmer and Raisch 2006) and of the moving pivot (Kumar and Ramkrishna 1997).

3.2.2.1 Code for Nanoparticle Aggregation

We used the method presented by Kumar and Ramkrishna (1997) to solve the PBE. The PBE was discretized and each finite size range was represented by a corresponding length, the pivot, and its lower and upper boundary. We used the arithmetic mean of the size of the lower and upper boundaries as the size of the pivot. The moving pivot technique overcomes the problem of numerical diffusion and instability, while providing the applicability to an arbitrary grid to guarantee accuracy and reasonable computational effort. In Figure 3.3, the evolution of the PSD of a nanoparticle suspension due to aggregation is shown. The corresponding MATLAB® code (The MathWorks 2015) is given in Figure 3.4.

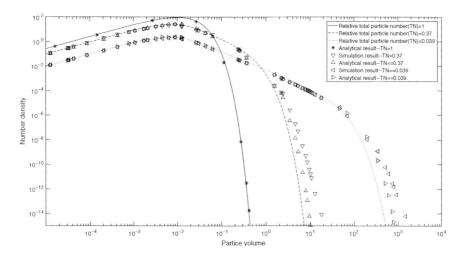

FIGURE 3.3
Evolution of the PSD of a nanoparticle suspension due to aggregation.

3.2.3 Importance of Entropic Repulsion for Reservoir Application

One of the major issues in the application of nanoparticles in the reservoir is the formation of clusters which can be mechanically retained within the pore-throats of the reservoir porous medium. Moreover, having a large ratio of surface area to volume, NPs tend to agglomerate quickly to minimize high surface energy due to the van der Waals attractive forces along with the strong magnetic dipole-dipole attractions for NPs with a magnetic core (Gupta and Gupta 2005). Two major forces are considered to impact of nanoparticles stability: Van der Waals forces and Coulombic (Entropic) forces. Entropy force is fairly ubiquitous in nature, but it is not practically beneficial for most cases, thus, how to reduce the entropic force of the system is very important. The entropy force also depends on the distance between two nanoparticles (Hua Yun-Feng 2017). Nanoparticles in a dispersion medium always show Brownian motion and hence collide with each other frequently. Upon collision, the formation of clusters may occur, and it is controlled by the forces of interaction. If attraction forces dominate, the particles will adhere with each other and finally the entire dispersion may coalesce. If repulsion dominates, the system will be stable and remain in a dispersed state (Sato and Ruch 1980). Therefore, only if sufficiently strong repulsive force counteracts the van der Waals attraction, the nanoparticle suspension will remain stable. Stability can be attained by, *e.g.*,

- electrical double layer (electrostatic or charge stabilization);
- grafted polymeric molecules (steric stabilization); and
- free polymer in the dispersion medium (depletion stabilization).

```
% Reference: Kumar, S., and Ramkrishna, D. (1997) On the solution of population balance equations
by discretization—III. Nucleation, growth and aggregation of particles. Chemical Engineering Science,
52(24), 4659–4679
% Parameters
M = 170; %
dt = 1;
tmax = 3000;
v0 = 1e-2;
N0 = 2.5;
s = 1.15;
% Initialize the size distribution
for i = 1:M
  v(i) = v0/2000*s^i;
  n(i,1) = N0/v0*(v(i)/v0)*exp(-v(i)/v0);
end
N = zeros(M,tmax);
N(1,1) = n(1,1)*v(1);
for i=2:M
  N(i,1) = n(i,1)*(v(i) - v(i-1));
end
N0 = N(:,1);
RR = zeros(M,tmax);
RD = zeros(M,tmax);
RRD = zeros(M,tmax);
NS = zeros(tmax,1);
Nratio = zeros(tmax,1);
for i = 1:M
  NS(1) = NS(1) + N(i,1);
end
%Equation 43 in the reference
for t = 2:tmax

% Birth term due to aggregation,
  for i = 2:M-1
    Rab = 0; % Equation 36 in the reference
    for j = 1:i
      for k = 1:j
        vjk = v(j)+v(k);
        if (v(i-1) <= vjk) && (v(i+1) >= vjk)
          if j == k
            deltajk = 1;
          else
            deltajk = 0;
          end
      % Equation 39 in the reference
          if vjk >= v(i) && vjk <=v (i+1)
            yita = (v(i+1) - vjk)/(v(i+1) - v(i));
          end
          if vjk >= v(i-1) && vjk <= v(i)
            yita = (vjk - v(i-1))/(v(i) - v(i-1));
          end
```

FIGURE 3.4

Code for the description of the particle size distribution during agglomeration.

```
Qjk = v(j) + v(k);  %% Figure 6 in the refrence
          if t > 1
              Rab = Rab + (1 - 0.5*deltajk)*yita*Qjk*N(j,t - 1)*N(k,t - 1);
          else
              Rab = Rab + (1 - 0.5*deltajk)*yita*Qjk*N0(j)*N0(k);
          end
        end
      end
    end
    RR(i,t) = Rab;
  end

% Death term due to aggregation
  for i = 1:M
    RDa = 0;
    for k = 1:M  %M is the total number of sections
      Qik = v(i) + v(k);
      if t > 1
        RDa = RDa + Qik*N(k,t-1);
      else
        RDa = RDa + Qik*N0(k);
      end
    end
    if t > 1
      RDa = RDa*N(i,t-1);
    else
      RDa = RDa*N0(i);
    end
    RD(i,t) = RDa;
  end
  for i = 1:M
    RRD(i,t) = RR(i,t) - RD(i,t);
    N(i,t) = N(i,t-1) + (RR(i,t) - RD(i,t));
  end
  n(1,t) = N(1,t)/v(1);
  for i = 2:M
    n(i,t) = N(i,t)/(v(i) - v(i-1));
    if n(i,t) < 1e-20
      n(i,t) = 0;
    end
  end
  for i = 1:M
    NS(t) = NS(t) + N(i,t);
  end
  Nratio(t) = NS(t)/NS(1);
end
save('All_data.mat')
```

FIGURE 3.4   (CONTINUED)

Figure 3.5 shows schematic of the three methods to provide stability to a nanoparticle suspension by entropic repulsion. Electrical double layer stabilization is obtained by adding to the solution ionic groups that adsorb to the surface of the NP forming a charged layer. The mutual repulsion of the double layers surrounding particles provides stability. Steric stabilization of NPs is achieved by grafting macromolecules to the surfaces of the NPs. Generally, polymers with molecular weights larger than 10,000 D (Dalton) are applied because the brush length is comparable with or longer than the range of the van der Waals forces of attraction. Therefore, if the grafted polymers can generate repulsion, they impart nanoparticle stability (Napper 1983). Depletion stabilization is imparted by polymers that are free in solution. In practice, the electrostatic and steric stabilization are combined (*electrosteric stabilization*) with the electrostatic component being the net charge of the polymer attached to the NP surface. In highly concentrated suspensions, depletion and steric stabilization are applied.

Steric stabilization is divided in to (Shi 2002):

- A polymer molecule with a relatively high molecular weight consisting of chemically similar units (or monomer) connected by primary covalent bonds.

- Copolymer which is a polymer having two different monomers incorporated into the same polymer chain. Copolymers may be of either random (statistical), block, or graft type.

In random copolymers, monomers have no definite order or arrangement. Block polymers have a long segment or block of one monomer followed by a block of a second monomer. The result is that different homopolymer chains are joined in a head-to-tail configuration (Ellerstein and Ullman 1961). So, a block polymer is a linear arrangement of blocks of different monomer composition. A diblock copolymer is poly-A-*block*-poly-B, and a triblock copolymer is poly-A-*block*-poly-B-*block*-poly-A. If A is a hydrophilic group and B is hydrophobic group, the result can be regarded as polymeric surfactant (Piirma 1992). If two NPs with adsorbed

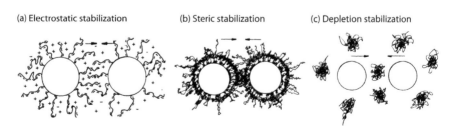

(a) Electrostatic stabilization (b) Steric stabilization (c) Depletion stabilization

FIGURE 3.5
Schematic of the three methods to provide stability to a nanoparticle suspension by entropic repulsion.

polymer layers reach a distance from each other which is less than twice the thickness of the adsorbed polymer layer, the two layers can interact. If the resulting Gibbs free energy upon interaction (ΔG) is negative, NPs will aggregate. Following Sato and Ruch (1980), the entropic stabilization theory assumes that a second surface approaching the adsorbed layer is impenetrable. Therefore, the adsorbed polymer layer of one of the interacting NP is compressed and the polymer in this compressed zone can occupy fewer configurations, hence, losing configurational entropy. Considering $\Delta G = \Delta H - T\Delta S$ and neglecting the change in enthalpy (ΔH), a reduction of entropy (ΔS) translates into an increase in ΔG and an overall effect of NP repulsion. Finally, the fundamental understanding of the entropic forces can help to control the self-assembly (Coalson et al. 2015), which can be used for applications in biomedical and process engineering.

3.2.4 Effect of Salinity, Hardness, and pH

One great disadvantage of NP suspensions stabilized through electrostatic or charge stabilization is its great sensitivity to the ionic strength (I) of the liquid phase where they are dispersed. The ionic strength is given by,

$$I = \frac{1}{2}\sum_i z_i^2 c_i \tag{3.16}$$

where z is the charge number of the ion i, and c the molar concentration of that ion. The thickness of the double layer depends, among others, on the ionic strength of the dispersion medium.

At low ionic strengths ($\sim 10^{-3}$ mole/kg), the thickness of the double layer is about 5–10 nm, which is the order of the distance of interaction of the van der Waals forces. Therefore, the NP suspension is stable due to the dominant effect of the repulsive forces. However, as the I increases, the thickness of the double layer is reduced significantly. For values larger than 10^{-1} mole/kg, the thickness of the double layer is less than 1 nm and the van der Waals forces prevail. Therefore, in a brine, NPs tend to aggregate. Kim et al. (2015) measured the change of the zeta potential of silica NPs with pH and ionic strength and they observed that as pH was reduced, the magnitude of zeta potential decreased and the point of zero charge dropped below 2 (Figure 3.6). In American Petroleum Institute (API) brine, i.e., 8% NaCl + 2% CaCl2, the silica NPs showed a zeta penitential approximately 0 mV. Due to the compression of the double layer or even the adsorption of cations, the repulsive forces were reduced and the interaction forces of attraction, i.e., van der Walls, would dominate and favor aggregation.

The van der Waals and the magnetic attraction forces cannot be altered easily, to induce agglomeration. The electrostatic, the osmotic, and the elastic-steric repulsion forces between the particles need to be controlled by NP

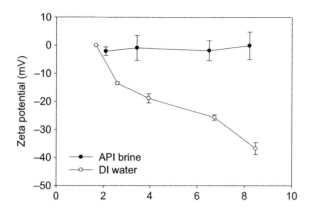

FIGURE 3.6
Zeta potentials of silica NPs suspended in deionized water and in API brine. Modified after Kim et al. (2015).

surface functionalization. This translates in to the manipulation of the model parameters Φ, d, ϕ, M_w, and ρ_p. If the values of M_w and ρ_p depend only on the type of polymer chosen for the coating, the values of d and ϕ depend on the techniques adopted to coat the nanoparticles. The surface potential, Φ, is instead a result of the polyelectrolyte layer (Allen et al. 2001, Shekar et al. 2012). Figure 3.7 shows how latex nanoparticles can aggregate when I exceeds certain values (Atmuri et al. 2013). Xue et al. (2014) studied the effect of I in conjunction with the type of salt on particle aggregation and straining. They observed that, for identical nanoparticle concentration and I, in the presence of $CaCl_2$, the retention of iron-oxide nanoparticle coated with poly (2-acrylamido-2-methyl-1-propanesulfonic acid-co-acrylic acid) (poly(AMPS-co-AA)) was 40% larger than in the presence of NaCl. Xue et al. (2014)

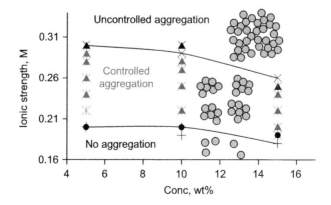

FIGURE 3.7
Aggregation of latex nanoparticle due to ionic strength. Modified after Atmuri et al. (2013).

attributed the attachment of NPs onto the porous medium wall to divalent cation (Ca^{2+}) mediated bridging and hydrophobic interactions.

Saleh et al. (2008) studied the transport of nanoscale zerovalent iron (NZVI) transport in porous media for their application in groundwater remediation. They observed that in order to make the NPs mobile in the subsurface their surface should be modified and the efficacy of the surface modification depends on groundwater ionic strength and cation type as well as physical and chemical heterogeneities of the aquifer material. They modified the NZVI surface using high molecular weight (MW) (125 kg/mol) poly(methacrylic acid)-*b*-(methyl methacrylate)-*b*-(styrene sulfonate) triblock copolymer (PMAA-PMMA-PSS), polyaspartate, which is a low MW (2–3 kg/mol) biopolymer, and the surfactant sodium dodecyl benzene sulfonate (SDBS, MW) 348.5 g/mol). Functionalizing the bare NZVI to triblock copolymer-modified NZVI provided the greatest mobility of the NPs by reducing the apparent ζ-potential approximately from −30 to −50 mV.

Kim et al. (2015) studied the aggregation of functionalized silica nanoparticles as a function of salinity and nanoparticle concentration. They observed that at alkaline conditions (pH = 8.5), aggregation kinetics of NPs was enhanced with increasing calcium and NP concentration. Contrarily, at acid conditions (pH = 3) the aggregation was negligible and it was not affected by either calcium or NP concentration. Single or aggregated NPs in a cluster smaller and an average diameter of 235 nm could be transported through a sandpack column without retention. Aggregated NPs in larger size clusters were lost due to retention and the retention was enhanced inversely to the average flow velocity.

Finally, Wang et al. (2017) studied the adsorption of calcium (Ca^{2+}) onto polyacrylic acid-functionalized iron-oxide magnetic nanoparticles (PAA-MNPs) to understand the adsorption behavior of alkaline earth elements at conditions typical of produced water from hydraulic fracturing. To evaluate the Ca^{2+} adsorption capacity by PAA-MNPs, a surface complexation model was developed which accounted for the Coulombic forces in the diffuse double layer and the competitive adsorption of protons (H^+) and Ca^{2+} onto the anionic carboxyl ligands of the PAA-MNPs. The measurements showed that Ca^{2+} adsorption is significant above pH 5 and decreases with the electrolyte concentration. Upon adsorption, the nanoparticle suspension destabilizes and creates large clusters, which favor an efficient magnetic separation of the PAA-MNPs, thus, helping their recovery and recycle.

3.2.5 Surface Coating Design for Long-Term Dispersion Stability

The major challenge in the application of NPs in petroleum engineering is the development of high mobility NPs for delivery in various porous media under high salinity conditions. One of the major reasons of reduced mobility is straining within the pore-throat upon formation of NP clusters in conjunction with irreversible attachment onto the porous medium walls. In order to mitigate the aggregation of NPs, the attractive forces of interaction must be counter-balanced. This can be achieved by providing a proper surface coating and protection of the nanoparticles for their long-term stability. Organic molecule stabilization, including small organic molecules, surfactants, and polymers, is often employed to maintain NP stability during or after synthesis. One of the major surface modifications of nanoparticles is based on the addition of polymer. Polymer has a distinct structure when grafted onto a particle surface. It is characterized by a polymer brush which is an array of polymer chains tethered at one end to a grafting surface (Coalson et al. 2015). A polymer molecule is defined as a molecule of relatively high molecular weight consisting of regularly repeating units, or chemically similar units (named monomer), connected by primary covalent bonds. Often, NPs are functionalized with copolymers (Stille 1962), as described earlier. The commonly used organic molecules are polyethylene glycol (PEG), polyvinylpyrrolidone (PVP), polyvinyl alcohol (PVA), polyacrylic acid (PAA), chitosan, dextra, and gelatin among polymeric materials and sodium oleates, dodecylamine, and sodium carboxymethylcellulose among surfactants (Gupta and Gupta 2005). Copolymer stabilizations, for examples, polyacrylamide-polyacryl acid (PAM-co-PAA) and poly (2-acrylamido-2-methyl-1-propanesulfonic acid)-polyacryl acid (PAMPSco-PAA), also provide great stability of NPs in a high salinity environment (Xue et al. 2014). Xue et al. (2014) used a series of poly(AMPS-co-AA) random copolymers to graft onto iron oxide nanoparticles to provide colloidal stability in American Petroleum Institute (API) standard brine (i.e., 8 wt/wt% NaCl and 2 wt/wt% CaCl2, anhydrous basis). The ratio of AMPS/AA was varied from 1:1 to 20:1 to balance the requirements of particle stabilization, low adsorption/retention, and permanent attachment of stabilizer. Transport experiments using the functionalized NPs through a porous medium containing sand showed that more than 91% were transported without retention and the fraction increased by approximately 6% when the AMPS to AA ratio was increased from 1:1 to 3:1. Worthen et al. (2016) investigated the behavior of ligands to provide steric stabilization of silica nanoparticles in concentrated brine containing divalent cations. They observed that a suspension of silica NPs remain stable for 30 days at temperature up to 80 °C when the NPs were grafted with low molecular weight ligands, namely a diol ether, [3-(2,3-dihydroxypropoxy)propyl]-trimethoxysilane, and a zwitterionic sulfobetaine, 3-([dimethyl(3-trimethoxysilyl)propyl]ammonio) propane-1-sulfonate. The ligand remains solvated also at high salinity providing a steric barrier against the van der Waals force of attraction for NPs as small as 10 nm. Electrosteric stabilization using, *e.g.*, triblock polymers (Saleh et al. 2008), can provide the best resistance to changing electrolyte conditions likely to be encountered in real groundwater aquifers providing long transport distances, up to 100s of meters in unconsolidated sandy aquifers.

3.3 Nanoparticle Transport in Porous Media

Deposition of NPs in porous media can be reversible or irreversible. Contrary to solutes, which may be adsorbed onto the porous medium surface and reach equilibrium with the liquid phase, particles attached onto the medium surface may not be detached unless the physico-chemical conditions change (Ryan and Elimelech 1996). The recent review paper by Babakhani et al. (2017) reports in detail the various modeling approaches (such as, abstract, mechanistic, and continuum), and goes to the details of mass transport equations and the constitutive equations of the continuum-scale models. Here, we will focus on this latter mathematical approach. Generally, *continuum model* is stays for a system with a spatial scale of the model domain of $\gg 1$ cm, in contrast to micro/porescale models (1–100 μm).

Two models generally describe particle transport, namely, the clean-bed colloid filtration theory (CFT) model (Yao et al. 1971) and the modified version of the CFT model (MFT), which considers a limited capacity of the porous medium walls for particle attachment (Elimelech and O'Melia 1990). Both models consider that particles are removed from a suspension through interception, diffusion, and sedimentation. Following these theories, the general mass conservation equation for NP transport in porous media accounts for advection, dispersion, and retention processes and it writes as (He et al. 2009)

$$\frac{\partial}{\partial t}\left(\rho c_{NP} + \theta S\right) + \rho v \frac{\partial}{\partial x} c_{NP} - \rho D \frac{\partial^2}{\partial x^2} c_{NP} = 0 \qquad (3.17)$$

where c_{NP} is the nanoparticle concentration (mole/m³), S is the retained concentration of the particles (mole/kg of porous medium), σ is the porosity (–), θ is the bulk density of the porous medium (kg/m³), D is the hydrodynamic dispersion (m/s²), and v is the interstitial velocity (m/s). Both models consider that the attachment process is kinetic controlled and that the detachment is of first-order. However, the CFT model considers an unlimited capacity of the porous medium wall for which the constitutive equation of particle attachment writes as

$$\frac{\theta}{\rho}\frac{\partial}{\partial t} S = k_{att} c_{NP} - \frac{\theta}{\rho} k_{det} S \qquad (3.18)$$

where k_{att} and k_{det} are the rate constants of attachment and detachment, and

$$k_{att} = \frac{3(1-\rho)}{2d_{50}} \alpha \eta_0 V \qquad (3.19)$$

where d_{50} is the median grain size (m), α is the attachment efficiency (–), η_0 is the single collector contact efficiency and is a measure of the frequency of particle collisions with porous medium collector surfaces. Following Tufenkji and Elimelech (2004), η_0 can be described as the combination of three collision mechanisms, i.e.,

$$\eta_0 = 2.4 A_s^{1/3} N_R^{-0.081} N_{Pe}^{-0.715} N_{vdW}^{0.052} + 0.55 A_s N_R^{1.675} N_A^{0.125} + 0.22 N_R^{-0.24} N_G^{1.11} \quad (3.20)$$

where the first term accounts for the Brownian diffusion, the second for the interception process, and the third one for the gravitational force. In Equation (3.20), As is the Happel factor ($A_s = \dfrac{2(1-\gamma^5)}{2-3\gamma+3\gamma^5-2\gamma^6}$, with $\gamma = (1-\rho)^{1/3}$) which depends on the porosity of the medium (–); N_R is the aspect ratio

($N_R = d_{NP}/d_{50}$, with d_{NP} the representative NP diameter) (–); $N_{Pe} = vd_{50}/D_{NP}$ where D_{NP} is the bulk diffusion coefficient (described by Stokes-Einstein

equation, i.e., $D_{NP} = \dfrac{kT}{6\pi\mu r}$, with k the Boltzmann constant and μ the absolute fluid viscosity) (–), $N_{vdW} = \dfrac{H_a}{k_{BT}}$ is the van der Walls number; $N_A = \dfrac{H_a}{3\pi d_{NP}^2 v}$ is

the attraction number (–); and $N_G = \dfrac{d_{NP}(\rho_{NP}-\rho_w)g}{18\mu v}$ is the gravity number,

with ρ_{NP} and ρ_w the density of the NPs and the fluid, respectively.

In Equation (3.17), which is written for one-dimensional uniform flow field, such as column experiments, transverse dispersion is neglected and thus the dispersion coefficient has only the longitudinal component which is function of the pore water velocity, i.e.,

$$D = \alpha_L v + D^* \quad (3.21)$$

where α_L is the longitudinal dispersivity (m) and D^* is the molecular diffusion coefficient (m/s²). The value of α_L can be determined through the inversion of the transport model on tracer breakthrough curves. However, recent studies have shown the dependence of the estimated α_L on the particle size (Chrysikopoulos and Katzourakis 2015). This suggests that in order to determine a representative value of α_L, tests with NPs should be performed instead of using values from colloidal experiments. However, such an approach would raise question on the knowledge of other transport phenomena which should be also accounted for. This issue creates new research questions in the field of the transport of NPs in porous media. NP diffusion has been rarely considered as a distinct parameter in the continuum models. However, the high NP diffusion has been considered as the reason of deposition behavior of NPs in porous media (Phenrat et al. 2009).

Equation (3.17) is coupled with the constitutive equations describing the processes of retention and adsorption of the NPs within and onto the porous medium.

One of the major questions in the oil industry when nanoparticles are used is the processes undergone by the NPs. For example, do the particles aggregate and then deposit onto the porous medium reducing thereby the reservoir permeability? The size, the concentration, the shape, and the coating of NPs as well as the fluid velocity and composition, and the presence of clay particles are the major factors affecting the NP deposition onto the rock wall. The work by Esfandyari Bayat et al. (2015) investigates the effect of the rock mineralogy on the fate of engineered aluminum oxide (Al_2O_3), titanium dioxide (TiO_2), and silicon dioxide (SiO_2). They performed column-flood tests using cleaned limestone, dolomite, and quartz. Retention of the particles within the porous medium changed with the type of NPs and the mineralogy of the porous medium. Esfandyari Bayat et al. (2015) observed that NPs transport and retention through various porous media is strongly dependent on NP stability in suspension against deposition, NP surface charge as well as porous media surface charge and roughness. When NPs have the same charge of the porous medium surface, they can be easily transported contrary to cases when NPs have opposite charge with which the medium and their retention is much more significant. Consequently, NPs made of Al_2O_3 and TiO_2 NPs tend to be strongly retained in quartz-sand porous medium. Stability of the nanoparticle suspension can then reduce even more the mobility of NPs.

For example, Esfandyari Bayat et al. (2015) observed that TiO_2-NPs recoveries through the carbonates porous media were lower than Al_2O_3-NPs recoveries. This was ascribed to the lower stability of TiO_2-NPs in DI water as compared with Al_2O_3- NPs. The formation of NP-clusters favors settling and therefore deposition of NPs within the porous medium. One important factor in the deposition of NPs in porous media is the size of an individual NP and of a formed cluster. The larger the size, the smaller the effect of Brownian motion, which favors the collisions between the particle and the porous medium wall. Therefore, smaller deposition of the particle will occur during transport (Nelson and Ginn 2011).

3.3.1 Code for Nanoparticle Transport Through Porous Media

In this section, a code developed in MATLAB® (The MathWorks 2015) to describe the transport of NPs is reported (Figures 3.8 and 3.9). The model was used to reproduce the data of experiment #7 in the work by Kim et al. (2015). Through minimization of the least-square difference between the measurements and the simulated concentrations of NPs, the model parameters, namely, S_{max}. k_{det}, and α_{PC} can be estimated (Figure 3.10).

```
% Nanoparticle transport
% --------------------

clear all
close all
clc

% Load data
% ---------
data=load('KIM15.txt');

% Geometry and conditions
% -----------------------
L=1; % m
T =273+25; % K, temperature

v=7.1; % m/day, flow velocity
    v=v/(60*60*24); % m/s
D=1e-8; % m2/s, longitudinal hydrodynamic dispersion
rho=0.35; % -, porosity
theta=1;
PVmax=5;  % -, pore volume
tmax=PVmax*L/v; % s

cL=1; % m, injected conditions
cR=0; % mole/kg, initial condition

% Discretization
% -------------
Nx =100;
Nt=100;
Dx=L/Nx;
Dt=0.05*Dx/v;
CFL1=Dt*v/Dx;
CFL2=D*Dt/Dx^2;
time=[0:Dt:tmax];
Nt=length(time);
xvec=linspace(0,L,Nx);
i=0;
t=0;

c=zeros(Nx,1);
c(1,1)=cL;
cnew=c;

%% Injection 1
% Nanoparticles dynamic
% --------------------
dp=5*1e-9; % m
dc=350*1e-6; % m
eta0=feta0(rho,dp,dc,v,T);
alphadc=3.77*10^(-7)*0.3;
katt=3/2*(1-rho)*v/dc*alphadc*eta0; % 1/hr
kdet=1e-15; % 1/h
```

FIGURE 3.8
MATLAB® code for coreflood data matching.

```
Smax=0.68; % g/kg
Sat=zeros(Nx,1);
Sat(1,1)=katt*cL/(theta/rho*kdet);

for i=1:length(time)
    cnew(2:Nx-1,1)=c(2:Nx-1,1)-CFL1*(c(2:Nx-1,1)-c(1:Nx-2,1))+CFL2*(c(3:Nx,1)-2*c(2:Nx-1,1)+c(1:Nx-2,1))-...
    katt*c(2:Nx-1,1)+theta/rho*kdet*Sat(2:Nx-1,1);
    cnew(Nx,1)=c(Nx,1)-CFL1*(c(Nx,1)-c(Nx-1,1))+CFL2*(cR-2*cR+c(Nx-1,1))-katt*c(Nx,1)+theta/rho*kdet*Sat(Nx,1);
    c=cnew;
    Satnew(1:Nx,1)=(katt*c(1:Nx,1)-theta/rho*kdet*Sat(1:Nx,1))*Dt*rho/theta+Sat(1:Nx,1);
    Sat=Satnew;
    cend(i)=cnew(Nx);
end

% Plotting
% --------
figure(1)
PV=v*time./L;
plot(data(:,1),data(:,2),'ok','MarkerSize',8,'MarkerFaceColor','w'); hold on
plot(PV,cend./cL,'-k','LineWidth',1.5);
ylabel('c/c_{inj}, -');
xlabel('PV, -');

%% Injection 2
PVmax=9;
tmax2=PVmax*L/v; % s
time2=[tmax:Dt:tmax2];

cL=0; % m, injected conditions
cR=1; % mole/kg, initial condition
c(1,1)=cL;
cnew=c;
Sat(1,1)=katt*cL/(theta/rho*kdet);
Satnew=Sat;

for j=1:length(time2)
    cnew(2:Nx-1,1)=c(2:Nx-1,1)-CFL1*(c(2:Nx-1,1)-c(1:Nx-2,1))+CFL2*(c(3:Nx,1)-2*c(2:Nx-1,1)+c(1:Nx-2,1))-...
    katt*c(2:Nx-1,1)+theta/rho*kdet*Sat(2:Nx-1,1);
    cnew(Nx,1)=c(Nx,1)-CFL1*(c(Nx,1)-c(Nx-1,1))+CFL2*(cR-2*cR+c(Nx-1,1))-katt*c(Nx,1)+theta/rho*kdet*Sat(Nx,1);
    c=cnew;
    Satnew(1:Nx,1)=(katt*c(1:Nx,1)-theta/rho*kdet*Sat(1:Nx,1))*Dt*rho/theta+Sat(1:Nx,1);
    Sat=Satnew;
    cend2(j)=cnew(Nx);
end

% Plotting
% --------
figure(1)
PV2=v*time2./L;
plot(PV2,cend2./cR,'--k','LineWidth',1.5);
ylabel('c/c_{inj}, -','FontSize',12);
xlabel('PV, -','FontSize',12); axis square;
ylim([0 1.01]);
xlim([0 8])
```

FIGURE. 3.9
MATLAB® code for coreflood data matching (cont. from Figure 3.7).

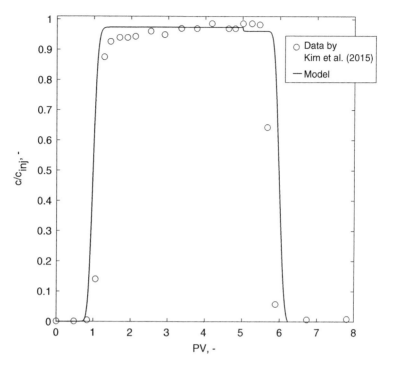

FIGURE. 3.10
Modeling the breakthrough dataset.

3.3.2 Nanoparticle Field-Scale Transport Simulation

In field applications, the mobility of the silica nanoparticles could be enhanced under the high flow rate. At high flow velocity, the hydrodynamic force is larger and this helps to prevent the desposition of aggregates of NPs formed. Kim et al. (2015) measured the transport of pre-aggregated silica NPs through a column. They observed that in experiments at the same conditions but at different velocities, the concentration of NPs measured at the outlet was initially much lower than in the case of no pre-aggregated NPs, but later increased if large average flow velocity (>10 m/day) was applied. The larger hydrodynamic forces present at large velocity might have facilitated the breakage of the NP clusters (Li et al. 2008).

References

Allen, E., Smith, P., Henshaw, J., and Morgan, M. (2001) *A Review of Particle Agglomeration*. US Department of Energy.

Atmuri, A. K., Henson, M. A., and Bhatia, S. R. (2013) A population balance equation model to predict regimes of controlled nanoparticle aggregation. *Colloids Surf. A: Physicochem. Eng. Asp.*, 436, 325–332.

Babakhani, P., Bridge, J., Doong, R., and Phenrat, T. (2017) Continuum-based models and concepts for the transport of nanoparticles in saturated porous media: A state-of-the-science review. *Adv. Colloid Interface Sci.*, 246, 75–104.

Chrysikopoulos, C. V., and Katzourakis, V. E. (2015) Colloid particle size-dependent dispersivity. *Water Resour. Res.*, 51(6), 4668–4683.

Coalson, R. D., Eskandari Nasrabad, A., Jasnow, D., and Zilman, A. (2015) A polymer-brush-based nanovalve controlled by nanoparticle additives: Design principles. *J. Phys. Chem. B*, 119(35), 11858–11866.

Derjaguin, B., and Landau, L. (1993) Theory of the stability of strongly charged lyophobic sols and of the adhesion of strongly charged particles in solutions of electrolytes. *Prog. Surf. Sci.*, 43(1), 30–59.

de Vicente, J., Delgado, A. V., Plaza, R. C., Durán, J. D. G., and González-Caballero, F. (2000) Stability of cobalt ferrite colloidal particles. Effect of pH and applied magnetic fields. *Langmuir*, 16(21), 7954–7961.

Elimelech, M., and O'Melia, C. R. (1990) Kinetics of deposition of colloidal particles in porous media. *Environ. Sci. Technol.*, 24(10), 1528–1536.

Ellerstein, S., and Ullman, R. (1961) The adsorption of polymethyl methacrylate from solution. *J. Polym. Sci.*, 55(161), 123–135.

Esfandyari Bayat, A., Junin, R., Shamshirband, S., and Tong Chong, W. (2015) Transport and retention of engineered Al_2O_3, TiO_2, and SiO_2 nanoparticles through various sedimentary rocks. *Sci. Rep.*, 5, 14264.

Gregory, J. (1977) Van der Waals interaction between mica surfaces: Comparison of theory and experiment. *J. Chem. Soc., Faraday Trans. 1: Phys. Chem. Condens. Phases*, 73, 1983–1987.

Gregory, J. (1981) Approximate expressions for retarded van der Waals interaction. *J. Colloid Interface Sci.*, 83(1), 138–145.

Gupta, A. K., and Gupta, M. (2005) Synthesis and surface engineering of iron oxide nanoparticles for biomedical applications. *Biomaterials*, 26(18), 3995–4021.

Hahn, M. W., Abadzic, D., and O'Melia, C. R. (2004) Aquasols: On the role of secondary minima. *Environ. Sci. Technol.*, 38(22), 5915–5924.

He, F., Zhang, M., Qian, T., and Zhao, D. (2009) Transport of carboxymethyl cellulose stabilized iron nanoparticles in porous media: Column experiments and modeling. *J. Colloid Interface Sci.*, 334(1), 96–102.

Hoek, E. M. V., and Agarwal, G. K. (2006) Extended DLVO interactions between spherical particles and rough surfaces. *J. Colloid Interface Sci.*, 298(1), 50–58.

Hogg, R., Healy, T. W., and Fuerstenau, D. W. (1966) Mutual coagulation of colloidal dispersions. *Trans. Faraday Soc.*, 62, 1638–1651.

Hotze, E. M., Phenrat, T., and Lowry, G. V. (2010) Nanoparticle aggregation: Challenges to understanding transport and reactivity in the environment. *J. Environ. Qual.*, 39(6), 1909–1924.

Hua Yun-Feng, Z. L.-X. (2017) Entropy forces of nanoparticles in self-propelled systems. *Acta Phys. Sin.*, 66(19), 190701.

Hunter, R. J., and White, L. R. (1987) *Foundations of Colloid Science*, Clarendon Press.

Israelachvili, J. N. (1992) *Intermolecular and Surface Forces*, 2nd ed. Academic Press, Inc.

Kim, I., Taghavy, A., DiCarlo, D., and Huh, C. (2015) Aggregation of silica nanoparticles and its impact on particle mobility under high-salinity conditions. *J. Pet. Sci. Eng.*, 133, 376–383.

Kobayashi, M., Juillerat, F., Galletto, P., Bowen, P., and Borkovec, M. (2005) Aggregation and charging of colloidal silica particles: Effect of particle size. *Langmuir*, 21(13), 5761–5769.

Kumar, S., and Ramkrishna, D. (1997) On the solution of population balance equations by discretization—III. Nucleation, growth and aggregation of particles. *Chem. Eng. Sci.*, 52(24), 4659–4679.

Li, Y., Wang, Y., Pennell, K. D., and Abriola, L. M. (2008) Investigation of the transport and deposition of fullerene (C60) nanoparticles in quartz sands under varying flow conditions. *Environ. Sci. Technol.*, 42(19), 7174–7180.

The MathWorks. (2015) Matlab 2014b. Available from www.mathworks.com/.

Mersmann, A. (2001) *Crystallization Technology Handbook*. Taylor & Francis. Available from https://books.google.com/books?id=BVJdDw59lDcC. Last access: 07/30/2018.

Napper, D. H. (1983) *Polymeric Stabilization of Colloidal Dispersions*. Academic Press Incorporated. Available from https://books.google.com/books?id=FQTwAA AAMAAJ. Last access: 07/30/2018.

Nelson, K. E., and Ginn, T. R. (2011) New collector efficiency equation for colloid filtration in both natural and engineered flow conditions. *Water Resour. Res.*, 47(5), 1–17.

Phenrat, T., Kim, H.-J., Fagerlund, F., Illangasekare, T., Tilton, R. D., and Lowry, G. V. (2009) Particle size distribution, concentration, and magnetic attraction affect transport of polymer-modified Fe0 nanoparticles in sand columns. *Environ. Sci. Technol.*, 43(13), 5079–5085.

Phenrat, T., Saleh, N., Sirk, K., Kim, H.-J., Tilton, R. D., and Lowry, G. V. (2008) Stabilization of aqueous nanoscale zerovalent iron dispersions by anionic polyelectrolytes: Adsorbed anionic polyelectrolyte layer properties and their effect on aggregation and sedimentation. *J. Nanopart. Res.*, 10(5), 795–814.

Piirma, I. (1992) *Polymeric Surfactants*. CRC Press.

Prieve, D. C., and Ruckenstein, E. (1976) Rates of deposition of brownian particles calculated by lumping interaction forces into a boundary condition. *J. Colloid Interface Sci.*, 57(3), 547–550.

Randolph, A. D., and Larson, M. A. (1988) *Theory of Particulate Processes*, 2nd ed. Academic Press.

Ryan, J. N., and Elimelech, M. (1996) Colloid mobilization and transport in groundwater. *Colloids Surf. A: Physicochem. Eng. Asp.*, 107, 1–56.

Sader, J. E., Carnie, S. L., and Chan, D. Y. C. (1995) Accurate analytic formulas for the double-layer interaction between spheres. *J. Colloid Interface Sci.*, 171(1), 46–54.

Saleh, N., Kim, H.-J., Phenrat, T., Matyjaszewski, K., Tilton, R. D., and Lowry, G. V. (2008) Ionic strength and composition affect the mobility of surface-modified Fe0 nanoparticles in water-saturated sand columns. *Environ. Sci. Technol.*, 42(9), 3349–3355.

Sato, T., and Ruch, R. (1980) *Stabilization of Colloidal Dispersions by Polymer Adsorption*. Dekker.

Shekar, S., Smith, A. J., Menz, W. J., Sander, M., and Kraft, M. (2012) A multidimensional population balance model to describe the aerosol synthesis of silica nanoparticles. *J. Aerosol Sci.*, 44, 83–98.

Shi, J. (2002) *Steric Stabilization*. The Ohio State University.

Stille, J. K. (1962) *Introduction to Polymer Chemistry*. John Wiley and Sons, Inc., New York, NY, USA.

Tufenkji, N., and Elimelech, M. (2004) Correlation equation for predicting single-collector efficiency in physicochemical filtration in saturated porous media. *Environ. Sci. Technol.*, 38(2), 529–536.

van Oss, C. J. (1993) Acid-base interfacial interactions in aqueous media. *Colloids Surf. A: Physicochem. Eng. Asp.*, 78, 1–49.

Verwey, E. J. W. (1948) *Theory of the Stability of Lyophobic Colloids*. Elsevier Publishing Company, Inc., Amsterdam.

Vollmer, U., and Raisch, J. (2006) Control of batch crystallization—A system inversion approach. *Chem. Eng. Process.: Process Intensif.*, 45(10), 874–885.

Wang, Q., Prigiobbe, V., Huh, C., and Bryant, S. L. (2017) Alkaline earth element adsorption onto PAA-coated magnetic nanoparticles. *Energies*, 10(2), 223.

Wiesner, M. R., and Bottero, J.-Y. (2007) *Environmental Nanotechnology: Applications and Impacts of Nanomaterials*. The McGraw-Hill Companies.

Worthen, A. J., Tran, V., Cornell, K. A., Truskett, T. M., and Johnston, K. P. (2016) Steric stabilization of nanoparticles with grafted low molecular weight ligands in highly concentrated brines including divalent ions. *Soft Matter*, 12(7), 2025–2039.

Xue, Z., Foster, E., Wang, Y., Nayak, S., Cheng, V., Ngo, V. W., Pennell, K. D., Bielawski, C. W., and Johnston, K. P. (2014) Effect of grafted copolymer composition on iron oxide nanoparticle stability and transport in porous media at high salinity. *Energy & Fuels*, 28(6), 3655–3665.

Yao, K.-M., Habibian, M. T., and O'Melia, C. R. (1971) Water and waste water filtration. Concepts and applications. *Environ. Sci. Technol.*, 5(11), 1105–1112.

4

Nanoparticles at Fluid Interfaces

4.1 Introduction

Ramsden (1904) and Pickering (1907) made the first observations of the remarkable ability of particles in stabilizing emulsions. For the first time, it was noticed that the accumulation of particles at the interfaces of fluids (liquid–gas or liquid–liquid) could allow separation of valuable products from a suspension, *e.g.*, applications of colloidal particles in froth flotation processes for the extraction of precious minerals (Gaudin 1957). Experiments have shown that the ability of particles to stabilize an interface is strongly dependent on their size and surface properties (Binks and Horozov 2006). In particular, nanoparticles (NPs) can stabilize droplets and bubbles for very long time, up to years, making them an attractive alternative to conventional surfactants (Binks 2002) in various engineering applications, *e.g.*, catalysis, optics, and biomedical and petroleum engineering. Therefore, the understanding of the dynamic adsorption of nanoparticles at fluid interfaces is important for the selection of the optimal type of nanoparticle and suspension composition to reach the desired process performance.

This chapter provides fundamental physico-chemical aspects of NPs at interfaces in combination with the models available to describe the adsorption process.

4.2 Mechanism of Adsorption of Nanoparticles at Fluid-Fluid Interfaces

Nanoparticles, similar to surfactant molecules, can spontaneously accumulate at the interface (adsorption) between two immiscible fluids providing outstanding stability to the interface. It is their small size that makes their behavior at the interface to depart from that of colloidal particles.

A key parameter for solid particles at fluid interfaces is the three-phase contact angle θ, which is the angle between the tangents to the solid surface and the liquid–liquid (or liquid–gas) interface measured through one of the liquids in each point of the three-phase contact line where the solid and two fluids meet (Figure 4.1) (Binks and Horozov 2006). In the case of an oil-water emulsion, the contact angle depends on the surface free energies (interfacial tensions) at the particle–water γ_{pw}, particle–oil γ_{po}, and oil–water γ_{ow}, interface according to Young's equation,

$$\cos\theta = \frac{\gamma_{po} - \gamma_{pw}}{\gamma_{ow}} \tag{4.1}$$

However, one of the main aspects that distinguishes the behavior of colloidal particles and nanoparticles at the interface of fluids is the effect of the thermal energy. In the case of nanoparticles, thermal energy can cause spatial fluctuations of the particles, comparable with the interfacial energy. The resulting energy balance results in a weak interfacial segregation of nanoparticles (Lin et al. 2003). Therefore, unlike micrometer-sized particles that are strongly held at the interface, smaller nanoparticles at an interface may be displaced with larger ones. It is possible however to modify the wettability of the particles through surface modification (grafting or functionalization) to favor adsorption onto the interface and, therefore, inhibit displacement.

There are two classes of particles that are used to stabilize interfaces between two fluids: homogeneous and Janus nanoparticles. The former ones are characterized by a surface of homogeneous chemical composition while the latter ones have an amphiphilic surface (i.e., possessing both hydrophilic and hydrophobic surface properties). Hydrophilic particles are preferentially wet by water ($\gamma_{po} > \gamma_{pw}$ in Equation 4.1), therefore $0° < \theta < 90°$, while hydrophobic particles are preferentially wet by oil in the case of an emulsion and air in the case of a foam ($\gamma_{po} < \gamma_{pw}$ in Equation 4.1), hence $90° < \theta < 180°$. Adsorption of nanoparticles with amphiphilic surface decreases the interfacial tension of a fluid interface and leads to the stabilization of foams and emulsions. As a particle adsorbs, the fluid-fluid interfacial area is replaced by

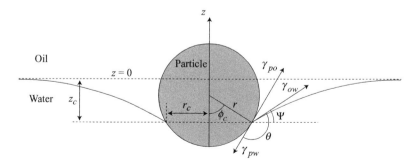

FIGURE 4.1
Spherical particle at equilibrium at the oil–water interface. The three-phase contact line with radius r_c is depressed at depth z_c below the zero level. From Binks and Horozov (2006).

particle-fluid interfaces. This area replacement causes a decrease of the free energy of the interface, ΔE_{ads} (per particle). In the case of a particle adsorbing at the oil-water/air-water interface, ΔE_{ads} writes (Binks 2002)

$$\Delta E_{ads} = -\pi r_{NP}^2 \gamma_{ij} \left(1 \pm \cos\theta_{ij}\right) \tag{4.2}$$

where

r_{NP} is the radius of the NP
γ_{ij} is the interfacial tension at the fluid-fluid interface
θ_{ij} is the contact angle at that interface.

The decrease in interfacial tension caused by the adsorption of a species to a fluid interface is a macroscopic mechanical response that is easily accessible experimentally. To compare the performance of various particles is of primary importance in measuring the interfacial tension. Measurements of interfacial tension at equilibrium state or during transient state have been employed (Pugh 2016), *e.g.*, the pendant drop method to determine the static surface tension (Song and Springer 1996; Zhang et al. 2017) and the bubble pressure to determine the dynamic surface tension (Du et al. 2010; Hua et al. 2018). These methods provide measurements that can be described employed in Equation (4.2).

4.2.1 Modeling Surface Tension at Stationary State

Functionalized particles with strong affinity to the interface, i.e., a wetting angle not extremely far from 90°, create favorable conditions for adsorption. Zhang et al. (2017) developed a simple model to predict interfacial tension under the assumption of negligible particle-particle interactions. The effective interfacial energy writes,

$$E^* = \gamma_{ij} A + N_s \Delta E_{ads} \tag{4.3}$$

where

A is the interfacial area
N_s is the number of adsorbed particles in the interface.

Considering the thermodynamic definition of surface tension as the change in free energy per unit surface area, the effective interfacial tension of the particle covered interface is given by,

$$\gamma_{ij^*} = \frac{\gamma_{ij} + N_s \Delta E_{ads}}{A} \tag{4.4}$$

Combining Equations (4.2) and (4.4) yields the effective interfacial tension (γ^*) of the interface between two fluids, *e.g.*, oil-water/air-water, in the presence of particles,

$$\gamma_{ij}^* = \gamma_{ij} \left[1 - \phi(1 - \cos\theta_{ij})\right] \tag{4.5}$$

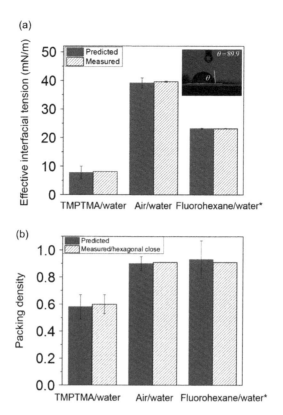

FIGURE 4.2
Comparison between measurements and prediction of (a) surface tension and (b) packing density. The inset reports an image of a trimethylolpropane trimethacrylate drop on a flat ethyl cellulose particle. Adapted with permission from Zhang et al. (2017). Copyright 2017 American Chemical Society.

where $\phi = N_s^2 / A$ is the packing density (area fraction) of the particles adsorbed at the interface. Figure 4.2 shows the comparison between measurements by Du et al. (2010) and the prediction using Equation (4.5) for the surface tension and packing density.

4.2.2 Modeling the Evolution of the Surface Tension

In contrast to surfactant, NPs because of their bigger size have smaller diffusivity that makes their adsorption on interfaces slower (Bizmark et al. 2014). Hua et al. (2018) proposed a diffusion-limited adsorption model that was validated with experiments run using NPs of average size equal to 5 and 10 nm to be adsorbed onto a water–toluene interface. The model combines the Ward-Tordai equations, commonly employed to describe the dynamics of surfactant adsorption at a fluid interface, combined with a Frumkin adsorption isotherm to model the diffusion-limited adsorption of NPs and with a wetting equation of state to characterize the dynamic interfacial tension during the adsorption of NPs to the interface. A schematic of the adsorption mechanisms is shown in Figure 4.3. The adsorbed concentration (z) of nanoparticles onto a droplet, for example, during the pendant drop test, can be described by the Frumkin adsorption isotherm, which incorporates non-ideal interactions between adsorbed species within the interface (Hua et al. 2018),

$$z = a \frac{\theta(t)}{1-\theta(t)} e^{K\theta(t)} \tag{4.6}$$

where

$\theta(t) = \Gamma(t)/\Gamma_\infty$	is the dynamic relative coverage of the nanoparticle with $\Gamma(t)$ the time-dependent surface coverage
Γ_∞	the maximum surface coverage at steady-state
a	is the affinity between the nanoparticle and the interface
K	accounts for the interactions between the nanoparticle within the interface relative to the thermal energy
kT	(k is the Boltzmann constant and T is the temperature)

Positive values of K represent net repulsion, and negative values represent attractive interactions.

The dynamic relative coverage is described by the Ward-Tordai equation (Hua et al. 2018),

$$\Gamma(t) = \sqrt{\frac{D}{\pi}} \left\{ 2c\sqrt{t} - \int_0^t \frac{c}{\sqrt{t-\tau}} d\tau \right\} + \frac{D}{x} \left\{ ct - \int_0^t c(\tau)d\tau \right\} \tag{4.7}$$

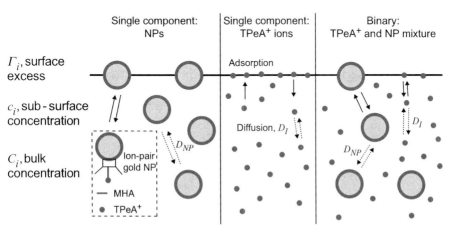

FIGURE 4.3
Schematic of adsorption dynamics of solely nanoparticles, surfactant, and nanoparticles with surfactant. The highlighted processes are: (1) adsorption and desorption indicated by the solid-line up and solid-line down arrows, respectively, between the subsurface region and the interface; (2) diffusion indicated by dotted-line arrows from bulk phase to the sub-surface. From Hua et al. (2018).

where
- c is the bulk concentration of the adsorbing species
- r is the radius of the pendent droplet
- D is the diffusivity of the adsorbing species, which can be estimated using Stokes-Einstein equation

$$D = \frac{kT}{6\pi\mu r_{NP}} \tag{4.8}$$

where
- μ is the fluid viscosity
- r_{NP} is the radius of the nanoparticle.

Generally, during the tests of surface tension determination, the surface pressure is measured, which writes as,

$$\gamma(t) = \gamma_{ow} - \Pi_{NP}(t) \tag{4.9}$$

where $\Pi_{NP}(t)$ is given by,

$$\Pi_{NP}(t) = \Delta E_{ads}z + kT\Gamma_{\infty}\left[\ln(1-\theta) - 0.5K\theta^2\right] \tag{4.10}$$

By fitting experimental data of surface tension, the adsorbed concentration of nanoparticles at the fluid-fluid interface can be determined. Figure 4.4 shows the comparison between measurements and predictions using the model outlined above.

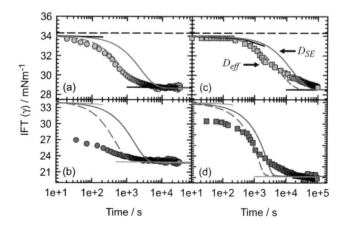

FIGURE 4.4
Comparison between measurements (points) and prediction (curves) of surface tension; 5 nm NP with concentration of 3.5×10^{11} 1/mL [(a) circles] and 1.5×10^{12} 1/mL [(b) circles] in 5 mM pH 11.0 and for 10 nm NP with concentration of 3.7×10^{11} 1/mL [(c) squares], and 9.0×10^{12} 1/mL [(d) squares]. Modified after Hua et al. (2018).

4.3 Stability of Thin Film between Bubbles within a Foam

Foam is injected into the subsurface to improve the gas mobility as it increases the effective gas viscosity and allows the gas to divert toward low permeability zones (Conn et al. 2014; Lake et al. 2014; Li et al. 2010). It has been successfully employed in gas-injection operations such as enhanced oil recovery (EOR) and remediation of contaminated sites to overcome problems of gravity segregation, viscous fingering, and early gas breakthrough (Hirasaki et al. 2000; Ransohoff and Radke 1988; Rossen 1996). In a wet foam, the gas phase is segregated into bubbles separated by thin films, approximately 30 nm thick, called lamellae (Figure 4.5). The number density of foam bubbles per unit volume of flowing gas gives the foam its texture (n_f) and its rheological properties (Kovscek and Radke 1994; Lake et al. 2014). The pressure difference (ΔP) across a lamella (i.e., the pressure difference within the two gas bubbles) resists the movement of the bubbles and it is given by

$$\Delta P = \frac{4\sigma_{l,g}}{r} \tag{4.11}$$

where

$\sigma_{l,g}$ is the interfacial tension between the liquid- and gas-phase
r is the radius of the mean curvature of the lamella.

Foam is thermodynamically metastable and its spontaneous decay can be mitigated using surface-active materials such as surfactants and/or particles, which make the foam stable within the time of an operation (Aronson et al. 1994; Elhag et al. 2018). Usually, surfactants are employed to stabilize bubbles in a foam. They adsorb on each of the two fluid-fluid interfaces of the lamella (Figure 4.5c), repelling the two interfaces by either electrical double-layer

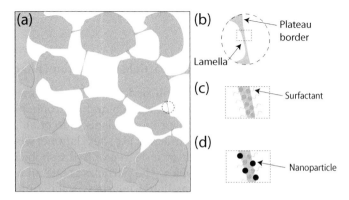

FIGURE 4.5
Schematic view of: (a) a foam within a porous medium; (b) lamella; (c) lamella with surfactant; and (d) lamella with surfactant and particles.

forces, steric effects, or both (Kovscek and Radke 1994; Lake et al. 2014; Pugh 2016). This repulsion is the "disjoining pressure." However, one of the major disadvantages is that surfactants adsorb onto the surface of the porous medium (Ma et al. 2013; Mannhardt et al. 1992) and also they are unstable at high temperature and salinity (Mannhardt et al. 1993; Pugh 2016; Zajac et al. 1997), therefore, corrupting the foam.

4.3.1 Stabilizing a Foam with Solely Nanoparticles

Particles at the nano- and micro-size have been observed to notably improve the stability of a foam by providing a barrier to gas mass transfer across the bubble wall and increasing the critical for bubble rupture (Aveyard et al. 2000; Azmin et al. 2012; Binks 2002; Binks et al. 2008; Carn et al. 2009; Cervantes Martinez et al. 2008; Kaptay 2006).

A comprehensive recent review by Yekeen et al. (2018) summarizes the achievements on understanding of the mechanism of foam/emulsion generation and stability in the presence of particles. Particle size, shape, hydrophobicity/wettability, and concentration, capillary attraction, and solution composition are the main factors for foam stabilization (Pugh 2016; Sun et al. 2016; Worthen et al. 2013b). For instance, nanoparticles can adsorb at the gas-bubble interface irreversibly, making the films very stable against rupture and inhibiting the coarsening (Binks and Horozov 2005). Moreover, Yu et al. (2012) showed that when nanoparticles are used to stabilize a foam, they have negligible adsorption onto sandstone, limestone, and dolomite rocks. Nanoparticles are not mechanically trapped within the pore bodies and pore throats, either. So, they do not cause plugging and blocking of the porous medium (Eftekhari et al. 2015). Therefore, nanoparticles are a promising alternative to surfactants in creating a foam, in particular in harsh environments (Alargova et al. 2004; Lotfollahi et al. 2017; Yekeen et al. 2018). But, to allow adsorption onto a gas-bubble (Liu et al. 2009; Pugh 2016; Worthen et al. 2013a), their surface must be modified through grafting techniques or by adding surfactants to the solution.

4.3.2 Addition of Surfactant for Nanoparticle Adsorption Enhancement

Recently, several studies have been published regarding the interaction of surfactants, particles, and bubbles. Ion exchange and hydrophobic interactions are considered to be the two major mechanisms (Sun et al. 2016). The study by Liu et al. (2009) shows that surfactant (*e.g.*, hexylamine) adsorbs onto particles (*e.g.*, Laponite), providing them a degree of hydrophobicity, which favors their adsorption onto gas bubbles. The particles create an armor that reduces gas diffusion through the gas-liquid interface, stabilizing the foam (Figure 4.5d). Flocculation of the particles between the armored bubbles and the un-adsorbed particles in the surrounding continuous phase, and the presence of particles packing at the plateau border, is actually beneficial for the foam, as it reduces water drainage and increases the interface rigidity, surface viscosity, and the strength of the foam structure (Liu et al. 2010). Sun et al. (2016) observed that the most stable foams are produced from a dispersion in which the silica particles in the presence of a surfactant are most aggregated. Similar observations were reported by other authors (*e.g.*, Carl et al. 2015; Du et al. 2003; Espinoza et al. 2010; Kim et al. 2016; Sun et al. 2015; Varade et al. 2011; Worthen et al. 2013b; Xu et al. 2018; Yu et al. 2014). In all cases, a synergistic effect of surfactant and particle on foam stability was reported.

For more detailed description and relevant literature on the nanoparticle-stabilized emulsions, Section 10.4.1, "Pickering Emulsion Re-Visited," is referred to.

References

Alargova, R. G., Warhadpande, D. S., Paunov, V. N., and Velev, O. D. (2004) Foam superstabilization by polymer microrods. *Langmuir*, 20(24), 10371–10374.

Aronson, A. S., Bergeron, V., Fagan, M. E., and Radke, C. J. (1994) The influence of disjoining pressure on foam stability and flow in porous media. *Colloids Surf. A: Physicochem. Eng. Asp.*, 83(2), 109–120.

Aveyard, R., Clint, J. H., and Nees, D. (2000) Small solid particles and liquid lenses at fluid/fluid interfaces. *Colloid Polym. Sci.*, 278(2), 155–163.

Azmin, M., Mohamedi, G., Edirisinghe, M., and Stride, E. P. (2012) Dissolution of coated microbubbles: The effect of nanoparticles and surfactant concentration. *Mater. Sci. Eng. C*, 32(8), 2654–2658.

Binks, B. P. (2002) Particles as surfactants—similarities and differences. *Curr. Opin. Colloid Interface Sci.*, 7(1), 21–41.

Binks, B. P., and Horozov, T. S. (Eds.) (2006) *Colloidal Particles at Liquid Interfaces.* Cambridge University Press, Cambridge, UK.

Binks, B. P., and Horozov, T. S. (2005) Aqueous foams stabilized solely by silica nanoparticles. *Angew. Chem. Int. Ed.*, 44(24), 3722–3725.

Binks, B. P., Kirkland, M., and Rodrigues, J. A. (2008) Origin of stabilisation of aqueous foams in nanoparticle–surfactant mixtures. *Soft Matter*, 4(12), 2373–2382.

Bizmark, N., Ioannidis, M. A., and Henneke, D. E. (2014) Irreversible adsorption-driven assembly of nanoparticles at fluid interfaces revealed by a dynamic surface tension probe. *Langmuir*, 30(3), 710–717.

Carl, A., Bannuscher, A., and von Klitzing, R. (2015) Particle stabilized aqueous foams at different length scales: Synergy between silica particles and alkylamines. *Langmuir*, 31(5), 1615–1622.

Carn, F., Colin, A., Pitois, O., Vignes-Adler, M., and Backov, R. (2009) Foam drainage in the presence of nanoparticle–Surfactant mixtures. *Langmuir*, 25(14), 7847–7856.

Cervantes Martinez, A., Rio, E., Delon, G., Saint-Jalmes, A., Langevin, D., and Binks, B. P. (2008) On the origin of the remarkable stability of aqueous foams stabilised by nanoparticles: Link with microscopic surface properties. *Soft Matter*, 4(7), 1531–1535.

Conn, C. A., Ma, K., Hirasaki, G. J., and Biswal, S. L. (2014) Visualizing oil displacement with foam in a microfluidic device with permeability contrast. *Lab Chip*, 14(20), 3968–3977.

Du, K., Glogowski, E., Emrick, T., Russell, T. P., and Dinsmore, A. D. (2010) Adsorption energy of nano- and microparticles at liquid–liquid interfaces. *Langmuir*, 26(15), 12518–12522.

Du, Z., Bilbao-Montoya, M. P., Binks, B. P., Dickinson, E., Ettelaie, R., and Murray, B. S. (2003) Outstanding stability of particle-stabilized bubbles. *Langmuir*, 19(8), 3106–3108.

Eftekhari, A.A., Krastev, R., and Farajzadeh, R. (2015) Foam stabilized by fly ash nanoparticles for enhancing oil recovery. *Ind Eng Chem Res*, 54(50), 12482–12491.

Elhag, A.S., Da, C., Chen, Y., Mukherjee, N., Noguera, J.A., Alzobaidi, S., Reddy, P.P., AlSumaiti, A.M., Hirasaki, G.J., Biswal, S.L., Nguyen, Q.P., and Johnston, K.P. (2018) Viscoelastic diamine surfactant for stable carbon dioxide/water foams over a wide range in salinity and temperature. *J Colloid Interf Sci*, 522, 151–162.

Espinoza, D.A., Caldelas, F.M., Johnston, K.P., Bryant, S.L., and Huh, C. (2010) Nanoparticle-Stabilized Supercritical CO_2 Foams for Potential Mobility Control Applications. SPE Improved Oil Recovery Symposium, Apr. 24–28, Tulsa, OK, USA. p. 13. doi:10.2118/129925-MS.

Gaudin, A.M. (1957) *Flotation*, 2nd edn. McGraw Hill Inc., New York, NY, USA.

Hirasaki, G., Miller, C., Szafranski, R., Tanzil, D.B., Lawson, J., Meinardus, H.W., Jin, M.T., Londergan, J., Jackson, R.A., Pope, G., and Wade, W.H. (2000) Field demonstration of the surfactant/foam process for aquifer remediation. In *NAPL Removal: Surfactants, Foams, and Microemulsions*. Lewis Publishers, Boca Raton. pp. 3–163.

Hua, X., Frechette, J., and Bevan, M.A. (2018) Nanoparticle adsorption dynamics at fluid interfaces. *Soft Matter*, 14(19), 3818–3828.

Kaptay, G. (2006) On the equation of the maximum capillary pressure induced by solid particles to stabilize emulsions and foams and on the emulsion stability diagrams. *Colloid Surface A*, 282–283, 387–401.

Kim, I., Worthen, A.J., Johnston, K.P., DiCarlo, D.A., and Huh, C. (2016) Size-dependent properties of silica nanoparticles for Pickering stabilization of emulsions and foams. *J Nanopart Res*, 18(4), 82.

Kovscek, A.R., and Radke, C.J. (1994) Fundamentals of foam transport in porous media. In *Foams: Fundamentals and Applications in the Petroleum Industry*. American Chemical Society. pp. 115–163.

Lake, L.W., Johns, R., Rossen, B., and Pope, G. (2014) *Fundamentals of Enhanced Oil Recovery*. Society of Petroleum Engineers, Texas, U.S.A, 2014.

Li, R.F., Yan, W., Liu, S., Hirasaki, G., and Miller, C.A. (2010) Foam mobility control for surfactant enhanced oil recovery. *SPE Journal*, 15(4), 928–942.

Lin, Y., Skaff, H., Emrick, T., Dinsmore, A.D., and Russell, T.P. (2003) Nanoparticle assembly and transport at liquid-liquid interfaces. *Science*, 299(5604), 226–229.

Liu, Q., Zhang, S., Sun, D., and Xu, J. (2009) Aqueous foams stabilized by hexylamine-modified Laponite particles. *Colloid Surface A*, 338(1), 40–46.

Liu, Q., Zhang, S., Sun, D., and Xu, J. (2010) Foams stabilized by Laponite nanoparticles and alkylammonium bromides with different alkyl chain lengths. *Colloid Surface A*, 355(1), 151–157.

Lotfollahi, M., Kim, I., Beygi, M.R., Worthen, A.J., Huh, C., Johnston, K.P., Wheeler, M.F., and DiCarlo, D.A. (2017) Foam generation hysteresis in porous media: Experiments and new insights. *Transport Porous Med*, 116(2), 687–703.

Ma, K., Cui, L., Dong, Y., Wang, T., Da, C., Hirasaki, G.J., and Biswal, S.L. (2013) Adsorption of cationic and anionic surfactants on natural and synthetic carbonate materials. *Colloid Interf Sci*, 408, 164–172.

Mannhardt, K., Schramm, L.L., and Novosad, J.J. (1992) Adsorption of anionic and amphoteric foam-forming surfactants on different rock types. *Colloid Surface*, 68(1), 37–53.

Mannhardt, K., Schramm, L.L., and Novosad, J.J. (1993) Effect of rock type and brine composition on adsorption of two foam-forming surfactants. *SPE Advanced Technology Series*, 1(1), 212–218.

Pickering, S.U. (1907) CXCVI.—Emulsions. *J Chem Soc*, 91, 2001–2021.

Pugh, R.J. (2016) *Bubble and Foam Chemistry*. Cambridge University Press, Cambridge, UK.

Ramsden, W. (1904) Separation of Solids in the Surface-Layers of Solutions and "Suspensions" (Observations on Surface-Membranes, Bubbles, Emulsions, and Mechanical Coagulation). Preliminary Account. Proceedings of the Royal Society of London, 72, pp. 156–164.

Ransohoff, T.C., and Radke, C.J. (1988) Mechanisms of foam generation in glass-bead packs. *SPE Reservoir Engineering*, 3(2), 573–585.

Rossen, W.R. (1996) Foams in enhanced oil recovery. In R.K. Prud'homme and S.A. Khan (Eds.), *Foams: Theory: Measurements: Applications*, 1st edn. Marcel Dekker, New York, NY, USA. pp. 413–464.

Song, B., and Springer, J. (1996) Determination of interfacial tension from the profile of a pendant drop using computer-aided image processing: 1. Theoretical. *J Colloid Interf Sci*, 184(1), 64–76.

Sun, Q., Li, Z., Wang, J., Li, S., Li, B., Jiang, L., Wang, H., Lü, Q., Zhang, C., and Liu, W. (2015) Aqueous foam stabilized by partially hydrophobic nanoparticles in the presence of surfactant. *Colloid Surface*, 471, 54–64.

Sun, Q., Zhang, N., Li, Z., and Wang, Y. (2016) Nanoparticle-stabilized foam for effective displacement in porous media and enhanced oil recovery. *Energy Technol-Ger*, 4(9), 1053–1063.

Varade, D., Carriere, D., Arriaga, L.R., Fameau, A.-L., Rio, E., Langevin, D., and Drenckhan, W. (2011) On the origin of the stability of foams made from catanionic surfactant mixtures. *Soft Matter*, 7(14), 6557–6570.

Worthen, A.J., Bagaria, H.G., Chen, Y., Bryant, S.L., Huh, C., and Johnston, K.P. (2013a) Nanoparticle-stabilized carbon dioxide-in-water foams with fine texture. *J Colloid Interfac Sci*, 391, 142–151.

Worthen, A.J., Bryant, S.L., Huh, C., and Johnston, K.P. (2013b) Carbon dioxide-in-water foams stabilized with nanoparticles and surfactant acting in synergy. *AIChE Journal*, 59(9), 3490–3501.

Xu, L., Rad, M.D., Telmadarreie, A., Qian, C., Liu, C., Bryant, S.L., and Dong, M. (2018) Synergy of surface-treated nanoparticle and anionic-nonionic surfactant on stabilization of natural gas foams. *Colloid Surface*, 550, 176–185.

Yekeen, N., Manan, M.A., Idris, A.K., Padmanabhan, E., Junin, R., Samin, A.M., Gbadamosi, A.O., and Oguamah, I. (2018) A comprehensive review of experimental studies of nanoparticles-stabilized foam for enhanced oil recovery. *J Petrol Sci Eng*, 164, 43–74.

Yu, J., An, C., Mo, D., Liu, N., and Lee, R.L. (2012) Study of Adsorption and Transportation Behavior of Nanoparticles in Three Different Porous Media. In SPE-153337-MS. Society of Petroleum Engineers, SPE. p. 13. doi:10.2118/153337-MS.

Yu, J., Khalil, M., Liu, N., and Lee, R. (2014) Effect of particle hydrophobicity on CO2 foam generation and foam flow behavior in porous media. *Fuel*, 126, 104–108.

Zajac, J., Chorro, C., Lindheimer, M., and Partyka, S. (1997) Thermodynamics of micellization and adsorption of zwitterionic surfactants in aqueous media. *Langmuir*, 13(6), 1486–1495.

Zhang, Y., Wang, S., Zhou, J., Zhao, R., Benz, G., Tcheimou, S., Meredith, J.C., and Behrens, S.H. (2017) Interfacial activity of nonamphiphilic particles in fluid–fluid interfaces. *Langmuir*, 33(18), 4511–4519.

5

Nanomagnetism

5.1 Introduction

Magnetic nanoparticles (typically in the form of stable dispersions, also known as ferrofluids) have a wide range of applications, from various engineering devices such as micro-electro-mechanical systems (Pérez-Castillejos et al. 2000) and lubrication and sealing of bearings, to biomedical applications (*e.g.*, magnetic nanoparticles are used in biomedicine as contrast agents to characterize human body tissues; Berkovsky et al. 1993). Their potential utility in subsurface environment and porous media was first recognized in the late 1990s (Moridis et al. 1998; Oldenburg et al. 2000). In their application in reservoir rocks, the size of the particles plays a critical role. Properly coated nanometer-size particles are capable of flowing through micron-size pores and throats without plugging or blocking them, and with minimal retention in rock (Yu et al. 2014), even at large particle concentrations (Rodriguez Pin et al. 2009) which opens up great potential for oil production applications.

Further, the magnetic minerals in Earth subsurface carry information on the past direction and magnetization of Earth's magnetic field, and as such, can be used for both characterization and correlation of subsurface formations as well as a geochronology tool.

We first review the basics of magnetism as relevant to magnetic solids and specifically superparamagnetic nanoparticles (SNPs) in Section 5.2, followed by dynamics of SNPs in liquid carriers in Section 5.3, as well as general magnetic behavior of their stable dispersions in bulk and in porous materials (Section 5.4). We continue with brief reviews of magnetic sensing (Section 5.5), high-gradient magnetic separation (Section 5.6), and hyperthermia heating (Section 5.7).

5.2 Nanomagnetism Basics

5.2.1 Brief Introduction to Electromagnetism

There are two sources of magnetism: the motion of electric charges (electric current) and the spin of elementary particles in the atoms forming solid matter. While the first refers to a condition that can be imposed experimentally, the latter characterizes an intrinsic property displayed by some materials. Three macroscopic fields are used to describe the magnetic effects: the magnetic field $\vec{H}$, the induction field $\vec{B}$, and the magnetization field $\vec{M}$. We will briefly introduce their inter-relationships.

The relation between the field $\vec{B}$ and an electric current is given by a volume integral known as the Biot-Savart law:

$$\vec{B}(\vec{r}) = \frac{\mu_o}{4\pi} \int_V \frac{\left(\vec{J}dV\right)\vec{r'}}{|\vec{r'}|^3} \tag{5.1}$$

where

$\vec{J}$	is the current density (current per unit of area)
$\mu_o = 4\pi \cdot 10^{-7} NA^{-2}$	is the magnetic permeability of vacuum,
$\vec{r'}$	is the displacement vector from a point where the integrand is being evaluated to the position $\vec{r}$ where the field $\vec{B}$ is being evaluated

$\vec{B}$ has the unit of Tesla (T), although is commonly measured in Gauss units ($G = 10^{-4}T$). As $\vec{B}$ can be generated by an electric current, it can conversely induce a force that will affect electric currents. This force is called Lorentz force and, in the absence of an external electrostatic field, is given by

$$\vec{F}_{\text{Lorentz}} = \int_V \vec{J}\vec{B}\,dV \tag{5.2}$$

Lastly, electric currents and the distribution of $\vec{B}$ must also obey to Ampere's circuit law:

$$\oint_C \vec{B}.dl = \mu_o \int_S \vec{J} \cdot dS \tag{5.3}$$

The left-hand side of Equation (5.3) consists of a line integral of $\vec{B}$ along a path C enclosing the surface S and the right-hand side represents the total current flowing through S. Whenever the distribution of $\vec{B}$ is well-determined around a closed-curve, known in this context as Amperian loop, Equation (5.3) provides a simple way of evaluating currents.

Equations (5.1) through (5.3) provide a framework for solving several problems involving magnetic effects and electric currents in the free space. However, these expressions alone are not sufficient for describing magnetic effects arising from the intrinsic magnetism of materials. In the context of ferrohydrodynamics (Section 5.4), for instance, this second class of magnetism is far more important than the first; ferrofluids are able to interact with magnetic fields even in the absence of electric currents and thus can be taken as magnetic materials.

The Amperian currents model is a way of modeling magnetic materials that postulates that magnetism emerges from the combined effect of electric charges moving in small loops contained by the material's volume. This approach provides the intuitive connection between the macroscopic fields and the microscopic processes that generate them, as the Amperian currents can be linked to the orbital motion of electrons around the positively charged nucleus. The currents forming these microscopic loops (Amperian currents or bound currents) face no resistance, however, and hence cannot be considered currents in a conventional sense. Yet, the magnetic effects they produce are completely analogous to the ones observed in macroscopic current loops.

We first define magnetic moment $\vec{m}$

$$\vec{m} = \int_S I_{\text{loop}} \hat{n} \, dS \tag{5.4}$$

where

I_{loop} is the intensity of the current loop
S is its area
$\hat{n}$ is the unit normal vector for the area S

The magnetic moment can be interpreted as the quantity that determines the strength of torque experienced by the circuit loop under an external field. The torque acts in the orientation of aligning $\vec{m}$ to $\vec{B}$. Given an external field $\vec{B}$ that intersects a current loop sufficiently small so that $\vec{B}$ can be taken as uniform within its area, it can be shown (Whitmer 1962) that the Lorentz force (that the loop undergoes due to the external field) and the associated torque are given by

$$\vec{F}_{\text{loop}} = \nabla \left(\vec{m} \cdot \vec{B} \right), \; \vec{\tau}_{\text{loop}} = \vec{m}\vec{B} \tag{5.5}$$

According to the Amperian current model, the current loop considered above is representative of an infinitesimal volume of magnetic matter if we take its current intensity as the intensity of the Amperian current. In order to generalize the analysis to finite volumes, the force and torque obtained for the loop should be written per unit of volume, which can be accomplished by defining the magnetization $\vec{M}$:

$$\vec{M} = \frac{d\vec{m}}{dV} \tag{5.6}$$

It can be shown (Whitmer 1962) from Equation (5.6) and the definition of $\vec{M}$ that

$$\nabla \vec{M} = \vec{J}_{\text{amp}} \tag{5.7}$$

where $\vec{J}_{amp}$ is the Amperian current density. The relevance of the magnetization (or magnetic moment per unit volume) comes from being directly linked to the microscopic currents that generate magnetism. Therefore, it serves as a measure of the degree of magnetic polarization of a material.

Differentiating Lorentz force in Equation (5.5) with respect to volume leads to the density of the body force acting on a medium with magnetization $\vec{M}$ when subjected to a field $\vec{B}$, known as Kelvin force:

$$\vec{f}_{kelvin} = \nabla\left(\vec{M}\cdot\vec{B}\right) \tag{5.8}$$

We further apply Equation (5.3) in its differential form, $\nabla\vec{B} = \mu_o\vec{J}$, to a magnetic material with no moving electric charges other than the Amperian currents and substitute the current density by the magnetization through Equation (5.8) to obtain

$$\nabla\left(\frac{\vec{B}}{\mu_o} + \vec{M}\right) = 0 \tag{5.9}$$

The curl-free field defined by the expression above is the magnetic field $\vec{H}$. It follows from the equation above that

$$\vec{B} = \mu_o\left(\vec{M} + \vec{H}\right) \tag{5.10}$$

Thus, the two fields that arise in macroscopic magnetism are related by the magnetization of the medium. In a free space, the fields are simply distinguished by a constant. For soft magnetic materials and the magnetic fields of moderate strength, $\vec{M}$ and $\vec{H}$ are co-linear, $\vec{M} = \chi\vec{H}$. The proportionality constant χ is dimensionless and is called magnetic susceptibility. In this case, we have $\vec{B} = \mu\vec{H}$ where $\mu = \mu_0(1+\chi)$.

5.2.2 Magnetostatic Equations

All the governing relations for magnetism presented throughout this chapter implicitly assumed steady state conditions. In addition, effects of an electric nature that can influence magnetic fields were disregarded. Given these assumptions, Equation (5.9) determines that $\vec{H}$ is irrotational. Combining this assertion with the Gauss's law of magnetism yields the magnetostatic formulation:

$$\nabla\times\vec{H} = 0$$
$$\nabla\cdot\vec{B} = 0 \tag{5.11}$$

It is easily verifiable that the two equations above can be derived directly from the macroscopic Maxwell equations by taking the time-derivatives and the free currents as zero. The resulting formulation is ideal for modeling ferrohydrodynamics problems in which the magnetic effects arise entirely due to a permanent external field. Applying Equation (5.11) to a magnetized ferrofluid surrounded by media that can be either non-magnetic or display a different kind of magnetic behavior creates important interface effects. In order to better understand these effects, the concept of magnetic permeability is introduced:

$$\mu = \mu_o \left(\left| \frac{M}{H} \right| + 1 \right) \tag{5.12}$$

It should be noticed that μ is defined such that $\vec{B} = \mu\vec{H}$ and thus it is a practical way of expressing magnetic behaviors in which $\vec{H}$ and $\vec{M}$ are parallel. In the non-magnetic media, the magnetic permeability is reduced to the permeability in free space $\mu = \mu_o$. Another case of constant μ occurs in materials that display linear magnetization, i.e., $\vec{M}$ varies linearly with $\vec{H}$ according to a constant χ known as magnetic susceptibility, which gives

$$\mu = \mu_o (1 + \chi) \tag{5.13}$$

Porous media are characterized by the presence of at least two phases (fluid/void and solid) of different mechanical and magnetic properties. Thus, it is useful to consider what happens to above continuous field definitions near an interface between the media (+) and (−) with different magnetic permeability, as shown in Figure 5.1. Satisfying both magnetostatic equations (Equation 5.11) when applied at the interface vicinity requires the tangential component of $\vec{H}$ and the normal component of $\vec{B}$ to be the same on both sides. As the sketch shows, in order to accommodate these conditions, the two fields change in both direction and magnitude across the interface. The following relations must hold where subscripts n and t refer to normal and tangential components respectively:

$$\frac{H_n^+}{H_n^-} = \frac{\mu^-}{\mu^+}$$
$$\frac{B_t^+}{B_t^-} = \frac{\mu^+}{\mu^-} \tag{5.14}$$

It is convenient to rewrite the magnetostatic formulation in a way that makes explicit the influence of the magnetic permeability distribution to the solution. This can be achieved by defining a potential function ψ such that $\vec{H} = -\nabla\psi$.

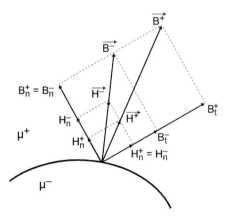

FIGURE 5.1
Magnetic and induction field being disturbed by the presence of an interface between two regions with different magnetic permeability. From Saint-Martin de Abreu Soares (2015).

Being described as the gradient of a scalar function guarantees $\vec{H}$ is irrotational and hence removes one of the equations in Equation (5.11). By substituting the potential function into the remaining equation:

$$\nabla \cdot \left(-\mu \nabla \psi\right) = 0 \qquad (5.15)$$

In the special case of a uniform magnetic behavior, the expression above is reduced to a Laplace equation. The interface relations (5.14) can also be written in terms of the potential:

$$\frac{\nabla \psi^+ \cdot \hat{n}}{\nabla \psi^- \cdot \hat{n}} = \frac{\mu^-}{\mu^+} \qquad (5.16)$$

A numerical solution for Equation (5.15) would then solve for ψ with the boundary condition given by Equation (5.16). The domain of the solution is a region with continuous μ and its boundaries reflect transitions to adjacent domains with different magnetic behaviors. The set of all domains must be solved simultaneously because the boundary condition employed has a coupling effect.

5.2.3 Magnetic Solids

In general, the magnetization of a natural material can be described in terms of the magnetic moments associated to the elementary particles in its atoms. The magnetic moments are often arranged in a way such that they cancel out and the material yields no magnetization. In other cases, some magnetization exists either permanently or by influence of external factors that disturb the arrangement of the magnetic moments, such as a change of temperature or an

imposed magnetic field. The different propensities of magnetic moments align-
ment and reactions to external factors characterize different kinds of magnetism. *Ferromagnetism* is the magnetic behavior displayed by most permanent
magnets. In a ferromagnetic material, magnetic moments align parallel to each
other spontaneously and hence magnetization exists even in the absence of an
external field. See Figure 5.2. Permanent magnetization also occurs in **ferrimag-
netism**. Ferrimagnetic materials, however, have atoms with magnetic moments
in opposite directions and each direction has a different moment magnitude.
The imbalance between the magnitudes is responsible for the magnetization.
Paramagnetism is the behavior in which the magnetic moments of a medium
are arranged in a way that produces no net magnetization when undisturbed
(because thermal motion randomizes the spin orientations), but are free to
align in the direction of an applied external field. Consequently, the magnetization of a paramagnetic material grows with the strength of the applied field.
Paramagnetism results from the presence of unpaired electrons in the material
(unpaired electrons spin and thus have a magnetic dipole moment). All atoms
with incompletely filled atomic orbitals are paramagnetic, and some examples
include aluminum, oxygen, titanium, and iron oxide.

For sufficiently small ferromagnetic nanoparticles (each of which consists
of one magnetic domain), magnetization can randomly flip direction under
the influence of temperature without any external magnetic field influence.
The time between two flips is called the Néel relaxation time. However, over
much more longer times, the magnetization appears to be in average zero.
This is referred to as **superparamagnetism**. **Diamagnetism** consists of the
opposite behavior, i.e., magnetic moments tend to align against an external
field so that $\vec{M}$ and $\vec{H}$ are oriented in opposite directions. The magnetization
resulting from diamagnetic behavior tends to be weak.

A rigorous analysis on how the subatomic structure of the medium
relates to its magnetism is out of the scope of this text and can be found in
Chikazumi et al. (2009). Fortunately, the magnetic behaviors described above
can be represented mathematically at a macroscopic level through appropriate constitutive models.

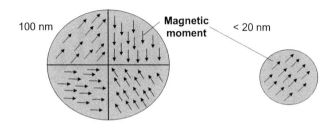

FIGURE 5.2
The effect of magnetic domain size on magnetization. Larger domains (100 nm, on the left)
are ferromagnetic materials and include multiple domains with randomly oriented magnetic
moments. Single domain particles (typically less than 20 nm, on the right) have uniform align-
ment and are referred to as superparamagnetic.

5.2.4 Unique Properties of Superparamagnetic Nanoparticles

As mentioned above, small ferromagnetic particles consisting of one domain act as small permanent magnets. Example visualization and size measurements are provided in Figure 5.3. If the particle is, say, suspended in fluid and can freely rotate, then the orientation of its magnetic field can both change due to physical rotation of the particle as part of Brownian motion (thermal fluctuation) and random flips of its field (Néel relaxation). When exposed to the external field, such particles can align and saturate at relatively low magnetic field, although the field still needs to be large enough to overcome the thermal motion that tends to counteract it. In other words, magnetic energy E_M per particle has to be larger than or equal to thermal energy E_T. Thus, in saturation we have

$$E_M = \mathrm{HVM}_s\mu_0 \gtrsim k_B T = E_T \tag{5.17}$$

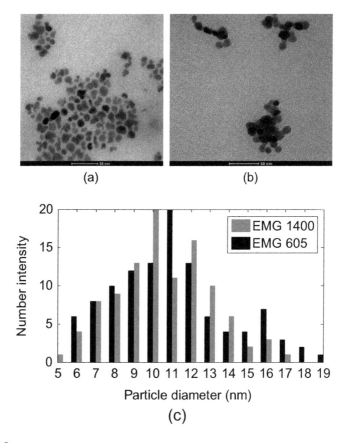

(a)   (b)

(c)

FIGURE 5.3
Transmission Emission Microscopy (TEM) images of two commercially available magnetite SNPs: (a) EMG 1400 and (b) EMG 605. The size distribution for both measured based on images is shown in (c). Figures from Mehta (2015).

where
  V is particle volume
  M_s is saturation magnetization of the particle material
  k_B is the Boltzmann constant
  T is absolute temperature.

The magnetization curve of a dispersion of small particles in thermal equilibrium is described by

$$\frac{M}{M_0} = L\left(\frac{HVM_s\mu_0}{k_BT}\right) \tag{5.18}$$

where

$L(x) = \coth(x) - \dfrac{1}{x}$ is the Langevin function

M_0 the saturation magnetization of the dispersion of the magnetic particles (Gleich 2014).

Figure 5.4 shows examples of how the magnetization changes depending on the particle size. While the Langevin function is not defined at 0, the magnetization curve is defined there as for $H = 0$ and $M = 0$ in case of SNPs.

Note that the slope of the curve in Figure 5.4 is magnetic susceptibility χ defined previously in Section 5.2.1. In the limit of small x, Langevin function is approximated as $L(x) = x/3$. It must be noted, however, that Langevin's theory is not particularly adequate in systems with high particle concentrations because it assumes no interaction between the particles.

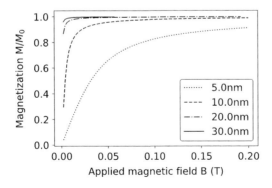

FIGURE 5.4
Example magnetization curve for superparamagnetic nanoparticles: the particles become magnetic only in the presence of external magnetic field and the magnetization curve exhibits no hysteresis. We used $M_s = 4.46 * 10^5$ for magnetite (Rosensweig 1997). The slope of the curves for small applied magnetic field is magnetic susceptibility χ defined in Section 5.2.1. For higher applied fields, all nanoparticles are oriented (which is referred to as magnetization saturation), as schematically shown in later sections in Figure 5.9b.

The settling of magnetic nanoparticles in a magnetic field is hindered by thermal energy inducing Brownian motion. The limit of $E_M = E_T$ also provides the minimum size of the particle that remains suspended under the given conditions (temperature and magnetic field):

$$d_{min} = \left(\frac{6k_BT}{\pi VM_s\mu_0} \right)^{1/3} \tag{5.19}$$

The above equations explain magnetization in thermal equilibrium. We next turn to how fast this equilibrium can be achieved. There are two ways a particle can change magnetization: (1) Brownian rotation, that is the (geometric) rotation of the single-domain particle, and (2) Néel rotation of the magnetization itself, without rotation of the crystal. It can be shown by a simple balance of hydrodynamic force/torque on a particle and the magnetic torque (Section 4.3 in Gleich [2014]) that the frequency of Brownian rotation is

$$\omega = \frac{\mu_0 MH}{\eta_f} \tag{5.20}$$

With η_f viscosity of the surrounding fluid. For $M = 1T\mu_0^{-1}$, $H = 1mT\mu_0^{-1}$, and $\eta_f = 1$ mPas, we get $\omega = 2\pi \cdot 100$ kHz. Since M chosen here is already a high value for a magnetic particle, in magnetic particle imaging, no relevant frequency components are expected to emerge for Brownian motion above 100 kHz, and imaging relying on capturing this needs to have much lower field frequency. The Neel rotation can be much faster than this (~30 MHz, and it relies on the coherent rotation of the coupled electron spins in the ferromagnet (Gleich 2014).

Some unique properties of SNPs can be exploited in various applications. Due to their small size, they can be moved in desired direction by applying the spatial gradient of magnetic field. Under applied high-frequency magnetic oscillation, SNPs can generate intense, highly localized heat. At the same time, under the applied field, SNP ensemble generates a magnetic induction field, allowing remote sensing. Finally, surface coating can be designed for SNPs to attach to desired locations, *e.g.*, for stable foam/emulsion generation or to carry out desired function(s). We will elaborate on the applications in the sections to follow.

5.3 Hydrodynamics of SNP Displacements

5.3.1 Settling and Separation of SNPs

There are a number of applications where the synergy of (1) magnetic nanoparticles' high surface area, (2) dispersivity related to their size, and (3) magnetization makes them very attractive as surface catalysts in various environmental and biomedical applications. Examples include softening of high salinity water (Xu et al. 2011, 2012), removal of oil droplets from water (Khushrushahi et al. 2013), imaging, and drug delivery (Kucheryavy et al. 2013; Tan et al. 2009). In particular, the nanoparticle surface could be functionalized to preferentially attach to different surfaces and ultimately aid in selective removal of certain chemicals, or the surface functionalization could allow certain chemical reactions, i.e., adsorption of aqueous ions, oxidation/reduction reactions, and dissolution/crystallization, to occur. In either case, naturally slow gravitational settling of the nanoparticles (due to their size) could be enhanced by the magnetic forces acting on the magnetic core.

Prigiobbe et al. (2015) performed a fundamental study on magnetic separation of superparamagnetic iron-oxide nanoparticles (that the authors refer to as SPIONs) from a brine. Experiments were carried out using suspensions of known concentrations settling under the effect of a magnetic field and continuously monitored. SNPs tend to aggregate to minimize high surface energy by strong magnetic dipole-dipole attractions and van der Waals force. A rapid separation of nanoparticles occurred, suggesting a non-classical settling phenomenon induced by magnetic forces which favor particle aggregation and therefore faster settling. In other words, the aggregation due to magnetic forces (which is in many cases considered undesirable and makes the process more difficult to model or quantify) is precisely what made the separation more effective. The rate of settling decreased with the concentration n of particles per volume, and an optimal condition for fast separation was found for an initial n of 120 g/L (see Figure 5.5). The model proposed in this work agrees well with the measurements in the early stage of the

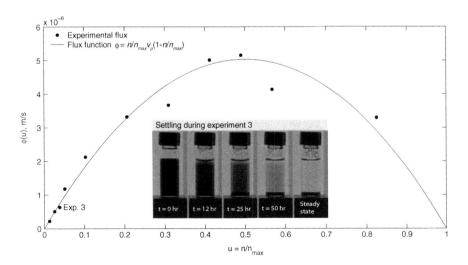

FIGURE 5.5
Graphical abstract figure reprinted from Prigiobbe et al. (2015). Complete settling is observed for all tested dispersions within five days. The range of concentrations n in the experiments was 3.15 to 202 g/L.

settling, but it fails to describe the upper interface movement during the later stage, probably because of particle aggregation induced by magnetization which is not accounted for in the model.

5.3.2 Derivation of a Single SNP Settling Velocity in a Brine

We summarize the model that explains the velocity of the particles in this case is an application of Newton's second law because it illuminates fundamental mechanisms. The details of the derivation are available in Prigiobbe et al. (2015).

Figure 5.6a shows the schematic of the forces acting on a magnetic particle in a fluid. The classic Newton's law applied for the case of the droplet in the figure. Let m_p refer to particle mass and v_p be the magnetic particle velocity. The forces acting on it are the force F_{drag}, magnetic force F_m, and the buoyancy force F_b. All forces act along z (gravity) direction so we omit the vector notation. Then the Newton's law requires:

$$m_p \frac{\partial v_p}{\partial t} = F_{drag} + F_m + F_b \tag{5.21}$$

The buoyant force is given by:

$$F_b = V_d \left(\rho_p - \rho_f \right) g \tag{5.22}$$

and the drag force by Stokes formula (assuming Reynolds number smaller than 1)

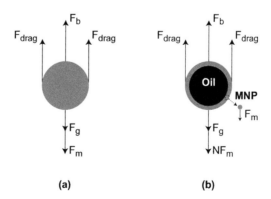

(a)   (b)

FIGURE 5.6
Schematic of forces acting on (a) single magnetic nanoparticle and (b) oil-in-water droplet stabilized with SNPs. F_{drag} is drag force, F_b is buoyant force, F_m is magnetic force, and F_g is gravity force on the particle (droplet). N in (b) refers to the number of SNPs at the oil/water interface stabilizing the droplet.

$$F_{\text{drag}} = -3\pi\eta_f d_p v_p \tag{5.23}$$

where V_p is particle volume, $\rho_p - \rho_f$ the density difference between the particle and the surrounding fluid (brine), g gravitational acceleration constant, η_f surrounding fluid (brine) viscosity, and ρ_p and d_p particle density and diameter. The magnetic force on the particle is given as

$$F_m = \frac{V_p\left(\chi_p - \chi_f\right)}{2\mu_0} \nabla B^2 \tag{5.24}$$

where

V_p	is particle volume
χ_p and χ_f	are particle and fluid (brine) magnetic susceptibilities
μ_0	is magnetic permeability of air/vacuum
B	is magnetic field.

χ_f is estimated as a simple volumetric mixture law of brine and particles. Units for all of the terms are SI and can be found in Nomenclature at the end of this chapter. The magnetic field was presumed produced by parallelepipedic magnets (so that analytic solution can be used). Assuming the left-hand side in Equation (5.21) is negligible, the velocity in z-direction can be found as:

$$v_p(z) = \frac{1}{3\pi\eta_f d_p}\left(\frac{V_p\left(\chi_p - \chi_f\right)}{2\mu_0} \nabla B_z^2 + V_d(\rho_d - \rho_f)g\right) \tag{5.25}$$

This velocity is incorporated into a suspension velocity $v(z) = v_p(z)(1 - n/n_{\max})$, where n is particle density and $n_{\max}$ the maximum particle density in settled suspension. The suspension velocity is then incorporated into a 1D conservation law that is solved over $0 \le z \le L, t > 0$ with L being the initial suspension height:

$$\frac{\partial n}{\partial t} + \frac{\partial (nv)}{\partial z} - D\frac{\partial^2 n}{\partial^2 z} = 0 \tag{5.26}$$

where diffusion coefficient is given by Stokes-Einstein formula $D = k_B T / 3\pi\eta_f d_p$

5.3.3 Displacement of SNP-Attached Dispersions and Chemicals

Ko et al. (2016) followed up this work for oil droplet removal from oil-in-water emulsion. The synthesized SNPs of approximately 10 nm of individual particle size (and approx. 60 nm when agglomerated) were functionalized to attach to surfaces of oil droplets. The mechanism of attachment was electrostatic: oil-in-water emulsions are negatively charged and the particle surface was positively charged. The schematic of the procedure is

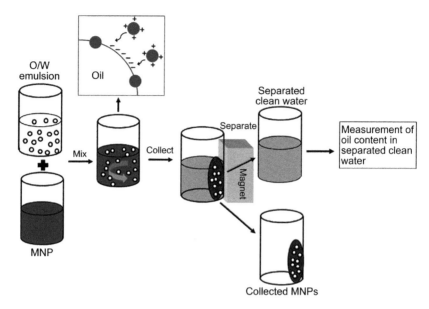

FIGURE 5.7
Schematic of the use of SNPs for separation of oil from oil-in-water emulsions. Reprinted from Ko et al. (2016). Copyright 2016, Society of Petroleum Engineers. Reproduced with permission of SPE. Further reproduction prohibited without permission.

shown in Figure 5.7. The magnetic force acting on the SNPs (see Figure 5.7) sped up the agglomeration of both SNPs and oil droplets, and sped up the separation process. The schematic of forces acting on a single SNP, analyzed in previous section, is contrasted with forces on an oil droplet with attached SNPs in Figure 5.6. The total magnetic separation when aggregation occurs was reduced to 1 second, as opposed to 36–72 hours in the case of individual SNPs. Crude oil content in the separated brine was reduced by 99.9% and the average SNP concentration required was 0.04% by weight.

5.3.4 Magnetic-Enhanced Flow Fractionation

Magnetic split-flow thin fractionation (SF) is a family of SF techniques for separating magnetically susceptible particles (Fuh et al. 2003). Particles with different degrees of magnetic susceptibility can be separated into two fractions by adjusting perpendicularly applied magnetic forces and flow rates at inlets and outlets—for example schematic, see Figure 5.8. Fuh et al. (2003) reported throughput of 3 g/h run over 8 h. The minimum difference in magnetic susceptibility required for complete separation was about $3 \times 10-6$ cgs. The throughput can be scaled up by using longer channel lengths, broader channel breadths, and stronger magnetic fields. This work orients the experimental setup so that gravity does not affect separation (unlike applications in the previous sections), but gravity could be combined.

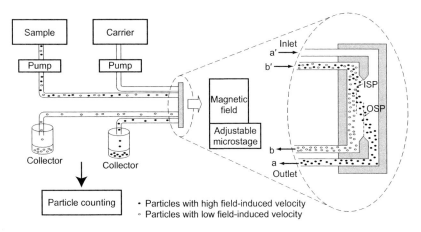

FIGURE 5.8
Schematic of the magnetic SF system. From Fuh et al. (2003).

5.4 Ferrofluids

Ferrofluids are stable colloidal suspensions of ferromagnetic particles in carrier fluids. They are suspended by Brownian motion and typically coated with a layer (see Figure 5.9 for schematic) that prevents magnetic domains from clumping together. Ferrohydrodynamics (FHD) is the science that deals with the mechanics of ferrofluid in a magnetic field (Rosensweig 1997). Ferrofluids are dielectric (non-conductive) and paramagnetic (do not retain magnetization in the absence of magnetic field). The study of FHD implies ferrofluids display sufficient magnetization so that the magnetic forces resulting from its exposure to an external field are strong enough to compete with other forces of fluid dynamics.

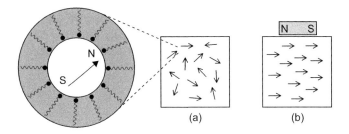

FIGURE 5.9
Schematic of nanoparticles with the single domain ferro- or ferri-magnetic core and coated with the adsorbed dispersant molecules. The magnetic core radius is typically about 80% of the particle radius. (a) With no external magnetic field applied; (b) in the presence of an external magnetic field, the nanoparticles become oriented. At the saturation magnetization, all nanoparticles are oriented (not that the N magnetic pole of the particle is attracted to the S magnetic pole of the applied field).

A common way of producing fluids that are controllable by magnetic fields is by applying an electric current through some conductive fluid, such as a liquefied metal, and resorting to the Lorentz force of Equation (5.2) that emerges when an external field intercepts an applied current at an angle. The fluids controlled by Lorentz forces are the subject of magnetohydrodynamics (MHD), which should be distinguished from FHD. The limitations of MHD are the constant need for an electric current and the fact that the Lorentz force is typically small except for extremely strong currents. Ferrohydrodynamics' control of ferrofluids was meant to overcome these limitations by relying on the magnetic force that occurs when magnetic matter interacts with a magnetic field.

There is a reason why ferrofluids are dispersions of nanoparticles as opposed to liquefied metal. Melting natural ferromagnetic solids is not a suitable procedure for obtaining ferrofluids because the Curie temperature of a material is always well below its melting point (Kittel 2005). This means that the ferromagnetic solids would lose their permanent magnetism before becoming liquids. Another approach that could potentially lead to ferrofluids would be to take advantage of the paramagnetism displayed by several water-soluble salts. Unfortunately, paramagnetism tends to be a very weak kind of magnetic behavior and thus the dynamics of the salt solutions is not expected to be significantly affected by a magnetic force.

The desired ferrofluid properties can be achieved with ferri- or ferromagnetic nanoparticles coated with some dispersant and suspended in a liquid carrier. This concept emerged in the mid-1960s, when the first stable dispersions were obtained (Papell Solomon 1965). Nowadays, a wide variety of techniques (Charles 2002) can be employed to produce spheroid particles with diameters in the range of 3–15 nm and stabilizing surfactant coatings. In most of the practical applications, the nanoparticles are made of ferrite compounds, usually a mix of magnetite and maghemite (Charles 2002). This kind of nanoparticle tends to be reliable and durable because the oxidation of magnetite to maghemite (Fe_2O_3) has a relatively small effect on the magnetic moment of the nanoparticle. Metal nanoparticles, such as the ones made of iron and cobalt, yield ferrofluids with higher saturation magnetizations, but also suffer greater loss of magnetic properties with oxidation.

5.4.1 Ferrofluid Applications

Early research on ferrofluids took advantage of the stresses emerging from magnetization for various mechanical applications, such as the lubrication and sealing of bearings (Berkovsky et al. 1993). The ferrofluid magnetization response is also useful for applications related to detection and imaging. In the biomedical field, for instance, magnetic nanoparticles designed to attach to specific body tissues can be used for enhancing MRI (Oh et al. 2006) or to burn off the cancer cells (see Pollert and Zaveta 2012).

5.4.1.1 Large Scale Ferrofluids Through Porous Media

In subsurface applications, one must account for processes occurring on a broad range of scales; each length scale (and its underlying physics) requires different modeling or measurement approaches. The pore scale is at the scale of individual porous medium grains (with grain size ranging from μm to mm). Any larger scale of interest ("macroscale," "Darcy scale") assumes a sufficiently large medium for continuous (upscaled) definition of properties such as porosity (void fraction) and permeability (the ability of medium to transmit fluids). This is the scale where prediction of flow properties is most needed and ranges from the scale of samples analyzed in laboratory (cm to m, "core" scale) to the size of the geologic body of interest (aquifer, bedding layer, reservoir).

Promising applications of ferrofluids in subsurface environments were first identified by Moridis et al. (1998). Numerical analysis of ferrofluid flows through porous media at Darcy-scale was pioneered by Oldenburg et al. (2000) and Lavrova et al. (2006), who added the magnetic force to the pressure gradient and gravitational body force terms in Darcy's law and calculated the resulting transport of the ferrofluid. These approaches treat ferrofluid as a (stable) continuum.

More recently, research effort recognized that stability of nanoparticle dispersions in a natural porous medium is not a given and focused on characterizing transport properties of nanoparticles in real sedimentary rocks. It has been verified (Zhang 2012) that the propagation of surface-treated silica (non-magnetic) nanoparticles through micron-sized pores occurs mostly without blocking or plugging the pore throats. In an experiment with the injection of magnetic nanoparticles with a coating designed to reduce adsorption to rocks surfaces (Yu et al. 2014), there was very little particle retention and insignificant permeability reduction in reservoir cores, even for high particles concentration. A number of researchers use traditional transport models for nanoparticle transport (Zhang 2012); however, those approaches do not incorporate magnetic forces.

Presence of multiple fluids further complicates the modeling and upscaled models of two-phase flow displacement in a magnetic field where one phase is ferrofluid exposed to the magnetic field **do not exist**. While Darcy flow equations could be modified to include magnetic forces, the closure relationship in multiphase flow on large (Darcy) scale involve knowing capillary pressure and relative permeability as a function of fluid saturation. Such relationships can be obtained via either experiments (micromodels proposed in this work are the first step) or pore scale modeling on a representative elementary volume of porous medium (as proposed here), and neither is currently available in the literature.

5.4.1.2 Microfluidics and Ferrofluids

The behavior of ferrofluid droplets exposed to magnetic field in microfluidic devices is a well studied problem: microfluidic systems can generate, handle, and manipulate many individual droplets which elongate when exposed to uniform magnetic field. Since the beginning study by Tarapov (1974), a great number of theoretical (Lavrova et al. 2004; Rosenzweig 1997), experimental (Berkovsky et al. 1987), and numerical (Afkhami et al. 2008) studies have been performed. Mathematical formulation for ferrofluid droplets is based on the coupled Maxwell's and Young-Laplace equations (Rosenzweig 1997). Since the interest is in low frequency magnetic fields (magnetostatic limit), and in ferrofluids, there is no free electric charge/current, the Maxwell's equations simplify to:

$$\nabla \times \vec{H} = 0 \text{ and } \nabla \cdot \vec{B} = 0 \qquad (5.27)$$

An application of these equations in the vicinity of the interface between the media with different magnetic permeabilities shows that the normal component of $\vec{B}$ and the tangential component of $\vec{H}$ must be the same at both sides of the interface. In order to ensure the condition $\vec{B} = \mu\vec{H}$, discontinuities must exist in these fields. In the absence of a magnetic field, according to the Young-Laplace equation, the interface at equilibrium supports a pressure difference (so-called capillary pressure $p_c = p_{nw} - p_w$). It does so by adopting a curved surface whose mean curvature κ satisfies $p_c = \sigma\kappa$, where σ is interfacial tension between fluids.

Extension of the Young-Laplace equation to ferrofluids yields an important result for modeling the motion of ferrofluid-fluid interfaces: the magnetic field manifests itself as several pressure-like terms, magneto-strictive pressure p_s, magnetic-normal traction p_n, and fluid-magnetic pressure p_m (details available in Rosenzweig [1997]). The magneto-strictive pressure is only relevant in cases of considerable fluid compressibility or occurrence of physicochemical phase change, so we will ignore it. A modified Young-Laplace equation that accounts for these stresses is then

$$p_{nw}v(p_m + p_n) + \sigma\kappa = p_w \qquad (5.28)$$

where magnetic pressures are subtracted from non-wetting pressure in case the non-wetting phase is ferrofluid (as is the case in microfluidic literature to date) and they are added otherwise (as is the case in this proposal). The situation where wetting fluid is magnetized is shown in Figure 5.9.

Non-wetting and wetting phase thermodynamic pressures p_w, p_{nw} and magnetic pressure $p_m = \mu_0 \int_0^H M dH$ are volume averaged (and in equilibrium assumed constant throughout each phase). Magnetic normal traction $p_n = \frac{1}{2}\mu_0(\vec{n} \cdot \vec{M})^2$, however, spatially varies based on the angle between the interface normal $\vec{n}$ (which considerably changes as one moves along the interface in capillarity dominated applications) and externally imposed magnetization (which is assumed of locally constant direction). Thus, the interface, in response, has spatially varying curvature.

In extensively studied manipulation of ferrofluid droplets (Berkovsky et al. 1993), these stresses cause (elliptical) extension of the droplet in direction parallel to the magnetic field which is a deviation from the spherical interface in a straight channel in the absence of magnetic forces. Most of the available studies assume linearized magnetization regime (which simplifies solving Maxwell's equations, and allows assumption of ellipsoidal shapes). Our group has recently improved on this knowledge for a ferrofluid blob in cylindrical channel in non-linear magnetization regime using a combined numerical and theoretical study (Rahmani et al. 2012).

Note that none of the microfluidics literature deals with irregular geometry of pore spaces which makes the studies in a channel of constant cross-section inapplicable; we next move on to discussing irregular pore geometries in more detail.

5.4.1.3 Non-Wetting Phase Trapping and Potential for Enhanced Oil Recovery

In subsurface applications, there is interest in studying residual non-wetting fluid mobilization in natural porous materials by magnetized ferrofluids, i.e., investigating ferrofluid potential for enhanced oil recovery. One of the main difference between microfluidic devices described above and natural porous media is that the cross-section of the pore pathways are constantly changing, and that contributes not only to more trapping of the non-wetting phase but also to more difficult mobilization of the blobs once trapped. Simple "snap-off" model represents the blob resulting from the rupture of a retreating non-wetting phase at a narrow pore throat due to strong capillary forces (see Figure 5.10, figure borrowed from Peters [2012]). The pressure difference required to overcome capillary forces and mobilize the blob in Figure 5.11 is

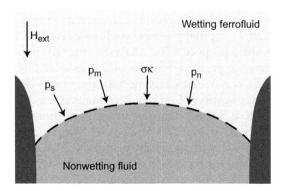

FIGURE 5.10
Schematic of pressure-like terms acting on a ferrofluid-fluid interface and arising from an externally imposed magnetic field; in this example, ferrofluid is wetting fluid. Thermodynamic pressures not shown.

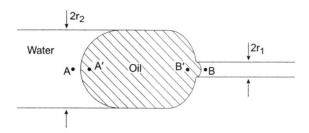

FIGURE 5.11
Simple pore (larger cross-sectional radius r_2) and throat (smaller radius r_1) arrangement. Mean curvature of the left and right interface is $2/r_2$ and $2/r_1$, respectively.

$\dfrac{P_A - P_B}{L} \geq 2\sigma\cos\theta \left(\dfrac{1}{r_1} - \dfrac{1}{r_2} \right)$, where we assume constant fluid-fluid-solid con-

tact angle θ on both ends. Another trapping mechanism would be that of bypassing of non-wetting fluid in a large pore by the wetting phase percolating through nearby small pores, so-called pore doublet model.

Note that realistic oil blobs in reservoirs after water flooding can fill up tens of individual pores (Payatakes 1982) and present complex spatial configurations. Nevertheless, image analysis of experimental investigations (Chatzis et al. 1983; Prodanovic et al. 2007) suggests that about 80% of the residual oil in the sample is found in single-pore snap-off geometries and the above models serves as a good basis for development of a pore-scale model. Quantitative understanding of fundamental trapping mechanisms can be translated to representative elementary volumes of porous material (Lake 1989).

5.4.1.4 Preliminary Modeling of Ferrofluid Surrounding Trapped Oil Blob

Mobilizing trapped non-wetting phase requires a driving force sufficient to overcome the effect of capillary forces, such as described in the previous section. In the traditional pore scale analysis, this driving force is related to a pressure difference produced by viscous stresses acting on the wetting fluid flow. However, the stresses arising with the magnetization of the ferrofluid under the application of an external magnetic field create an additional contribution to this force. The goal of the preliminary modeling study (Soares et al. 2014) was to evaluate how these magnetic stresses compare with the capillary stresses in the given geometry. The model provided an equilibrium blob shape for a given combination of external magnetic field, ferrofluid magnetic properties and pore geometric parameters in a snap-off pore model. The authors assumed incompressibility, linear magnetization, and quasi-static conditions and couple a level set method (Osher and Sethian 1988) for the quasi-static displacement of the fluid interface until equilibrium, and a second-order accurate immersed interface method (Li and Ito 2006) for the magnetic field distribution.

The preliminary simulations from the method described above were restricted to 2D pores occupied by a discrete perfectly non-wetting blob. Figure 5.12 shows the results for different magnetic field orientations and ferrofluid magnetic susceptibilities in a sinusoidal (diverging-converging) pore geometry. Note how changing the magnetic field orientation has a strong impact on the equilibrium configuration. Whereas the field parallel to the flow path pushes the blob interface away from the pore walls and extends the blob along the flow path, the perpendicular field slightly increases the length of the blob contact to the wall. This difference can be explained by the different kinds of magnetic field distribution in the three-phase contact region on each case: the parallel field is greatly amplified in the cusp while the perpendicular one is reduced. The result suggests that the interface of a wetting ferrofluid inside a constrained space is mostly determined by the local interactions of the magnetic field with the geometry of three-phase zones, rather than the pore geometry and magnetic field distribution as a whole. In Soares et al. (2014), it was shown that a small increase in magnetic susceptibility leads to much larger interface displacements (up to a saturation point, of course), and could potentially be optimized by appropriate ferrofluid synthesis.

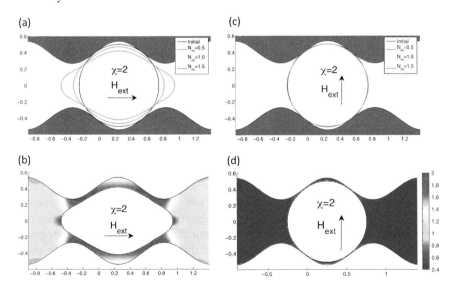

FIGURE 5.12
Each colored line in (a) and (c) is an equilibrium interface shape obtained with a different ratio of magnetic to capillary strength N_m and the dark line is the original interface of the blob (absence of magnetic field); (b) and (d) show corresponding normalized magnetic field H/H_{ext} for two different orientation of the external magnetic field. Ferrofluid magnetic susceptibility is 2 in all cases shown, for different values and more details see Soares et al. (2014).

5.4.1.5 Summary: Ferrofluid Potential for Enhanced Oil Recovery

No experimental studies have been reported that investigate the pore scale behavior of a non-wetting fluid surrounded by wetting ferrofluid phase in a magnetic field, either in a simple (microfluidic) geometry nor in more complicated pore structures resembling those in natural porous materials (*e.g.*, micromodels). The preliminary numerical investigation described above shows potential for reducing residual non-wetting phase (contaminant or residual oil) and thus a multitude of practical applications are contemplated. In the particular case of ferrofluids in natural porous media and when the ferrofluid is the wetting phase, large magnetic field gradients are expected in the ferrofluid regions occupying the sharp cusp formed by a highly non-wetting phase and the porous medium surface. Predicting how these gradients will affect the geometry of the interface is difficult without an appropriate model. A complex geometry also poses difficulty in solving and coupling both fluid displacement and magnetic field distributions, and efficient 3D models and methods are lacking. Finally, in natural porous media, one is typically interested in averaged behavior (ferrofluid/fluid saturations, displacement efficiency, permeability, and so on, as a function of saturation) in a representative volume of individual pores and this would be the only way to show overall enhanced EOR in applications. In the case of ferrofluid-fluid displacement in a magnetic field, such upscaling is lacking due to the absence of fundamental knowledge about pore scale behavior. Finally, in all applications, it remains a problem to deliver a strong field in practical subsurface applications.

5.5 Magnetic Sensing

5.5.1 Brief Introduction to Magnetometers

Magnetometers (Jackson Kimball and Budker 2013), the simplest of them being a compass, measure the magnetization of a magnetic material or the direction/strength/relative change of a magnetic field in a location. They can also be used as metal detectors, which is not the focus in subsurface operations. Carl Friedrich Gauss in 1833 created the first magnetometer for measurement of absolute magnetic density. Magnetometers are classified based on the field strength into magnetometers (<1 mT) and gaussmeters (>1 mT). Scalar magnetometers measure the total strength of the magnetic field to which they are subjected, but not its direction. Vector magnetometers have the capability to measure the component of the magnetic field in a particular direction, relative to the spatial orientation of the device. Table 5.1 lists the instrument types with their field strength range, resolution, and bandwidth. The most sensitive of them all is superconducting quantum interference

TABLE 5.1

Field Measurement of Magnetic Field Characteristics, After Macintyre (2000)

Instrument	Range (mT)	Resolution (nT)	Bandwidth (Hz)	Comment
Induction coil	10^{-10} to 10^6	Variable	10^{-1} to 10^6	Cannot measure static fields
Fluxgate	10^{-4} to 0.5	0.1	dc to 2×10^3	General-purpose vector magnetometer
SQUID	10^{-9} to 0.1	10^{-4}	dc to 5	Highest sensitivity magnetometer
Hall effect	0.1 to 3×10^4	100	dc to 10^8	Best for fields above IT
Magnetoresistance	10-3 to 5	10	dc to 10^7	Good for mid-range applications
Proton precession	0.02 to 0.1	0.05	dc to 2	General-purpose scalar magnetometer
Optically pumped	0.01 to 0.1	0.005	dc to 5	Highest resolution scalar magnetometer

The Hall effect device is a gaussmeter, all others in the table are magnetometers. The magneto-resistive sensors cover the middle ground between the low- and high-field sensors.

device (SQUID) magnetometers. However, these sensors operate at temperatures near absolute zero and require special thermal control systems. Thus, they are not robust for field use.

5.5.2 Electromagnetic Tomography

Electromagnetic (EM) tomography images the contrast in electromagnetic properties of different geologic features present in the subsurface. Electromagnetic properties include electrical conductivity, magnetic permeability, and dielectric permittivity of the rock/fluid system. Crosswell EM measurements are gaining importance in the petroleum industry as important reservoir characterization and monitoring tools, adding to the understanding of reservoir heterogeneity and fluid front evolution (Marion et al. 2011; Sanni et al. 2007). Note that traditional well log data provides high resolution measurements of the formation very close to the wellbore and surface-based methods provide a large volume of investigation but with very coarse resolution. Thus, crosswell measurements bridge this information gap by imaging the inter-well region at the reservoir scale. Crosswell EM typically employs low frequencies (tens of kHz and less) to increase the depth of investigation.

A typical EM system consists of a transmitter that broadcasts a time-varying magnetic field and multiple receivers that detect the field at some distance away. For conventional EM tomography, the recorded magnetic fields are a combination of the primary field of the transmitter and the secondary fields produced by currents induced in the electrically conductive formation.

The ratio of scattered, or secondary field, to primary magnetic field depends upon conductivity, frequency, and borehole separation (Wilt et al. 1995). Whenever possible, the source and receiver are placed at regularly spaced intervals below, within, and above the depth of interest (Alumbaugh et al. 2008). The process starts with creating an initial model (mainly through logs and background geology). The EM data is collected and interpreted via an inverse technique whereby the inversion procedure adjusts the initial model until the observed and calculated data fit within a given tolerance. This process results in building an image of the inter-well conductivity structures (Montaron et al. 2007).

5.5.2.1 Combining SNP Flooding and Cross-well EM Tomography

As of now, the only EM property of interest for deep EM readings has been electrical conductivity. Surface-coated (superpara)magnetic nanoparticles are capable of flowing through micron-size pores across long distances in a reservoir with modest retention in rock. At the same time, they change magnetic permeability of the flooded region. Tracing these contrast agents using the current electromagnetic tomography technology could potentially help track the flood-front in waterflood and EOR processes and characterize the reservoir. Once the induction responses are measured at the observation well, a set of Maxwell equation solutions will be obtained with the assumed locations/configurations of the nanoparticle dispersion zone; and an inversion algorithm is employed to identify the exact location/configuration. Flooding the reservoir with ferrofluids has a potential of improving the contrast in electromagnetic tomography between the flooded region and the rest. One key advantage is that the magnetic nanoparticles provide high resolution measurements at very low frequencies where the conductivity contrast is hardly detectable and thus the sensitivity of magnetic measurements can be improved. However, for that to work, the effect of wellbore casing (which uses iron) has to be manageable.

   Rahmani (2013) quantified the change in magnetic permeability in standard reservoir conditions and investigated the potential of SNPs to create sufficient resultant contrast for improved EM imaging. We point the reader to Chapter 9 for more details.

5.5.3 Paleomagnetism: Its Use for Petrophysical Characterization

Paleomagnetism is the study of the Earth's magnetic field of rocks, sediment, or archaeological materials (Tauxe 2010). Magnetic minerals in rocks carry the information on the direction and intensity of the magnetic field when they form. This record provides information on the past behavior of Earth's magnetic field and allows reconstruction of the past position of tectonic plates. The record of geomagnetic field reversals found in volcanic and sedimentary rock sequences (also known as magnetostratigraphy) provides a timescale and is used as a geochronology tool.

Aluminum, sodium, magnesium, and most of the clay minerals are paramagnetic. The important paramagnetic minerals in marine sediments include clay minerals (*e.g.*, illite, chlorite, and smectite), ferromagnesian silicates (*e.g.*, biotite, tourmaline, pyroxene, and amphiboles), iron sulfides (*e.g.*, pyrite and marcasite), iron carbonates (such as siderite and ankerite), and other iron- and magnesium-bearing minerals. They can be used for regional or global correlation (*e.g.*, see Ellwood et al. 2000). The study of susceptibility in contrast with rock magnetic analyses provides quantitative and qualitative information about the paramagnetic and ferromagnetic materials in the examined rock. Low-field magnetic and mass-specific susceptibility measured on mini-cores can be cross-correlated with all other measurements in well logging.

5.5.4 Magnetic Particle Imaging

Magnetic particle imaging (MPI) is a new tomographic imaging technique which measures the spatial distribution of superparamagnetic nanoparticles (Biederer et al. 2009; Gleich 2014; Gleich and Weizenecker 2005). MPI is a quantitative imaging modality, providing high sensitivity and sub-millimeter spatial resolution. Furthermore, the acquisition time is short, allowing for real time applications.

The basic idea of magnetic particle imaging is detection of magnetic material using its non-linear magnetization. The superparamagnetic nanoparticles, whose properties are described in Section 5.2.3 are the ideal candidates for imaging. The zero coercive field and the unstable magnetic moment of these particles makes them more sensitive to the presence of an applied magnetic field compared with the non-superparamagnetic single-domain nanoparticles. In other words, the magnetization (the magnetic moment) of the superparamagnetic nanoparticles responds immediately to the magnetic field (and magnetically rotates). For single-domain (non-superparamagnetic) nanoparticles, the magnetization will respond to the magnetic field in a weaker sense because the magnetization will not respond to the magnetic field until the field exceeds the coercive field (H_c), effectively reducing sensitivity. Practical review of MPI principles is available in Gleich (2014).

The fundamental principle of MPI is illustrated in Figure 5.13. To determine the spatial distribution of magnetic nanoparticles, a time varying, sinusoidal magnetic field is applied to the nanoparticles (Figure 5.13a). Due to their non-linear magnetization curve (Figure 5.13a), the magnetization response contains the excitation frequency f_0 as well as harmonics (i.e., integer multiples) of this frequency (Figure 5.13c). In a receive coil, a signal is induced, which is directly proportional to the time derivative of the particle magnetization (Figure 5.13d). By Fourier transformation of the induced signal, the harmonics can be determined (Figure 5.13e).

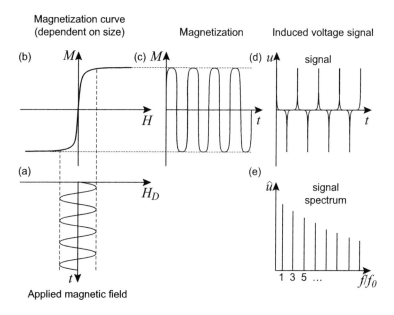

FIGURE 5.13

Physical effect exploited in MPI: (a) a sinusoidal magnetic field oscillation is applied to SPM-NP which has its unique magnetization curve, (b). The anharmonic magnetization, shown in (c), induces a voltage signal, $u(t) \propto dM(t)/dt$, in a receiver coil, as shown in (d). Due to the non-linear magnetization curve, the spectrum of the acquired signal contains the excitation frequency f_0 as well as higher harmonics, (e). From Huh et al. (2005).

5.5.4.1 Detection of Fluid-Ferrofluid Interface Motion Detection

SNPs are currently being used to enhance magnetic resonance imaging (MRI) of targeted human body parts (Oh et al. 2006). A potential application of SNPs for remote detection of the presence of oil in subsurface formations has been explored based on the successful application in these biomedical applications. The idea is to exploit the imbalance of forces at the ferrofluid (wetting fluid, presumably water) and non-magnetic fluid (presumably oil). When exposed to the external magnetic field oscillation, there is an interface fluctuation which can be detected (see Figure 5.14). Investigation so far has been limited to the lab and it is hypothesized that one could detect acoustic response (and even resonance) of interface between gas or oil and magnetized water (ferrofuid) in the field. The measurements of interfacial motion by phase-sensitive optical coherence tomography (as well as theoretical predictions) show frequency doubling of the interface movement whose amplitude ranges from 50 to 3300 nm depending on the SNP magnetic properties and applied magnetic field (Ryoo et al. 2012). The frequency doubling arises due to super-paramagnetism of the nanoparticles: they can rotate freely yet are bound by ferrofluid phase boundary, and at investigated frequencies 5–50 Hz they have "enough time" to relax. In Chapter 9, we provide more

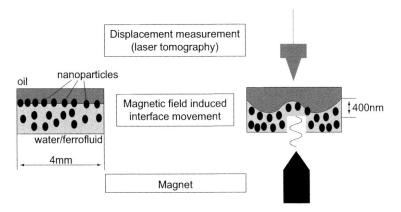

FIGURE 5.14
Schematic of the fluid-ferrofluid interface movement measurement in the lab (not to scale).
From Ryoo et al. (2012).

detail on how the fundamental mechanisms behind MPI are employed in a remote reservoir sensing application.

5.6 Hyperthermia Heating

5.6.1 Brief Introduction to Medical Hyperthermia

The superparamagnetic nature of the SPM-NPs means that, when an external magnetic field is applied, the nanoparticles generate intense, localized heating. This targeted hyperthermia is currently used to treat cancerous tumors in vivo without damaging surrounding tissue (Laurent et al. 2011; Nedyalkova et al. 2017).

5.6.2 Hyperthermia Theory

SNPs can generate localized heat when exposed to a magnetic field. This is because of the energy released by the magnetization of each particle attempting to return to equilibrium (see Section 5.2.3). The specific absorption rate (SAR), which is defined as $c(dT / dt)$ where dT / dt is the change in temperature with time and c is the specific heat capacity (Chou 1990; Jordan et al. 1993), is used to describe the heating. SAR may be calculated for a material with imbedded SNPs as

$$\text{SAR} = \pi\mu_0^2 \frac{\varphi M_s^2 V}{k_B T} H_0^2 \nu \frac{2\pi\nu\tau}{1+(2\pi\nu\tau)^2} \tag{5.29}$$

where μ_0 is the magnetic permeability of free space, φ is the volume fraction of SNPs in dispersion, M_s is the specific magnetization of the SNPs, V is the volume of an individual SNP, is the magnetic field intensity, v is the frequency of the magnetic field, τ is the relaxation time (time constant for the return of the particle's magnetization to equilibrium), k_B is Boltzmann's constant, and T is absolute temperature (Laurent et al. 2008, 2011). Equation (5.29) yields a maximum when $\tau = 1/(2\pi v)$ (Rosensweig 2002). In medicine, magnetic field frequencies on the order of hundreds of kHz are usually used, since higher frequencies tend to create hot spots because of local variations in tissue properties (Andrä and Nowak 2007; Johnson et al. 1993). For a frequency of 500 kHz, the maximum SAR should be obtained with SNPs with τ values around 0.3 μs, indicating that Néel relaxation is the dominant mechanism in which only the magnetization vector is perturbed (in contrast with Brownian relaxation, in which the entire particle rotates and relaxes viscously as explained in Section 5.2.3). The Néel relaxation time τ_N scales with the particle radius r as $\ln \tau_N \sim r^3$, and $\tau_N = 0 \left(10^{-1} \mu s\right)$ occurs for SNPs when r is roughly 6–7 nm. SAR values of up to 400 W/g have been reported in the literature for SNPs of this size dispersed as ferrofluid (Hergt et al. 2004).

5.6.3 Potential for Oilfield Applications

Flow assurance is a critical problem in the oil and gas industry, as an increasing number of wells are drilled in deep water and ultra-deep water environments. High pressures and temperatures as low as 5°C in these environments hinder flow of hydrocarbon-based fluids by formation of methane hydrate and wax deposits on the inner surface of pipelines. Two main issues arise in applying the SNPs and localized heating for this application: (1) delivering nanoparticles to the desired location and having them effectively raise the local temperature; and (2) delivering magnetic field to the location. Preliminary experimental and numerical work was done on (1) and is briefly described in Chapter 8. The preliminary work described therein paves the way for additional, detailed experimental and modeling study of the potential of this technology for flow assurance in deepwater environments.

List of Abbreviations

EM	Electromagnetic
EOR	Enhanced oil recovery
FHD	Ferrohydrodynamics
MHD	Magnetohydrodynamics
NP	Nanoparticle

SAR   Specific absorption rate
SNP   Superparamagnetic nanoparticle
TEM   Transmission Electron Microscopy

Nomenclature

$\vec{B}$	Magnetic induction field (T)
F_b	Buoyant force (N)
F_{drag}	Drag force (N)
F_g	Gravity force (N)
F_m	Magnetic force (N)
g	Gravitational acceleration (9.81 m/s^2)
$\vec{H}$	Magnetic field (A/m)
$\vec{M}$	Magnetization
n	Particle density in a suspension (g/m^3)
n_{max}	Maximum particle density in a settled suspension (g/m^3)
V_p, V_d	Particle and droplet volumes respectively (m^3)
η_f	Fluid viscosity (Ns/m)
μ	Magnetic permeability (dimensionless)
μ_0	Magnetic permeability of vacuum (= $4\pi \cdot 10^{-7}\,\text{NA}^{-2}$)
ρ	Density (fluid, particle density in the text) (kg^3/m)
χ	Magnetic susceptibility (dimensionless)

References

Afkhami, S., Renardy, Y., Renardy, M., Riffle, J.S., and St Pierre, T. (2008) Field-induced motion of ferrofluid droplets through immiscible viscous media. *J. Fluid Mech.*, 610, 363–380.

Alumbaugh, D. L., Donadille, J. M., Gao, G., Levesque, C., Nalonnil, A., Reynolds, L., Wilt, M., and Zhang, P. (2008) Multi-Scale Data Integration in Crosswell EM Imaging and Interpretation. Proceeding of the Society of Exploration Geophysics (SEG) Annual Meeting, November 9–14, Las Vegas, NV, USA.

Andrä, W., and Nowak, H. (Eds.) (2007) *Magnetism in Medicine: A Handbook*, 2nd completely rev. and enl. ed. Wiley-VCH, Weinheim, Germany.

Berkovsky, B., Bashtovoi, V., Mikhalev, V., and Rex, A. (1987) Experimental study of the stability of bounded volumes of magnetic fluid with a free surface. *J. Magn. Mater.*, 65, 239–241.

Berkovsky, B. M., Medvedev, V. F., and Krakov, M. S. (1993) *Magnetic Fluids: Engineering Applications*, 1st ed. Oxford University Press, Oxford; New York, NY, USA.

Biederer, S., Knopp, T., Sattel, T. F., Lüdtke-Buzug, K., Gleich, B., Weizenecker, J., Borgert, J., and Buzug, T. M. (2009) Magnetization response spectroscopy of superparamagnetic nanoparticles for magnetic particle imaging. *J. Phys. Appl. Phys.*, 42, 205007.

Charles, S. W. (2002) The Preparation of Magnetic Fluids. In Odenbach, S. (Ed.), *Ferrofluids*, Springer, Berlin Heidelberg, Germany, pp. 3–18.

Chatzis, I., Morrow, N. R., and Lim, H. T. (1983) Magnitude and detailed structure of residual oil saturation. *Soc. Pet. Eng. J.*, 23, 311–326.

Chikazumi, S., Graham, C. D., and Chikazumi, S. (2009) *Physics of Ferromagnetism*, 2nd ed., *International Series of Monographs on Physics*. Oxford University Press, Oxford; New York, NY, USA.

Chou, C.-K. (1990) Use of heating rate and specific absorption rate in the hyperthermia clinic. *Int. J. Hyperthermia*, 6, 367–370.

Ellwood, B. B., Crick, R. E., Hassani, A. E., Benoist, S. L. and Young, R. H. (2000) Magnetosusceptibility event and cyclostratigraphy method applied to marine rocks: Detrital input versus carbonate productivity. *Geology*, 28, 1135–1138.

Fuh, C. B., Tsai, H., and Lai, J. (2003) Development of magnetic split-flow thin fractionation for continuous particle separation. *Anal. Chim. Acta.*, 497, 115–122.

Gleich, B. (2014) *Principles and Applications of Magnetic Particle Imaging, Aktuelle Forschung Medizintechnik*. Springer Vieweg, Wiesbaden, Germany.

Gleich, B., and Weizenecker, J. (2005) Tomographic imaging using the nonlinear response of magnetic particles. *Nature*, 435, 1214–1217.

Hergt, R., Hiergeist, R., Zeisberger, M., Glöckl, G., Weitschies, W., Ramirez, L., Hilger, I., and Kaiser, W. (2004) Enhancement of AC-losses of magnetic nanoparticles for heating applications. *J. Magn. Magn. Mater.*, 280, 358–368.

Huh, C., Nizamidin, N., Pope, G. A., Milner, T. E., and Wang, B. (2015) Hydrophobic paramagnetic nanoparticles as intelligent crude oil tracers, US Patent Appl. 2015/0376493 A1.

Jackson Kimball, D. F., Budker, D. (2013) *Optical Magnetometry*. Cambridge University Press, Cambridge, UK.

Johnson, R. H., Robinson, M. P., Preece, A. W., Green, J. L., Pothecary, N. M., and Railton, C. J. (1993) Effect of frequency and conductivity on field penetration of electromagnetic hyperthermia applicators. *Phys. Med. Biol.*, 38, 1023–1034.

Jordan, A., Wust, P., Fähling, H., John, W., Hinz, A., and Felix, R. (1993) Inductive heating of ferrimagnetic particles and magnetic fluids: Physical evaluation of their potential for hyperthermia. *Int. J. Hyperth. Off. J. Eur. Soc. Hyperthermic Oncol. North Am. Hyperth. Group*, 9, 51–68.

Khushrushahi, S., Zahn, M., and Hatton, T. A. (2013) Magnetic separation method for oil spill cleanup. *Magnetohydrodynamics*, 49, 546–551.

Kittel, C. (2005) *Introduction to Solid State Physics*, 8th ed. Wiley, Hoboken, NJ, USA.

Ko, S., Kim, E. S., Park, S., Daigle, H., Milner, T. E., Huh, C., Bennetzen, M. V., and Geremia, G. A. (2016) Oil Droplet Removal from Produced Water Using Nanoparticles and Their Magnetic Separation. Society of Petroleum Engineers. SPE Annual Technical Conference and Exhibition, September 26–28, Dubai, UAE. https://doi.org/10.2118/181893-MS

Kucheryavy, P., He, J., John, V. T., Maharjan, P., Spinu, L., Goloverda, G. Z., and Kolesnichenko, V. L. (2013) Superparamagnetic iron oxide nanoparticles with variable size and an iron oxidation state as prospective imaging agents. *Langmuir*, 29, 710–716.

Lake, L. W. (1989). *Enhanced Oil Recovery.* Prentice Hall, Englewood Cliffs, NJ, USA.

Laurent, S., Dutz, S., Häfeli, U. O., and Mahmoudi, M. (2011) Magnetic fluid hyperthermia: Focus on superparamagnetic iron oxide nanoparticles. *Adv. Colloid Interface Sci.,* 166, 8–23.

Laurent, S., Forge, D., Port, M., Roch, A., Robic, C., Vander Elst, L., and Muller, R. N. (2008) Magnetic iron oxide nanoparticles: Synthesis, stabilization, vectorization, physicochemical characterizations, and biological applications. *Chem. Rev.,* 108, 2064–2110.

Lavrova, O., Matthies, G., Mitkova, T., Polevikov, V., and Tobiska, L. (2006) Numerical treatment of free surface problems in ferrohydrodynamics. *J. Phys. Condens. Matter,* 18, S2657.

Lavrova, O., Matthies, G., Polevikov, V., and Tobiska, L. (2004) Numerical modeling of the equilibrium shapes of a ferrofluid drop in an external magnetic field. *PAMM,* 4, 704–705.

Li, Z., and Ito, K. (2006) *The Immersed Interface Method: Numerical Solutions of PDEs Involving Interfaces and Irregular Domains, Frontiers in Applied Mathematics.* SIAM, Society for Industrial and Applied Mathematics, Philadelphia, PA, USA.

Macintyre, S. A. (2000) *Magnetic Field Measurement,* CRC Press LLC.

Marion, B. P., Zhang, P., Safdar, M., Wilt, M., Loh, F. F. H., and Nalonnil, A. (2011) Crosswell Technologies: New Solutions for Enhanced Reservoir Surveillance. SPE Enhanced Oil Recovery Conference. Presented at the SPE Enhanced Oil Recovery Conference, Society of Petroleum Engineers, July 19–21, Kuala Lumpur, Malaysia. https://doi.org/10.2118/144271-MS

Mehta, P. (2015) Application of Superparamagnetic Nanoparticle-Based Heating for Non-Abrasive Removal of Wax Deposits from Subsea Oil Pipelines. Master's Thesis, The University of Texas at Austin, Austin, TX, USA. https://doi.org/10.15781/T2HD7NZ3C

Montaron, B. A., Bradley, D. C., Cooke, A., Prouvost, L. P., Raffn, A. G., Vidal, A., and Wilt, M. (2007) Shapes of Flood Fronts in Heterogeneous Reservoirs and Oil Recovery Strategies. SPE/EAGE Reservoir Characterization and Simulation Conference. Presented at the SPE/EAGE Reservoir Characterization and Simulation Conference, Society of Petroleum Engineers, October 28–31, Abu Dhabi, UAE. https://doi.org/10.2118/111147-MS

Moridis, G. J., Borglin, S. E., Oldenburg, C. M., and Becker, A. (1998) *Theoretical and Experimental Investigations of Ferrofluids for Guiding and Detecting Liquids in the Subsurface (No. LBL-41069).* Lawrence Berkeley National Laboratory, Berkeley, CA, USA.

Nedyalkova, M., Donkova, B., Romanova, J., Tzvetkov, G., Madurga, S., and Simeonov, V. (2017) Iron oxide nanoparticles—In vivo/in vitro biomedical applications and in silico studies. *Adv. Colloid Interface Sci., Recent Nanotechnol. Colloid Sci. Devel. Biomedical Appl.,* 249, 192–212.

Oh, J., Feldman, M. D., Kim, J., Condit, C., Emelianov, S., and Milner, T. E. (2006) Detection of magnetic nanoparticles in tissue using magneto-motive ultrasound. *Nanotechnology,* 17, 4183–4190.

Oldenburg, C. M., Borglin, S. E., and Moridis, G. J. (2000) Numerical simulation of ferrofluid flow for subsurface environmental engineering applications. *Transp. Porous Media,* 38, 319–344.

Osher, S., and Sethian, J. A. (1988) Fronts propagating with curvature-dependent speed: Algorithms based on Hamilton-Jacobi formulations. *J. Comput. Phys.,* 79, 12–49.

Papell Solomon, S. (1965) Low viscosity magnetic fluid obtained by the colloidal suspension of magnetic particles. United States Patent number 3215572.

Payatakes, A. C. (1982) Dynamics of oil ganglia during Immiscible displacement in water-wet porous media. *Annu. Rev. Fluid Mech.*, 14, 365–393.

Pérez-Castillejos, R., Plaza, J. A., Esteve, J., Losantos, P., Acero, M. C., Cané, C., and Serra-Mestres, F. (2000) The use of ferrofluids in micromechanics. *Sens. Actuators Phys.*, 84, 176–180.

Peters, E. J. (2012) *Advanced Petrophysics*. Live Oak Book Company, Austin, TX, USA.

Pollert, E., and Zaveta, K. (2012) Nanocrystalline Oxides in Magnetic Fluid Hyperthermia. In Thanh, N.T.K. (Ed.), *Magnetic Nanoparticles: From Fabrication to Clinical Applications*, CRC Press, Boca Raton, Fl, USA, 449–477.

Prigiobbe, V., Ko, S., Huh, C., and Bryant, S. L. (2015) Measuring and modeling the magnetic settling of superparamagnetic nanoparticle dispersions. *J. Colloid Interface Sci.*, 447, 58–67.

Prodanovic, M., Lindquist, W. B., and Seright, R. S. (2007) 3D image-based characterization of fluid displacement in a Berea core. *Adv. Water Resour.*, 30, 214–226.

Rahmani, A. R. (2013) Modeling of Recovery Process Characterization Using Magnetic Nanoparticles. PhD Thesis. The University of Texas at Austin, Austin, TX, USA. http://hdl.handle.net/2152/28732.

Rahmani, A. R., Prodanović, M., Bryant, S. L., and Huh, C. (2012) Quasi-static analysis of a ferrofluid blob in a capillary tube. *J. Appl. Phys.*, 111, 074901.

Rodriguez Pin, E., Roberts, M., Yu, H., Huh, C., and Bryant, S. (2009) Enhanced Migration of Surface-Treated Nanoparticles in Sedimentary Rocks. Proceedings of SPE Annual Technical Conference and Exhibition. Presented at the SPE Annual Technical Conference and Exhibition. SPE Annual Technical Conference and Exhibition, 4–7 October, New Orleans, LA, USA. https://doi.org/10.2118/124418-MS

Rosensweig, R. E. (2002) Heating magnetic fluid with alternating magnetic field. *J. Magn. Magn. Material. Proc Int. Conf. Magn. Fluids*, 252, 370–374.

Rosenzweig, R. E. (1997) *Ferrohydrodynamics*. Dover Publications, Inc., Mineola, NY.

Ryoo, S., Rahmani, A. R., Yoon, K. Y., Prodanović, M., Kotsmar, C., Milner, T. E., Johnston, K. P., Bryant, S.L., and Huh, C. (2012) Theoretical and experimental investigation of the motion of multiphase fluids containing paramagnetic nanoparticles in porous media. *J. Pet. Sci. Eng.*, 81, 129–144.

Sanni, M. L., Yeh, N., Al-Afaleg, N. I., Al-Kaabi, A. U., Levesque, C., Donadille, J.-M., and Ma, S. M. (2007) Cross-Well Electromagnetic Tomography: Pushing the Limits. SPE Middle East Oil and Gas Show and Conference. Presented at the SPE Middle East Oil and Gas Show and Conference, Society of Petroleum Engineers, March 11–14, Manama, Bahrain. https://doi.org/10.2118/105353-MS

Saint-Martin de Abreu Soares, F. (2015) A Pore Scale Study of Ferrofluid-Driven Mobilization of Oil. Master's Thesis, The University of Texas at Austin, Austin, TX, USA. http://hdl.handle.net/2152/32139

Soares, F. S.-M. de A., Prodanovic, M., and Huh, C. (2014) Excitable Nanoparticles for Trapped Oil Mobilization. Presented at the SPE Improved Oil Recovery Symposium, Society of Petroleum Engineers, April 12–16, Tulsa, OK, USA, Paper number 169122. https://doi.org/10.2118/169122-MS

Tan, M. C., Chow, G. M., and Ren, L. (2009) *Nanostructured Materials for Biomedical Applications*. Transworld Research Network, Kerala, India.

Tarapov, I. E. (1974) Hydrodynamics of magnetizable and polarizable media. *Fluid Dyn.*, 9, 806–810.

Tauxe, L. (2010) *Essentials of Paleomagnetism.* University of California Press, Berkeley, CA, USA.

Whitmer, R. M. (1962) *Electromagnetics*, 2nd revised ed., Prentice Hall, Englewood Cliffs, NJ, USA.

Wilt, M. J., Alumbaugh, D.L., Morrison, H. F., Becker, A., Lee, K. H., and Deszcz-Pan, M. (1995) Crosswell electromagnetic tomography: System design considerations and field results. *Geophyics*, 60, 871–885.

Xu, M., Zhang, Y., Zhang, Z., Shen, Y., Zhao, M., and Pan, G. (2011) Study on the adsorption of Ca2+, Cd2+ and Pb2+ by magnetic Fe3O4 yeast treated with EDTA dianhydride. *Chem. Eng. J.*, 168, 737–745.

Xu, Y.-Y., Zhou, M., Geng, H.-J., Hao, J.-J., Ou, Q.-Q., Qi, S.-D., Chen, H.-L., and Chen, X.-G. (2012) A simplified method for synthesis of Fe3O4@PAA nanoparticles and its application for the removal of basic dyes. *Appl. Surf. Sci.*, 258, 3897–3902.

Yu, H., Yoon, K. Y., Neilson, B. M., Bagaria, H. G., Worthen, A. J., Lee, J. H., Cheng, V., Bielawski, C. W., Johnston, K. P., Bryant, S. L., and Huh, C. (2014) Transport and retention of aqueous dispersions of superparamagnetic nanoparticles in sandstone. *J. Pet. Sci. Eng.*, 116, 115–123.

Zhang, T. (2012) Modeling Nanoparticle Transport in Porous Media. PhD Thesis, University of Texas at Austin, Austin, TX, USA.

6

Nanotechnology and the Environment

6.1 Introduction

While nanotechnology has the potential to provide benefits to oil and gas operations, some of the unique properties of the nanoparticles that make them beneficial raise questions about potential negative impacts on the environment and human health. Several types of nanoparticles are employed in petroleum engineering and they are organic as well as inorganic. The most common nanoparticles used are made of silica (SiO_2), magnetic nanoparticles (*e.g.*, magnetite, Fe_3O_4), and metallic nanoparticles (*e.g.*, copper oxide, CuO, nickel oxide, NiO, titanium oxide, TiO_2, zinc oxide, ZnO), and carbon-based (Bennetzen and Mogensen 2014). They are applied in reservoir characterization to enhance image contrast (Al-shehri et al. 2013; Rahmani et al. 2015), to provide thermal information (Alaskar et al. 2011), and to identify the presence of hydrocarbons (Berlin et al. 2011). Nanoparticles are also used for enhanced oil recovery to affect viscosity, interfacial tension, and wettability of the residual oil (Hendraningrat et al. 2013; Zhang et al. 2010) and to stabilize foam (Espinoza et al. 2010; Yu et al. 2014). Moreover, nanoparticles, particularly magnetic nanoparticles, are used for water treatment to remove metals and residual oil droplets. These types of nanoparticles have a strong magnetic response and when functionalized they can selectively adsorb contaminants and then be separated from the solution through a magnetic field (Ko et al. 2014; Wang et al. 2017). Application of nanoparticles has also been reported in hydraulic fracturing to increase the viscosity, in the stability of drilling fluid (Al-Muntasheri et al. 2017), to reduce the flow of water into the shale in order to improve wellbore stability (Sharma et al. 2012), and to stabilize foam (Emrani et al. 2017). It is the latter application that raises the major concern regarding the release of nanoparticles into the environment (Amarfio and Mohammed 2014). The open arenas of the drilling sites are the locations where exposure to the environment is high and therefore to living organisms. Figure 6.1 shows a scheme of the possible exposure routes of nanoparticles to living organisms, reporting the processes that nanoparticles can undergo between the three environmental compartments, i.e., soil, water, and air.

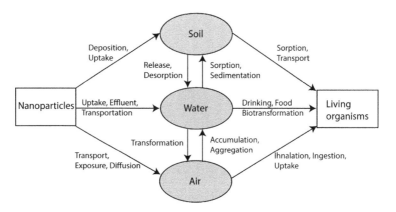

FIGURE 6.1
Exposure routes of nanoparticles to living organisms. From Shan et al. (2009).

6.2 Release of Nanoparticles into the Environment

Releases of nanoparticles can occur during any stage from their production through manufacturing, use/application, and ultimate end-of-life management/disposal (Salieri et al. 2018). Life cycle assessment (LCA) is the method used to assess the potential environmental impacts of nanoparticles across their entire life cycle. The recent paper by Salieri et al. (2018) summarizes the efforts done in this field of engineering and science and provides recommendation for future works. The life cycle of nanoparticles involves three major components: source, media, and receptors. The link between the former and the latter is named pathway or route. The main challenges in the LCA are (Hischier et al. 2017): (1) description of the processes the nanoparticles are subjected to; (2) the knowledge of the nanoparticles in use; and (3) release into the environment. The former point seems to have the larger effect on the uncertainty of the LCA. Nanoparticles can enter into the environment through several routes, specifically for the oil and gas sector: (1) wastewater from industrial application; (2) runoff from sites; (3) atmospheric transport; (4) spills from wells or storage pools; and (5) deliberate releases into the subsurface, *e.g.*, for the remediation of contaminated sites or for the enhanced recovery of petroleum (Babakhani et al. 2017). Once in the environment, nanoparticles will go to different environmental compartments (Zhang et al. 2009). Based on the evidence that nanoparticles are released to the environment, several authors created models to predict their environmental concentration. The review by Gottschalk et al. (2013) reports a comprehensive summary of all the models developed to identify processes of nanoparticles in the environment and analytical methods for their detection. The models distinguish in two classes, deterministic and stochastic, and the analytical methods divide in filtration, microscopic examination (*e.g.,* scanning electron microscopy/electron disper-sive X-ray microanalysis, SEM/EDX), spectroscopic analysis (*e.g.,* inductively coupled plasma-optical emission spectroscopy, ICP-MS), and liquid chroma-tography. The models and methods aim at predicting the distribution of the nanoparticles at scales of difference size, and their spatial distributions from local to regional. Comparing the results from several modeling and experi-mental studies, Gottschalk et al. (2013) derived the diagrams of the distribu-tion of different types of nanoparticles, namely, nanosized TiO_2, ZnO, Ag, fullerenes, carbon nanotube (CNT), and CeO_2. Figure 6.2[1] reports the distri-bution of these nanoparticles in three environmental compartments, i.e., soil, surface water, and air. This comparison shows that, within a certain range, there is some understanding of the expected environmental concentrations of nanoparticles commonly used in several products and industrial activities.

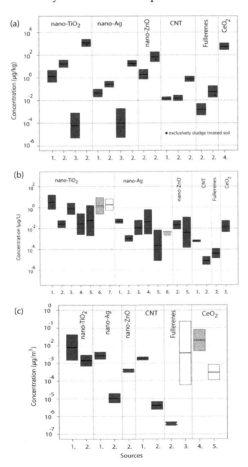

FIGURE 6.2
Modeled (dark gray) and measured (white) concentrations of nanoparticles in (a) soils, (b) water, and (c) air. The light gray boxes combine modeling and measurements. After Gottschalk et al. (2013).

6.3 Human Health Risk Assessment

Smita et al. (2012) developed a causal diagram to handle the complexity of issues on nanoparticle safety, from their exposure to the effects on the environment and health. Figure 6.3 gives an overview of the common sources of NPs and their interactions with various environmental processes that may pose threats to both human health and the environment. A person can potentially be exposed to nanoparticles through various routes such as inhalation, ingestion, skin contact, and injection. The exposure to nanoparticles and their accumulation in microbiota, plants, and humans may result in various adverse effects. In a person, the effects can change between respiratory discomforts and dermatitis, as well as being carcinogenic. Nanotoxicology is the sub-discipline at the interface of toxicology and nanomaterial science which studies the possibility of the direct interaction and interference of nanoparticles with vital cellular processes (Shekar et al. 2012).

Life cycle impact assessment (LCIA) aims at calculating the potential environmental and human health impacts of a certain product (Hristozov

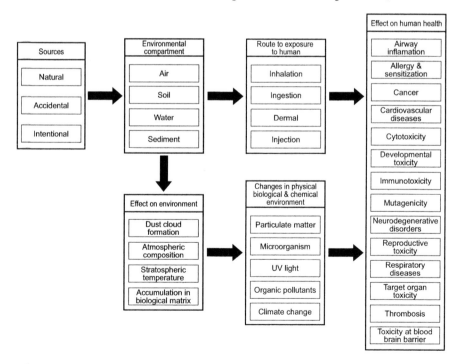

FIGURE 6.3
Proposed cause-effect diagram of NPs on environment and human health. From Smita et al. (2012).

et al. 2016). It is largely used as a tool for risk assessment of manufactured nanomaterials. One of the quantitative approaches applied to nanoparticles allows us to determine an impact score (*IS*) (Hauschild and Huijbregts 2015),

$$IS = \sum_{i,j} m_{i,j} CF_i \qquad (6.1)$$

where
$m_{i,j}$ is the mass of nanoparticles emitted (m^3), with *i* a specific nanoparticle emitted, and *j* a certain stage in the nanoparticle life cycle,
CF_i is the corresponding characterization factor (m^3d/kg).

To determine the value of CF_i information on the environment, the extent of human and environment exposure and (eco)toxic effects are required. A tool has been developed by the United Nations Environment Program (UNEP) and the Society for Environmental Toxicology and Chemistry, which is named USEtox R (Rosenbaum et al. 2008). The tool allows the determination of CF as:

$$CF = FF \cdot EF \cdot XF \qquad (6.2)$$

where *FF* is the fate factor, i.e., the persistence of a chemical in a given media (days); *XF* is the exposure factor for human toxicity, which relates the concentration of a substance to its intake (days^{-1}); and *EF* is the effect factor for human toxicity (cases of intake/kg emitted), which relates to the amount of a substance taken in by the population via inhalation or ingestion to the probability of adverse effect (carcinogenic/non-carcinogenic). The work by Salieri et al. (2018) summarizes the values of *CF* calculated using USEtox® for some types of nanoparticles. One of the recent models to determine *FF* is SimpleBox4Nano (SB4N) (Meesters et al. 2014). It is a fate modeling tool that accounts for the release of nanoparticles in various environmental compartments, such as soil, water, air, and sediment. The model accounts for processes such as agglomeration/aggregation, attachment, and dissolution. Figure 6.4 reports a scheme of the physical processes the nanoparticles undergo in the three environmental compartments accounted for. The output of the model is the PECs of free, aggregated, and attached species as a function of time and at steady state. From this, the values of *FF* are determined. The work by Ettrup et al. (2017) combines USEtox® and SB4N to develop the LCIA of TiO_2 nanoparticles, providing the *CF* values of 1.19, 1.55×10^3, and 6.05×10^2 m^3d/kg emitted, correspondingly, to soil, water, and air.

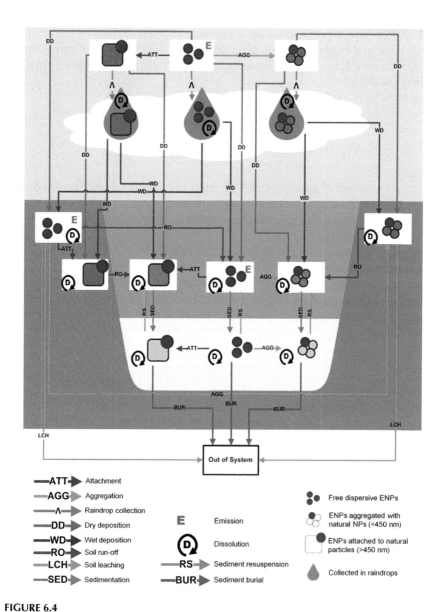

FIGURE 6.4
Overview of model concept SimpleBox4nano (SB4N). Reprinted with permission from Meesters et al. (2014). Copyright 2014 American Chemical Society.

6.4 Environmental Regulation

In the United States, nanoscale materials are regulated under the Toxic Substances Control Act (TSCA) (EPA 2018b). The Environmental Protection Agency (EPA) is pursuing a comprehensive regulatory approach under TSCA including an information gathering rule on new and existing nanomaterials and pre-manufacture notifications for new nanomaterials. Therefore, each nanoscale material will be evaluated on the basis of the specific nanoscale material properties and those of any structural analogs. All the information is available in ChemView (EPA 2018a). Upon characterization, EPA will determine whether further action, including additional information collection, is needed for that specific nanoscale material to facilitate assessment of risks and risk management, examination of the benefits and costs of further measures, and making future decisions based on available scientific evidence. In the European Union (EU), nanoscale materials are covered by the regulation concerning the Registration, Evaluation, Authorisation and Restriction of Chemicals (REACH) (EU 2006) and Classification, Labelling and Packaging of Substances and Mixtures (CLP) (EU 2009). REACH is the most comprehensive legislative provision for chemicals (substances) in the EU and applies to chemicals in whatever size, shape, or physical state. Nanomaterials are substances at the nanoscale covered by the REACH definition of substances as laid down in its article 3(1) (Rauscher et al. 2017). Most information requirements in REACH are triggered by the quantity of a substance produced or imported per year per manufacturer or importer. Substances that are marketed in the EU in volumes larger than 1 t/y have to be registered, and for more than 10 t/y, a chemical safety assessment is required. The CLP regulation requires that suppliers of a hazardous substance (on its own or mixture) notify the European Chemicals Agency (ECHA) of its classification and labeling within one month of placing the substance on the market for the first time. Suppliers need to decide on the hazard class and hazard category of a substance or mixture by gathering available data and referring to chemical classification criteria given in the CLP regulation.

Note

1. In part a of the figure, sources are: (1) Predicted environmental concentrations (PEC) for a current and high exposure scenario at regional level (Switzerland) covering diffusive engineered nanomaterials (ENM) emissions from a comprehensive spectrum of ENM applications. (2) PECs (15–85% quantiles range of modeled probability distributions) at regional level (Europe) reflecting diffusive ENM emissions from a comprehensive spectrum of ENM applications. (3) Range of local PECs (Johannesburg Metropolitan City, South Africa) for a minimum and maximum ENM release scenarios from cosmetic products via WWTPs. (4) PECs for soils near

highways considering the ENM accumulated during a 40-year period. In part b of the figure, sources are: (1) PECs for a current and high exposure scenario at regional level (Switzerland) covering diffusive ENM emissions from a comprehensive spectrum of ENM applications. (2) PECs (15–85% quantiles range of modeled probability distributions) at regional level (Europe) reflecting diffusive ENM emissions from a comprehensive spectrum of ENM applications. (3) Range of minimal and maximal values of 90% confidence intervals for mean PECs modeled at regional level (Ireland) covering selected applications. (4) Range of local PECs (Johannesburg Metropolitan City, South Africa) for a minimum and maximum ENM release scenarios from cosmetic products via WWTPs and a dilution factor of 1. (5) PECs at local resolution for river water (Switzerland) based on a conservative scenario without any ENM degradation/deposition and an optimistic one assuming rapid ENM removal. (6) PECs of sunscreen ENM TiO_2 for river water of the Thames and Anglian regions in southern England and different water flow conditions. (7) Measured environmental concentrations (MEC) (filtered Ti <0.45 mm) for river water that drains different land types (urban, industrial, rural agricultural land) (UK). (8) PECs in the Trent and the Thames basins derived from long-term average water flows and two scenarios for the ENM removal efficiency in WWTPs. In part c of the figure, sources are: (1) PECs for a current and high exposure scenario at regional level (Switzerland) covering diffusive ENM emissions from a comprehensive spectrum of ENM applications. (2) PECs (15–85% quantiles range of modeled probability distributions) at regional level (Europe) reflecting diffusive ENM emissions from a comprehensive spectrum of ENM applications. (3) MEC for the Mediterranean Sea atmosphere. (4) PECs in different areas of street canyons. (5) MECs of a Newcastle city center monitoring site after the introduction of the nanoparticulate (CeO_2) diesel fuel additive.

References

Alaskar, M., Ames, M., Liu, C., ... Cui, Y. (2011) Smart nanosensors for in-situ temperature measurement in fractured geothermal reservoirs. *Geoth Res*, 35 2(2006), 1371–1381.

Al-Muntasheri, G. A., Liang, F., and Hull, K. L. (2017) Nanoparticle-enhanced hydraulic-fracturing fluids: A review. *SPE Prod Oper*, 32(2), 186–195.

Al-shehri, A. A., Ellis, E. S., Servin, J. M. F., Kosynkin, D. V., Kanj, M. Y., and Schmidt, H. K. (2013) Illuminating the Reservoir: Magnetic NanoMappers. In *SPE Middle East Oil and Gas Show and Conference, Mar. 10-13, Manama, Bahrain*, pp. 40–47.

Amarfio, E. M., and Mohammed, S. (2014) Environmental and Health Impacts of Nanoparticles Application in Our Oil Industry. Society of Petroleum Engineers. SPE Nigeria Annual International Conference and Exhibition, Lagos, Nigeria. p. 4.

Babakhani, P., Bridge, J., Doong, R. A. , and Phenrat, T. (2017) Continuum-based models and concepts for the transport of nanoparticles in saturated porous media: A state-of-the-science review, *Adv Colloid Interfac*, 246, 75–104.

Bennetzen, M. V., and Mogensen, K. (2014) Novel Applications of Nanoparticles for Future Enhanced Oil Recovery, Kuala Lumpur, Malaysia: International Petroleum Technology Conference. doi:10.2523/IPTC-17857-MS.

Berlin, J. M., Yu, J., Lu, W., and Tour, J. M. (2011) Engineered nanoparticles for hydrocarbon detection in oil-field rocks. *Energ Environ Sci*, 4(2), 505–509.

Emrani, A. S., Ibrahim, A. F., and Nasr-El-Din, H. A. (2017) Mobility Control using Nanoparticle-Stabilized CO2 Foam as a Hydraulic Fracturing Fluid. SPE Europec featured at 79th EAGE Conference and Exhibition, Paris, France. p. 16.

EPA. (2018a) *Chemview*. Retrieved from https://chemview.epa.gov/chemview. Last access: 07/31/2018.

EPA. (2018b) Reviewing New Chemicals Under the Toxic Substances Control Act (TSCA). Retrieved from www.epa.gov/reviewing-new-chemicals-under-toxic-substances-control-act-tsca/control-nanoscale-materials-under. Last access: 07/31/2018

Espinoza, D. A., Caldelas, F. M., Johnston, K. P., Bryant, S. L., and Huh, C. (2010) Nanoparticle-Stabilized Supercritical CO2 Foams for Potential Mobility Control Applications. SPE Improved Oil Recovery Symposium, Tulsa, OK, USA. p. 13.

Ettrup, K., Kounina, A., Hansen, S. F., Meesters, J. A. J., Vea, E. B., and Laurent, A. (2017) Development of comparative toxicity potentials of TiO 2 nanoparticles for use in life cycle assessment. *Environ Sci Techno*, 51(7), 4027–4037.

EU (2006) Regulation (EC) No 1907/2006 of the European Parliament and of the Council of 18 December 2006 Concerning the Registration, Evaluation, Authorisation and Restriction of Chemicals (reach), Establishing a European Chemicals Agency, Amending Directive 1999/4.

EU (2009) Regulation (EC) no 1272/2008 on Classification, Labelling and Packaging of Substances and Mixtures. Available at: https://echa.europa.eu/documents/10162/13643/questions_and_answers_clp_20090526_en.pdf. Last access: 07/31/2018.

Gottschalk, F., Sun, T., and Nowack, B. (2013) Environmental concentrations of engineered nanomaterials: Review of modeling and analytical studies. *Environ Pollut*, 181, 287–300.

Hauschild, M. Z., and Huijbregts, M. A. J. (2015) *Introducing Life Cycle Impact Assessment*, Springer, Dordrecht, the Netherlands.

Hendraningrat, L., Li, S., and Torsater, O. (2013) A Coreflood Investigation of Nanofluid Enhanced Oil Recovery in Low-Medium Permeability Berea Sandstone. SPE International Symposium on Oilfield Chemistry, The Woodlands, TX, USA. p. 14.

Hischier, R., Salieri, B., and Pini, M. (2017) Most important factors of variability and uncertainty in an LCA study of nanomaterials—Findings from a case study with nano titanium dioxide. *NanoImpact*, 7, 17–26.

Hristozov, D., Gottardo, S., Semenzin, E.…Marcomini, A. (2016) Frameworks and tools for risk assessment of manufactured nanomaterials. *Environ. Int.*, 95, 36–53.

Ko, S., Prigiobbe, V., Huh, C., Bryant, S. L., Bennetzen, M. V., and Mogensen, K. (2014) Accelerated Oil Droplet Separation from Produced Water Using Magnetic Nanoparticles. SPE Annual Technical Conference and Exhibition, Amsterdam, the Netherlands. p. 14.

Meesters, J. A. J., Koelmans, A. A., Quik, J. T. K., Hendriks, A. J., and Van De Meent, D. (2014) Multimedia modeling of engineered nanoparticles with simpleBox4nano: Model definition and evaluation. *Environ Sci Techno*, 48(10), 5726–5736.

Rahmani, A. R., Bryant, S., Huh, C., … Wilt, M. (2015) Crosswell magnetic sensing of superparamagnetic nanoparticles for subsurface applications. *SPE Journal*, 20(5), 1067–1082.

Rauscher, H., Rasmussen, K., and Sokull-Klüttgen, B. (2017) Regulatory aspects of nanomaterials in the EU. *Chemie-Ingenieur-Technik*, 89(3), 224–231.

Rosenbaum, R. K., Bachmann, T. M., Gold, L. S., … Hauschild, M. Z. (2008) USEtox - The UNEP-SETAC toxicity model: Recommended characterisation factors for human toxicity and freshwater ecotoxicity in life cycle impact assessment. *Int J Life Cycle*, 13(7), 532–546.

Salieri, B., Turner, D. A., Nowack, B., and Hischier, R. (2018) Life cycle assessment of manufactured nanomaterials: Where are we? *NanoImpact*, 10, 108–120.

Shan, G. Bin, Surampalli, R. Y., Tyagi, R. D., Zhang, T. C., Hu, Z., and Yan, S. (2009) Environmental Risks of Nanomaterials. In *Nanotechnologies for Water Environment Applications*, American Society of Civil Engineers (ASCE). pp. 591–618.

Sharma, M. M., Zhang, R., Chenevert, M. E., Ji, L., Guo, Q., and Friedheim, J. (2012) A New Family of Nanoparticle Based Drilling Fluids. SPE Annual Technical Conference and Exhibition, San Antonio, TX, USA. p. 13.

Shekar, S., Smith, A. J., Menz, W. J., Sander, M., and Kraft, M. (2012) A multidimensional population balance model to describe the aerosol synthesis of silica nanoparticles. *Journal of Aerosol Science*, 44, 83–98.

Smita, S., Gupta, S. K., Bartonova, A., Dusinska, M., Gutleb, A. C., and Rahman, Q. (2012) Nanoparticles in the environment: Assessment using the causal diagram approach, *Environ Health*, 11(1), S13.

Wang, Q., Prigiobbe, V., Huh, C., and Bryant, S. L. (2017) Alkaline earth element adsorption onto PAA-coated magnetic nanoparticles. *Energies*, 10(2), 223.

Yu, J., Khalil, M., Liu, N., and Lee, R. (2014) Effect of particle hydrophobicity on CO2foam generation and foam flow behavior in porous media. *Fuel*, 126, 104–108.

Zhang, T. C., Surampalli, R., and Lai, K. C. K. (2009) Fate and Transport of Nanomaterials in Aquatic Environments. In *Nanotechnologies for Water Environment Applications* (Chap. 15). American Society of Civil Engineers (ASCE).

Zhang, T., Davidson, D., Bryant, S. L., and Huh, C. (2010) Nanoparticle-Stabilized Emulsions for Applications in Enhanced Oil Recovery. SPE Improved Oil Recovery Symposium, Tulsa, OK, USA. p. 18.

7

Drilling and Completions

7.1 Introduction

Drilling and completions represent major capital investments from oil companies, since these are essentially sunk costs that must be committed before any revenue is generated from sales of produced products. Because of this, there is tremendous interest in the oil and gas industry in developing new technologies to minimize the time required to drill and complete wells, as well as the complexity of drilling and completion design (*e.g.*, Gilje et al. 2017; Lukawski et al. 2014). In addition to financial consideration, operators are always seeking to reduce environmental and safety hazards during drilling in completion, particularly by mitigating downhole pressure control issues, improving cement quality and formation-cement bonds, and reducing the amount of water needed for hydraulic fracturing operations. Various nanotechnology solutions are being developed to meet these needs. Most of these applications rely on using nanoparticles or other nanomaterials as additives in drilling fluids, fracturing fluids, and cement to alter their properties. In this chapter, we will discuss these applications, as well as nanotechnology applications in drilling and completion hardware and cuttings analysis.

7.2 Drilling Fluids

When drilling an oil or gas well, drilling fluid must be circulated down through the interior of the drill pipe and back up through the annulus between the drill pipe and formation. This drilling fluid serves four purposes: to carry out the cut pieces of rock (cuttings) and keep the hole clean; to equalize the formation pore pressure and prevent flow of formation fluids into the wellbore; to prevent the hole from caving; and to keep the drill bit cooled and lubricated (Laik 2018). As such, the rheology, density, thermal conductivity, and chemical properties of the drilling fluid must be carefully controlled. This is typically accomplished through the use of chemical additives, including fluid loss additives, clay stabilizers, lubricants, bactericides, corrosion inhibitors, and scale inhibitors (Fink 2015). As we will see, nanoparticles can often be substituted for these chemicals, thus reducing the environmental risk associated with the chemicals.

7.2.1 Rheology Improvement

The term *rheology* refers generally to the study of the flow of fluids and soft matter in response to an applied force. More specifically, rheology usually describes the relationship between shear stress and shear rate. Newtonian fluids are fluids in which strain rate is linearly proportional to shear stress, with the viscosity being the constant of proportionality (Bird et al. 2007). Any fluid that displays a non-linear relationship between shear stress and strain rate is non-Newtonian. Most drilling fluids are non-Newtonian (*e.g.*, Melrose and Lilienthal 1951), with their rheology often described by the Bingham model (Bingham 1922; Melrose et al. 1958) or the Herschel-Bulkley (power-law) model (Herschel and Bulkley 1926; Kelessidis et al. 2006). The Bingham model is

$$\dot{\gamma} = \begin{cases} 0, & \tau < \tau_0 \\ \dfrac{\tau - \tau_0}{\mu_B}, & \tau \geq \tau_0 \end{cases}' \tag{7.1}$$

where
 $\dot{\gamma}$ is the shear rate
 τ is the shear stress
 τ_0 is the yield stress
 μ_B is the Bingham viscosity.

The Herschel-Bulkley model is usually expressed in terms of shear rate and is given by

$$\tau = \tau_0 + I\dot{\gamma}^n, \tag{7.2}$$

where I is the consistency index and n is an exponent that describes whether the fluid is shear thinning ($n < 1$), Newtonian ($n = 1$), or shear thickening ($n > 1$).

 Drilling fluid rheology is important to control as it affects the efficiency of cuttings transport and hole cleaning, and the friction between the drilling fluid and bit, drillstring, and formation can cause wear and tear on pumps and downhole equipment and decrease efficiency of bit cooling. Theoretical and experimental work has demonstrated that better hole cleaning occurs when the velocity profile in the annulus between the drill pipe and formation is flatter and when the yield stress is higher (*e.g.*, Becker et al. 1991; Iyoho and Azar 1981; Okrajni and Azar 1986). The velocity profile for a Herschel-Bulkley fluid with zero yield stress in a centered annulus with outer radius r_1 and inner radius r_2 is given by

$$v(r) = \frac{n}{n+1}\left(\frac{1}{I}\frac{dP}{dl}\right)^{1/n}\left[\left(\frac{r_1-r_2}{2}\right)^{\frac{n+1}{n}} - \left|r - \frac{r_1+r_2}{2}\right|^{\frac{n+1}{n}}\right], \tag{7.3}$$

where dP/dl is the pressure gradient in the direction of flow (Iyoho and Azar 1981). Smaller values of n yield flatter velocity profiles, and clearly it is advantageous for drilling fluids to be shear thinning ($n<1$) for more efficient hole cleaning. Similarly, Founargiotakis et al. (2008) show that the friction factor f for flow of a Herschel-Bulkley fluid with zero yield stress through an annulus is given by

$$f = \frac{24}{N_{Re}} = 24\left(\frac{r_1-r_2}{12\bar{v}}\right)^{1-n}\frac{I}{\rho\bar{v}(r_1-r_2)}, \tag{7.4}$$

where

N_{Re} is the Reynolds number
$\bar{v}$ is the average flow velocity in the annulus
ρ is the fluid density.

It can be seen from inspection of Equation (7.4) that the friction factor is minimized when $n < 1$, again corresponding to a shear-thinning fluid. A larger yield stress has a similar effect in reducing the friction factor. Therefore, we may conclude that any additives to drilling fluid that give it more shear-thinning behavior and/or a larger yield stress will be advantageous for its performance.

Several different types of nanoparticle additives have been investigated to improve the rheology of drilling fluids. Vryzas and coworkers have tested iron oxide and silica nanoparticles for rheology control. They showed that adding both types of nanoparticles to an aqueous bentonite solution retained the shear-thinning behavior of the solution, but that iron oxide nanoparticles tended to increase the yield stress while silica nanoparticles tended to decrease it (Vryzas et al. 2015) (Figure 7.1a,b). This result was also found by Barry et al. (2015). Subsequent work showed that iron oxide nanoparticles improved the rheology of bentonite slurry at high temperatures, particularly after aging at 177°C for 16 hours (Vryzas et al. 2016). The superparamagnetic nature of the iron oxide nanoparticles has been demonstrated to allow control of fluid rheology through application of an external magnetic field. Vryzas et al. (2017) demonstrated that the yield stress of a bentonite slurry amended with 1 wt% iron oxide nanoparticles increased over 6-fold when a magnetic field of 0.7 T was applied (Figure 7.1c). They attributed this behavior to the formation of nanoparticle chains within the fluid as the magnetic moments of the nanoparticles aligned with the applied field, although another possibility may lie in magnetization of the bentonite particles themselves upon sorption

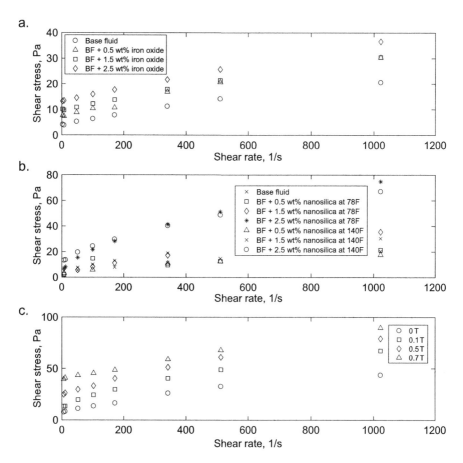

FIGURE 7.1

(a) Rheology plot for different iron oxide nanoparticle concentrations in 7 wt% aqueous bentonite suspension. All data measured at 78°F (26°C). (b) Rheology plot for different nanosilica concentrations in 7 wt% aqueous bentonite suspension. (c) Rheology plot for 7 wt% aqueous bentonite suspension containing 1 wt% iron oxide nanoparticles under different magnetic field strengths. Measurements were performed at 25°C. Figures generated using data from Vryzas et al. (2017).

of iron oxide nanoparticles as demonstrated by Galindo-Gonzalez et al. (2009). Vipulanandan et al. (2017) demonstrated similar behavior in water-based mud amended with bentonite and iron oxide nanoparticles exposed to a magnetic field. Wahid et al. (2015) tested the rheology of synthetic-based drilling fluid amended with silica nanoparticles at elevated temperatures (up to 177°C). They found that the plastic viscosity and yield point could be reduced nearly by half with the addition of 1.05 wt% nanosilica in combination with gilsonite. Abdo and Haneef (2013) found similar results using palygorskite nanoparticles. William et al. (2014) showed that CuO and ZnO nanoparticles in very low concentration (0.1–0.3 wt% base nanofluid added to water-based mud at a concentration of 1 vol%) improved the rheology of water-based mud at elevated temperatures by reducing the effect of pressure on viscosity and maintaining good shear-thinning behavior. While many of the physical processes leading to this behavior remain to be worked out, these results are promising for the development of a new class of smart drilling fluids.

Nanoparticles and nanocomposites, including both metal oxides and single- and multi-walled carbon nanotubes, have also shown promise as friction reducers in drilling fluids. These results mirror advancements made in the materials science community with applications in automotive and mechanical engineering (*e.g.*, Ettefaghi et al. 2013; Hwang et al. 2011; Ji et al. 2011; Liu et al. 2011; Ma et al. 2013; Peng et al. 2007; Wu et al. 2007). The mechanism behind nanoparticle-based friction reduction is generally agreed to be formation of a film of nanoparticles at the fluid-solid interface, allowing easier slip of fluid past the solid, combined with nanoparticles filling in pits and etches that may develop as a result of wear (Ji et al. 2011; Liu et al. 2011; Peng et al. 2007) (Figure 7.2). Nanoparticle surface chemistry may additionally be manipulated through surface functionalization or dispersion chemistry to

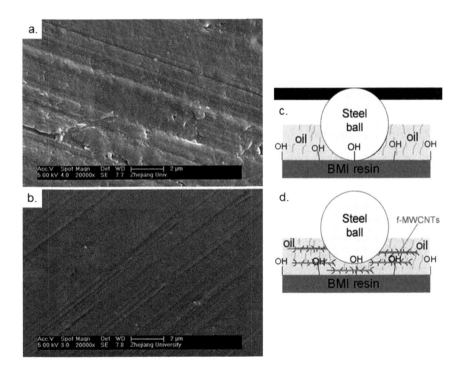

FIGURE 7.2
(a) Scanning electron microscope image of worn resin surface lubricate with pure paraffin oil. (b) Same as (a), but lubricant had functionalized multi-walled carbon nanotubes. (c) Illustration of wear in the presence of pure paraffin oil. (d) Carbon nanotubes bind to the resin surface and reduce contact between steel ball and resin, reducing wear. Figures from Liu et al. (2011).

promote this behavior (*e.g.*, Peng et al. 2007). Aftab et al. (2016) showed a 25% increase in lubricity with a nano-sized zinc oxide-coated acrylamide composite dispersed in water-based drilling fluid, which they attributed to the nanocomposite particles rotating during shear, acting like ball bearings. Similar results were found by Mao et al. (2015). and Ismail et al. (2016) and are similar to the results and analysis of Wu et al. (2007), who added nanoparticles to engine oil. Laboratory tests of carbon nanotubes in both water-based and oil-based drilling fluids have shown similar success, with slightly better friction reduction compared with metal oxide nanoparticles and nanocomposites (Ismail et al. 2016; Taha and Lee 2015).

Taha and Lee (2015) presented the results of a field test in Myanmar using graphene nanoparticles as a friction reducer. With 3% graphene nanoparticles by volume in water-based mud, reaming torque was reduced by 20% and the severe bit wear that had historically been a problem when drilling in that particular field was greatly mitigated. Good mud lubricity was observed at high temperatures up to 176°C. Krishnan et al. (2016) also performed a field trial in Myanmar using a water-based mud amended with a borate ester-based nanocomposite. Their laboratory results indicated up to 80% torque reduction with 5 wt% nanocomposite addition in 10 lbm/gal drilling fluid, and >50% torque reduction with 5 wt% addition in 13.5 lb/gal drilling fluid. The field test indicated that downhole torque was reduced by 44% at a temperature of 176°C. The results of these field tests are quite promising, and future work should expand the application of nanoparticles as friction reducers in drilling fluid.

7.2.2 Fluid Loss Control

When a new section of open hole is drilled, some fluid exchange between the wellbore and formation will occur, depending on the pressure differential between the two. In the case of overbalanced drilling, which is a typical practice to prevent uncontrolled fluid flow into the wellbore, drilling fluid will be driven into the formation. This process, known as invasion, can have deleterious effects caused by chemical interactions between the drilling fluid and formation, also known as formation damage, which can permanently alter the permeability of the near-wellbore region and ultimately limit production. Invasion can also inhibit the driller's ability to control wellbore pressure, which may be a serious safety concern. Solids that are added to the drilling fluid for rheology and density control and other purposes are typically filtered at the borehole wall since these solid particles are usually too large to penetrate into the formation. As invasion progresses, a filter cake of these solid particles builds on the borehole wall, and when the filter cake reaches some critical thickness, invasion can be arrested (Figure 7.3). A great deal of work has been devoted to designing drilling fluid additives known as fluid loss reducers that allow invasion to stop as early as possible. It has been known for some time that fibrous and

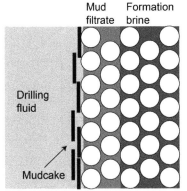 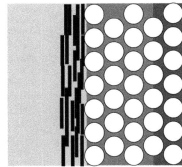

Early stage mudcake buildup

Thick mudcake formation and cessation of invasion

FIGURE 7.3

Left: Mudcake buildup at the borehole wall. Solids (black) are filtered out of the drilling fluid while the mud filtrate can flow into the formation. Right: After a sufficiently thick mudcake forms, no more fluid exchange is possible between the borehole and formation.

sheet-like additives make very effective fluid loss reducers (*e.g.*, Dyke 1949) because they form very low-permeability filter cakes (Jaffal et al. 2017). Recent developments in nanoparticle-based drilling fluids have indicated that nanoparticles accomplish similar results, either by acting synergistically with clay additives or by specifically targeting and clogging the pores in the formation.

Two types of experiment are used to evaluate the efficiency of fluid loss control. The first is a filtration test described in American Petroleum Institute (API) Recommended Practice 13B (American Petroleum Institute 2016, 2018). In this test, drilling fluid is placed in a pressure vessel with filter paper on the outlet. The upstream side of the vessel is pressurized and held for 30 min, and the volume of fluid passing through the filter paper as well as the amount of material filtered out are recorded (Figure 7.4). In practice, the filtrate and filter cake volumes may be measured periodically throughout the duration of the test. This test is typically performed at low temperatures and low pressures (LTLP; ~700 kPa and 25°C) or high temperatures and high pressures (HTHP; ~7 MPa and 200°C).

The second type of experiment is a pore pressure transmission test, which measures the drilling fluid's ability to prevent mass transfer between the wellbore and formation. This test draws on the principle of the pressure transient decay permeability test (Brace et al. 1968) and was described with respect to drilling fluid testing by van Oort (1994). In this test, a sample of rock is placed in a core holder with confining pressure applied. The upstream end of the core holder contains drilling fluid while the downstream end contains brine (Figure 7.5). To test, the pressure at the upstream end is raised and the pressure response at the downstream end is recorded. Assuming

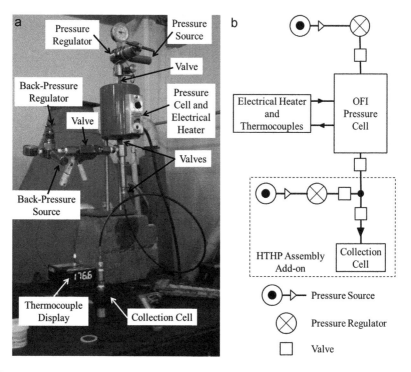

FIGURE 7.4
(a) HTHP API filtration test setup. (b) Schematic of equipment for both LTLP and HTHP tests. Figure from Barry et al. (2015).

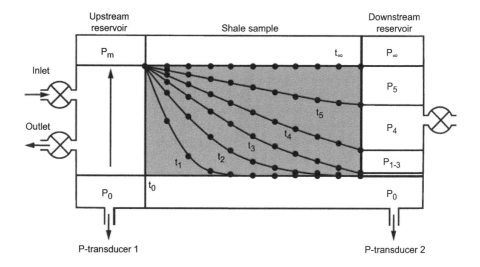

FIGURE 7.5
Schematic of pressure transmission test setup showing pressure profiles within the sample at different times. Figure from van Oort (1994).

one-dimensional flow that obeys Darcy's law, the pressure pulse decay is described by

$$\frac{\partial^2 P}{\partial x^2} = \frac{1}{K}\frac{\partial P}{\partial t},$$ (7.5)

where P is the pressure at position x along the axis of the sample and K is given by

$$\frac{1}{K} = \frac{\mu\beta}{k}\left(\varphi + \frac{\beta_r}{\beta} - \frac{[1+\varphi]\beta_s}{\beta}\right),$$ (7.6)

where
  μ is the fluid viscosity
  β is the fluid compressibility
  β_r is the rock compressibility
  β_s is the compressibility of the rock matrix
  k is the permeability
  φ is the interconnected porosity (Trimmer 1981).

Following Carslaw and Jaeger (1959), if l is the length of the sample and A its cross-sectional area, then Equation (7.5) may be solved with the boundary conditions

$$P = P_m, \qquad\qquad x = 0,\ t > 0$$
$$\frac{\partial P}{\partial x} = -\frac{\mu V\beta}{Ak}\frac{\partial P}{\partial t}, \qquad x = l,\ t > 0,$$ (7.7)

where
  P_m is the pressure in the upstream reservoir
  V is the volume of the downstream reservoir.

This treatment assumes that the upstream reservoir is infinite. The solution to Equation (7.5) is

$$\frac{P(x,t) - P_0}{P_m - P_0} = 1 - 2\sum_{n=1}^{\infty}\frac{\exp\left(-\frac{\psi_n^2 Kt}{l^2}\right)\sin\left(\frac{x\psi_n}{l}\right)}{\cos\psi_n\sin\psi_n + \psi_n},$$ (7.8)

where P_0 is the initial pore pressure and ψ_n is the nth solution to

$$\psi_n\tan\psi_n = \frac{Al}{V}\left(\varphi + \frac{\beta_r}{\beta} - \frac{[1+\varphi]\beta_s}{\beta}\right).$$ (7.9)

According to Trimmer (1981), when the ratio of sample pore volume to downstream reservoir volume is smaller than 0.25, Equation (7.8) may be approximated as

$$\frac{P(x,t)-P_0}{P_m - P_0} = 1 - \exp\left(-\frac{Akt}{\mu\beta Vl}\right).$$
(7.10)

The apparent permeability reduction as a result of mudcake formation and associated delay of pressure pulse propagation may thus be assessed from this measurement.

Nanoparticle applications in fluid loss reduction generally fall into two areas: applications in which nanoparticles block the pores of the rock formation and thus prevent fluid leakoff, and applications in which nanoparticles accelerate mudcake buildup and enhance the properties of the mudcake. Applications in the first category tend to be effective in reducing fluid infiltration in shales since shale pores are roughly the same size as nanoparticles. Work by Sensoy et al. (2009), Cai et al. (2012), Hoelscher et al. (2012), Ji et al. (2012), Sharma et al. (2012), Zakaria et al. (2012), and Hoxha et al. (2017) demonstrated that the addition of silica nanoparticles to both water-based and oil-based drilling fluid reduced the apparent permeability of shales by up to two orders of magnitude in pore pressure transmission tests, a fact that the authors all attributed to nanoparticles plugging the pores of shales (Figure 7.6). Gao et al. (2016) obtained similar results using aluminum oxide and magnesium oxide nanoparticles. The exact mechanism behind these results is not clear. Shale typically has a very low matrix permeability, and it is difficult to envision advection of

FIGURE 7.6
Silica nanoparticles plugging a pore throat in Atoka shale as imaged by scanning electron microscopy. Image from Sensoy et al. (2009).

nanoparticles to the shale surface in a process analogous to mudcake formation because the rate of fluid flow into the shale would be very low, except in the case of a fractured shale. Hoxha et al. (2017) demonstrated the role of electrostatic interactions between nanoparticles and shale in dictating fluid loss prevention by testing a variety of nanoparticles with different surface charge; better results were obtained in cases with greater electrostatic attraction between shale and nanoparticles. Future work should focus on a better understanding of this mechanism.

Applications focused on improving mudcake quality while reducing the thickness of the mudcake and thus the amount of filtrate invasion generally exploit synergistic effects between nanoparticles and bentonite in the drilling fluid through electrostatic effects. Barry et al. (2015) performed a detailed study using iron oxide and hybrid alumina-silica nanoparticles to form clay-nanoparticle composites by precipitating the nanoparticles in an aqueous bentonite solution. These hybrid mixtures were added to water-based drilling fluid and yielded a reduction in fluid loss, but for different reasons (Figure 7.7): the iron oxide clay hybrids promoted formation of cross-linked and coagulated networks of bentonite particles due to the positive surface charge on the nanoparticles, while the alumina-silica clay hybrids repelled each other electrostatically. In both cases, the resulting filter cakes were thinner with lower permeability than in the case of bentonite alone. When iron oxide nanoparticles themselves were added to water-based drilling fluid containing bentonite, the results were poor in LTLP filtration tests because the clay particles coagulated and the resulting filter cakes were more permeable.

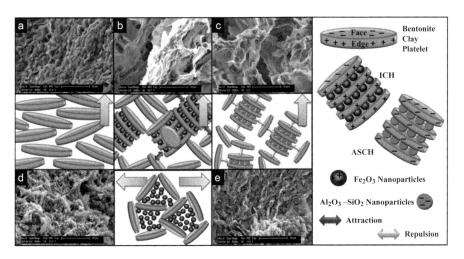

FIGURE 7.7
Scanning electron microscope images of different filter cakes prepared with bentonite and different types of nanoparticles. Interactions between nanoparticles and clay platelets are shown on the right. Figure from Barry et al. (2015).

However, in HTHP filtration tests, the results were better than the bentonite control because the nanoparticles replaced dissolved Na^+ in solution and caused deflocculation of bentonite. Salih and Bilgesu (2017) found similar results, with anionic metal oxide nanoparticles generally reducing fluid loss and yielding more compact, lower permeability mudcakes because they promote better structuring of bentonite platelets. They also found that cationic metal oxide nanoparticles performed more poorly. Wahid et al. (2015) found synergistic effects between nanosilica and gilsonite in synthetic-based drilling fluid, with filtrate volume decreased by 41.67% at 135°C and by 28.57% at 177°C with 0.7 wt% nanosilica in the drilling fluid.

A final, novel application of nanoparticles in reducing fluid loss and mudcake thickness was reported by Borisov et al. (2015). They used calcium- and iron-based nanoparticles in oil-based drilling fluid and reduced fluid losses by an average of 27% in concentrations of 0.5–2 wt%. Since these nanoparticles were hydrophilic, the authors designed the invert emulsion to carry the nanoparticles downhole in the aqueous phase (Figure 7.8).

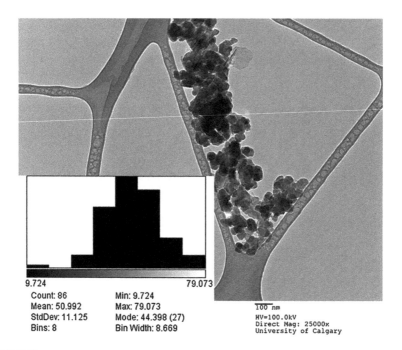

9.724	79.073
Count: 86	Min: 9.724
Mean: 50.992	Max: 79.073
StdDev: 11.125	Mode: 44.398 (27)
Bins: 8	Bin Width: 8.669

100 nm
HV=100.0kV
Direct Mag: 25000x
University of Calgary

FIGURE 7.8
Calcite nanoparticles encapsulated within the aqueous phase of a diesel-based invert emulsion imaged by scanning electron microscopy. The inset shows the particle size distribution for the particles pictured here. Image from Borisov et al. (2015).

7.2.3 Wellbore Stability and Strengthening

If the density of drilling fluid is too large, the pressure in the wellbore may fracture the surrounding formation. This presents a significant safety hazard because drilling fluid may leak readily into the fractures, making it difficult to control the pressure in the wellbore. The problem of preventing such fracturing, and mitigating it when it does occur, has been a long-standing concern in oil and gas drilling.

There are two conceptual models that explain fracture prevention and mitigation: tip resistance and stress caging. There is vigorous debate in the drilling community regarding which one of these models more accurately describes downhole processes (*e.g.*, Feng and Gray 2017), and it is not our purpose here to advocate one over the other, as results have been published showing the efficacy of nanoparticles in promoting both processes. The tip resistance model (Dupriest 2005) states that mudcake buildup inside a propagating fracture will tend to inhibit pressure propagation to the fracture tip, arresting the growth of the fracture (Figure 7.9). The result is a strengthened wellbore, because a borehole pressure exceeding the initial fracture initiation pressure would be required for a fracture to continue propagating. Tip resistance should not be as effective in low-permeability formations, since filtrate invasion and mudcake buildup are very slow in such cases.

The stress caging model (Alberty and McLean 2004) states that solids deposition at the mouth of a growing fracture will tend to isolate the fracture from the wellbore, allowing dissipation of excess fluid pressure within the fracture and eventual closure (Figure 7.10). After closure, the compression at the fracture aperture results in an increase in the tangential (hoop) stress at the borehole wall, yielding a strengthened wellbore. Again, this model will be most effective in higher-permeability formations where the pressure inside the fracture may dissipate rapidly. Stress caging in shales has been postulated to occur by filling the fracture with some sort of cementing material (Aston et al. 2007).

Contreras et al. (2014) used calcium- and iron-based nanoparticles combined with graphite in oil-based drilling fluid at concentrations up to 2.5 wt% nanoparticles to mitigate fracturing in sandstone cores. They observed

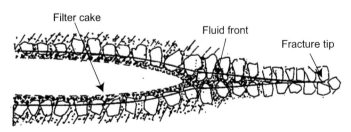

FIGURE 7.9
Illustration of the tip resistance model. Filter cake builds along the fracture surface and inhibits pressure propagation to the fracture tip. Figure modified from Feng and Gray (2017).

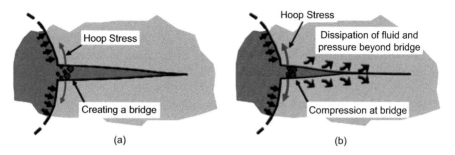

FIGURE 7.10

The stress caging model. (a) Bridged particles in the fracture aperture inhibit pressure propagation into the fracture, inhibiting its growth. (b) When the fracture closes, the compression on the bridge strengthens the wellbore by increasing the tangential (hoop) stress. Figure from Alberty and McLean (2004).

strengthening at concentrations as low as 0.5 wt%. Image analysis confirmed that the nanoparticles had plugged fractures, suggesting a tip resistance mechanism (Figure 7.11). The degree of strengthening observed was highly correlated with the results of HTHP filtration tests, suggesting that straining of nanoparticles was necessary for success. Better results were obtained with the calcium-based than with the iron-based nanoparticles. The presence of graphite probably prevented the nanoparticles themselves from moving into the relatively large sandstone pores since the nanoparticle sizes were reported as 30–60 nm.

In a complementary study, Hoxha et al. (2017) demonstrated that nanosilica could increase the apparent strength of shale. They used thick-walled cylinder (TWC) collapse testing, in which a cylindrical sample of rock has a bore drilled down its axis, which is filled with fluid and left under overbalanced conditions for about 24 hours. After this time, the pressure in the borehole is gradually increased until the sample fractures. Hoxha et al. (2017) showed that 5 wt% of various nanosilica particles in water-based drilling fluid resulted in strengthening in tests using Mancos shale. Most interestingly, they demonstrated a positive relationship between the degree of strengthening and the pressure transmission delay as measured in pore pressure transmission tests (Figure 7.12). This suggests that the electrostatically driven pore plugging responsible for the pressure transmission delay may also play a role in wellbore strengthening by reducing fluid infiltration or reducing the number of flaws on the wellbore wall that may act as nucleation sites for fractures. Boul et al. (2017) similarly linked electrostatic interactions to wellbore strengthening in tests with Mancos and Pierre shales. They tested two particles: bare silica (sodium-stabilized), and AEAPTS (2-aminoethyl-3-aminopropyl trimethoxysilane) functionalized silica. Better results were obtained with functionalized nanoparticles, probably because the positive charge promoted better bonding to the shale surface.

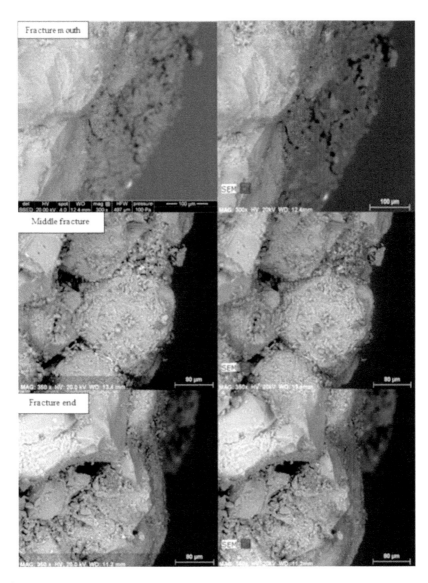

FIGURE 7.11

Scanning electron microscope images (left) and energy dispersive X-ray spectroscopy overlay (right) of the filter cake on a fracture surface after exposure to calcite nanoparticle-doped drilling fluid. The pink coloration at right indicates the presence of nanoparticles in the filter cake, suggesting a tip resistance mechanism in arresting fracture development. Figure from Contreras et al. (2014).

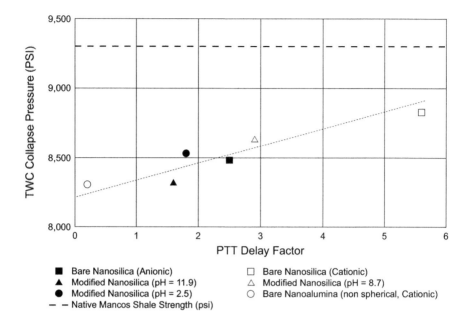

FIGURE 7.12

Comparison of thick-walled cylinder (TWC) collapse pressure and pore pressure transmission test (PTT) delay factor for Mancos shale samples exposed to water-based drilling fluid containing different nanoparticles. Figure from Hoxha et al. (2017). Copyright 2017, SPE International Conference on Oilfield Chemistry. Reproduced with permission of SPE. Further reproduction prohibited without permission.

7.2.4 Thermal Conductivity

As one of the principal roles of drilling fluid is to prevent the drill bit from overheating, the thermal conductivity of the drilling fluid must be controlled and optimized for efficient drilling. A few studies have investigated the degree to which metal oxide nanoparticles may be used to optimize thermal conductivity of drilling fluids. Ponmani et al. (2014) and William et al. (2014) both reported results of adding mixtures of CuO and ZnO nanoparticles in a xanthan gum solution to water-based drilling fluid. William et al. (2014) used 0.1–0.3 wt% nanoparticles in a base xanthan gum solution, added at 1 vol% to water-based drilling fluid. The nanoparticle-amended drilling fluids displayed higher thermal and electrical conductivity compared with the base case. CuO had better thermal conductivity (28–53% improvement) than ZnO (12–23% improvement). Ponmani et al. (2014) performed a detailed study of the structure of the nanoparticles in dispersion and found that the xanthan gum was a necessary additive for keeping the nanoparticles suspended and dispersed.

7.3 Fracturing Fluids

Hydraulic fracturing is widely used to allow oil and gas production from formations whose matrix permeability is insufficient to sustain economic flow rates. The fluids used in hydraulic fracturing must be designed to initiate and propagate fractures in the formation and transport proppant necessary to hold open the fractures that are generated. In addition, these fluids must not damage the wellbore in such a way that negatively affects subsequent production. As such, the physical and chemical properties of fracturing fluids must be carefully controlled through the use of various additives. These additives can include polymers, crosslinkers and breakers, emulsifiers, surfactants, foamers, gel stabilizers, pH control additives, and biocides (Fink 2015). Many applications use water-based fracturing fluids, but alternative fracturing fluids are available for use in formations that might be chemically sensitive to water (*e.g.*, Aderibigbe and Lane 2013) or to mitigate the environmental impact of hydraulic fracturing in areas with sparse water supplies (Middleton et al. 2015).

The efficacy of nanoparticles in addressing these concerns has been demonstrated by many studies. Nanoparticles can improve the rheological characteristics of fracturing fluids at elevated pressures and temperatures. They can stabilize foams and gels in alternative fracturing fluids, greatly reduce fluid leakoff into the formation, and assist in post-fracturing wellbore cleanup. In this section, we will describe these applications and the underlying mechanisms.

7.3.1 Rheology Improvement

Viscous, water-based fluids have been widely used in hydraulic fracturing for many years. High viscosity proved advantageous for generating wider, longer fractures, minimizing fluid leakoff into the formation, and efficiently transporting proppant (Barati and Liang 2014). In organic shales, fluid leakoff is less of a concern because of the extremely low matrix permeability, and slickwater, a low-viscosity water-based fracturing fluid, has been used extensively because it contains fewer additives that require post-fracturing cleanup. However, its low viscosity makes it less efficient at transporting proppant, and as a result there has been a trend away from slickwater recently toward slickwater that incorporates crosslinked gel for better proppant delivery (Al-Muntasheri 2014).

Viscous fracturing fluids generally contain a water-soluble polymer and a crosslinking agent, which allows efficient proppant transport in the crosslinked polymer. After the proppant has been pumped, the fracturing fluid must be removed from the fracture to allow oil or gas to flow into the fracture, and chemical breakers are typically used to break the crosslinked polymer and decrease the fluid viscosity (Fink 2015). More recently, viscoelastic surfactant (VES) fracturing fluids have been introduced. These fluids obviate the need for a breaker and flow more easily back to surface after proppant placement (Al-Muntasheri 2014). Nanoparticles have been shown to improve the rheology of polymer-based and VES fracturing fluids.

7.3.1.1 Polymer-Based Fluids

Several studies have been conducted on the effects of adding metal oxide nanoparticles to polymer-based fracturing fluid. Fakoya and Shah (2013) showed that 20 nm silica nanoparticles increased the viscosity of aqueous guar solutions as nanoparticle concentration increased to a certain point (between 0.24 and 0.4 wt% depending on the solution), but that larger nanoparticle concentrations could be detrimental. They attributed this behavior to formation of a gelled network at lower nanoparticle concentrations by adsorption at active sites on the nanoparticle surface, and interference with this process at higher nanoparticle concentrations. Vipulanandan et al. (2014) found a similar result with nanosilica in guar-based fracturing fluid. The mechanism of viscosity increase was investigated in detail by Hurnaus and Plank (2015), who showed that titania nanoparticles could act as crosslinkers in guar-based fracturing fluid. Their results indicated that crosslinking occurred through interactions between –OH groups on the polymer and those on the nanoparticle surfaces (Figure 7.13). A similar mechanism could be proposed for interactions with silanol groups in silica nanoparticle surfaces. The size and weight fraction of nanoparticles used in these applications can likely be optimized through consideration of active surface sites per unit mass of nanoparticles. Both Fakoya and Shah (2013) and Vipulanandan et al. (2014) found that viscosity decreased with increasing temperature in nanoparticle-amended polymer-based fracturing fluid, which is probably a result of weak hydrogen bonding being the primary driver of viscosification (*e.g.*, Cordier and Grzesiek 2002).

7.3.1.2 Viscoelastic Surfactant-Based Fluids

VES fracturing fluids have become popular recently, especially for use in low-permeability organic shale reservoirs. The VES typically has a hydrophobic tail and a hydrophilic head, so no additional crosslinking agent is necessary to form a gel. Also, since the VES molecules are smaller than the polymers used in polymer-based fracturing fluids, less residue is left in the fracture and the wellbore, simplifying cleanup.

Viscoelastic materials exhibit a combination of fluid-like and solid-like behavior. When sheared, these materials will strain at a rate that is proportional to the applied shear stress, like a fluid; however, they will also exhibit finite, reversible strain whose magnitude is proportional to the applied shear stress, like a solid. Many different constitutive models exist to describe the stress-strain behavior of viscoelastic materials. A widely used

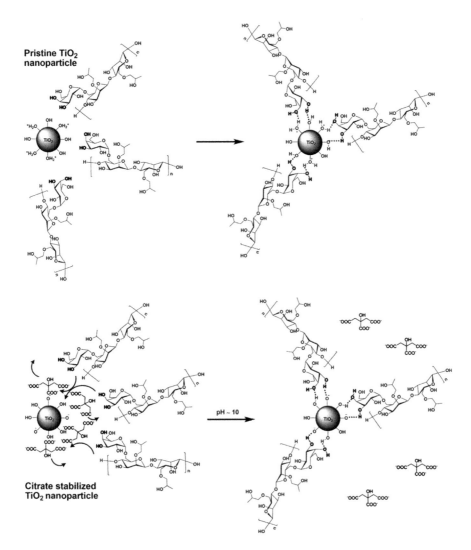

FIGURE 7.13

Crosslinking mechanism between Guar-derived biopolymers and titania nanoparticles with bare surfaces (top) and citrate stabilization (bottom). Reprinted with permission from Hurnaus and Plank (2015). Copyright 2015 American Chemical Society.

approach to characterize this behavior is oscillatory rheology experiments. In these experiments, an oscillating strain with maximum amplitude ε_0 is applied with angular frequency ω. The strain and stress responses in the material are described by

$$\varepsilon = \varepsilon_0 \sin(\omega t),$$ (7.11a)

$$\sigma = \sigma_0 \sin(\omega t + \delta),$$ (7.11b)

where

 ε and σ are the instantaneous strain and stress at time

 t, σ_0 is the maximum stress amplitude

 δ is the phase lag between the strain and the stress responses.

In the case of a shear oscillatory rheology test, the complex shear modulus G^* is defined as

$$G^* = G' + iG' \tag{7.12}$$

where the storage modulus G' is given by

$$G' = \frac{\sigma_0}{\varepsilon_0} \cos \delta, \tag{7.13}$$

and the loss modulus G'' is given by

$$G'' = \frac{\sigma_0}{\varepsilon_0} \sin \delta. \tag{7.14}$$

In purely elastic materials, $\delta = 0$, so G' is a measure of the elastic behavior of the material. In purely viscous materials, $\delta = \pi/2$, so G'' is a measure of the viscous behavior of the material. Viscoelastic surfactant solutions typically display complex rheology, with more viscous behavior at some strain frequency ranges and more elastic behavior at other frequency ranges (Fischer and Rehage 1997). Analysis of oscillatory rheology data can give some indication of the degree of internal ordering in the fluid.

Several studies have investigated the effects of nanoparticle addition on the rheology of VES fluids. Huang and Crews (2008) showed that 20 nm ZnO nanoparticles increased the viscosity of fracturing fluid containing amidoamine surfactant (the composition of the nanoparticles and VES were not disclosed by Huang and Crews [2008] but are identified as such by Al-Muntasheri et al. [2017]). They observed a nearly 10-fold increase in viscosity at low shear rates (< 0.01 s^{-1}) at room temperature, and a 5-fold increase at high shear rates (100 s^{-1}) and high temperature (121°C) with no degradation over time, unlike the base case. The viscosity enhancement was attributed to a pseudo-crosslinking mechanism whereby surfactant micelles experienced electrostatic attraction and chemisorption on the nanoparticle surfaces (Figure 7.14).

Maxey et al. (2008) added nanoparticles to aqueous solutions of nonionic amine oxide surfactant and performed oscillatory rheology experiments at different temperatures. They found that the surfactant by itself transitioned from primarily viscous behavior at room temperature to primarily elastic behavior at 77°C, while the surfactant solution amended

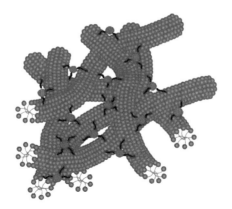

FIGURE 7.14
Surfactant micelles pseudo-crosslinked with nanoparticles (pink) forming a strong structure.
Figure from Huang and Crews (2008).

with nanoparticles exhibited elastic behavior at all temperatures tested
(Figure 7.15). In addition, they found that the storage modulus of the
nanoparticle-amended solution did not change appreciably after 2 hours at
77°C, while that of the surfactant solution without nanoparticles lost much
of its elastic behavior at this temperature. They interpreted their results
as evidence that nanoparticles enhanced micelle-micelle interactions.
Gurluk et al. (2013) presented similar results, showing that adding MgO
or ZnO nanoparticles to amidoamine solution increased the viscosity and
imparted greater elastic properties to the solution. They tested rheology
up to 135°C and found that nanoparticles greatly increased the elastic
behavior of VES solutions.

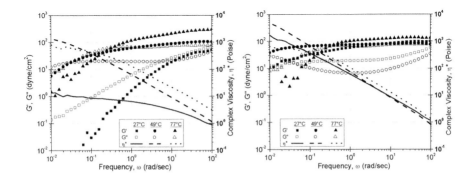

FIGURE 7.15
Oscillatory rheology data for 2 vol% non-ionic amine oxide VES in dense brine (left), and with
addition of 0.077 wt% nanoparticles (right). Note that the nanoparticles impart greater elastic
behavior as G was larger than G'' over most of the frequency and temperature range tested.
Figure from Maxey et al. (2008).

7.3.1.3 Physical Mechanism of Rheology Improvement

The rheology improvement seen in polymer- and VES-based fracturing fluids when nanoparticles are added is attributable to the interactions between organic molecules in solution and nanoparticle surfaces. In the case of polymers, interactions generally rely on hydrogen bonding. Typical biopolymers used in fracturing fluids include guar and cellulose and similar polymers (Al-Muntasheri 2014). These polymers usually contain hydroxyl functional groups, which may interact with other molecules. Metal oxides in aqueous solution will typically hydrate, coating the surface of a metal oxide particle with hydroxyl groups (Brown et al. 1999). This promotes hydrogen bonding between hydroxyls on metal oxide nanoparticle surfaces and those on polymer chains, as proposed by Hurnaus and Plank (2015). The result is a crosslinking phenomenon, which imparts a high viscosity and gel-like behavior to the fracturing fluid. However, since hydration of metal oxide surfaces depends on solution pH, and because hydrogen bonding is strongly affected by temperature (Brown et al. 1999; Cordier and Grzesiek 2002), nanoparticle-based viscosification of polymer-based fracturing fluids is only effective under certain conditions.

In the case of surfactants, electrostatic attraction and chemisorption are the dominant means of interaction between solutes and nanoparticles. When the surfactant and nanoparticle surfaces have opposite charges, electrostatic adsorption is readily achieved. However, even when the surfactant and nanoparticle have like charges, chemisorption may still be achieved when one end of the surfactant is attracted to a nanoparticle with the same wettability. For example, Nettesheim et al. (2008) showed that the hydrophobic endcaps of cetyltrimethylammonium bromide (CTAB) adsorbed on the surface of hydrophobic silica nanoparticles, which promoted the growth of wormlike micelles emanating from the nanoparticle surface. Helgeson et al. (2010) showed that this phenomenon is only effective when the nanoparticle surface is not strongly hydrophilic or hydrophobic, but even in the case of moderate wettability, the sorption of surfactant molecules on the nanoparticle is a lower-energy configuration relative to surfactant in aqueous solution (Figure 7.16). The result is a phenomenon called pseudocrosslinking in which the nanoparticles promote the formation of a micellar network. This network has pronounced elastic properties; Helgeson et al. (2010) showed that, even when the aqueous surfactant solution has Newtonian behavior, the addition of nanoparticles can impart viscoelasticity. Because the nanoparticle-surfactant interactions are driven by stronger bonds than in the case of nanoparticle-polymer interactions, the rheology is less sensitive to temperature.

7.3.2 CO_2 Foams for "Waterless" Fracturing Fluids

Foams have been used as fracturing fluids since the late 1960s (Grundmann and Lord 1983). They possess several advantages over water-based fracturing fluids, including less demand for water, easier wellbore and fracture

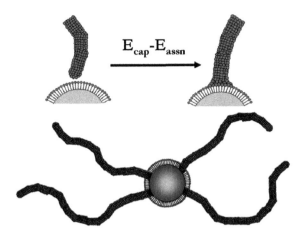

FIGURE 7.16
Attachment of a micelle to a nanoparticle. Surfactant molecules adsorb on the nanoparticle surface and the micellular endcap attached to these surfactant molecules because it represents a lower-energy configuration; the energy required to maintain the endcap of the micelle, E_{cap}, is larger than the energy required to associate with the adsorbed surfactant, E_{assn}. The bottom image shows the nanoparticle incorporated into the micellular structure. Reprinted with permission from Nettesheim et al. (2008). Copyright 2008 American Chemical Society.

cleanup following fracture stimulation, and improved proppant transport (Ribeiro and Sharma 2013). A few studies have probed the use of nanoparticle-stabilized CO_2 foams for hydraulic fracturing. Li et al. (2015) showed that using an aqueous nanoparticle dispersion as a pad fluid prior to injecting liquid CO_2 as a fracturing fluid improved the sweep efficiency of the fracturing fluid and mitigated viscous fingering, which is analogous to the advantages of CO_2 foams in enhanced oil recovery.

Prodanović and Johnston (2017) investigated the stability and proppant-carrying capacity of ultra-dry CO_2 and N_2 foams stabilized with nanoparticles, surfactant, and polymers. They used silica nanoparticles, lauramidopropyl betaine (LAPB) surfactant, and hydrolyzed polyacrylamide (HPAM) polymer. Their foams had water contents between 2% and 5% and displayed good stability at temperatures up to 120°C and pressures up to 20.7 MPa (3000 psi). In particular, the foam displayed the ability to suspend sand proppant for up to 1 day at a pressure of 13.8 MPa (2000 psi) with negligible change in foam quality. The generated foams were shear thinning and had apparent viscosities of hundreds to thousands of centipoises, which was much larger than the apparent viscosity of foams without nanoparticles. Emrani and Nasr-El-Din (2017) used an anionic alpha-olefin sulfonate (AOS) surfactant along with silica and iron oxide nanoparticles to stabilized CO_2 foams. They found that nanoparticles greatly improved foam stability relative to the base case of AOS alone and that silica nanoparticles formed more stable foam than polymers or viscoelastic surfactants. They additionally showed that mutual electrostatic repulsion between nanoparticles yielded stronger foams, and that for the silica nanoparticles, the optimal concentration was 0.1 wt%. Further studies are necessary to fully understand the effects of nanoparticle surface chemistry on foam stability, rheology, and strength.

Qajar et al. (2016) modeled the fracture propagation and subsequent cleanup processes with nanoparticle-stabilized CO_2 foam fracturing fluid. They showed that the large viscosity of the foam generated wider, shorter fractures than water-based fracturing fluid, and that cleanup of the fracturing fluid was as much as 100 times faster than in the case of a water-based fluid (10 days vs. 1000 days).

The exceptional stability and rheology of nanoparticle-stabilized foams for hydraulic fracturing gives them an advantage over surfactant-stabilized foams. However, more fundamental work is necessary to understand the controls on physical behavior and refine plans for field implementation.

7.3.3 Fluid Loss Control

Similar to the research using nanoparticles as drilling fluid additives to reduce fluid infiltration into the formation, nanoparticles have been shown to minimize fracturing fluid leakoff at the fracture face. This is advantageous, since fluid leakoff can arrest fracture propagation, and the near-fracture region of the formation can be adversely affected by invasion of fracturing fluid, particularly in the case of water-based fluids, to the extent that hydrocarbon production is seriously inhibited (the so-called water blockage effect).

Fluid loss mitigation during hydraulic fracturing can rely on either viscosification of the fluid, which tends to reduce the rate of infiltration by Darcy's law, or formation of a filter cake on the fracture surface. Any of the viscosifying treatments discussed earlier would tend to reduce fluid loss. Some notable work was performed by Huang et al. (2010) and Crews and Gomaa (2012), who showed that silica nanoparticles could promote the formation of a filter cake in VES-based fracturing fluid through the same pseudocrosslinking behavior that causes rheology improvement. This mitigates a pervasive problem of fluid loss when using conventional VES-based fracturing fluids. Barati (2015) performed a series of experiments using polymer-based fracturing fluid (guar) with a borate crosslinker and silica nanoparticles (SnowTex, 110 nm diameter). He found that the silica nanoparticles reduced fluid loss, but only in rocks with permeabilities lower than 0.1 mD. Fluid loss reduction was even observed in nanoparticle-amended fracturing fluid without polymer, which suggests a pore blockage mechanism. This would explain the poor results in higher-permeability rocks as well.

7.3.4 Post-Fracturing Wellbore Cleaning

After hydraulic fracture stimulation, the fracturing fluid(s) must be removed from the fracture and wellbore before hydrocarbon production can proceed. With conventional water-based fracturing fluids, this process can take weeks to months as water that imbibes into the formation at the fracture face is progressively drained. Even when steps are taken to minimize fluid loss, any viscous polymer gels or surfactant solutions must be pumped back to the surface. This is typically accomplished by reducing the viscosity of the fracturing fluids to allow easier flow. Polymer-based fluids include breaker chemicals that break the crosslinks in the gel, thus reducing the viscosity (Fink 2015). VES-based fluids leave minimal residue, greatly simplifying cleanup; this is one of their main advantages over polymer-based fluids (Barati and Liang 2014). Breaker chemicals called internal breakers are sometimes added to VES-based fluids to reduce what little residue might be left behind.

In polymer-based fluids, a main advantage of nanoparticle additives is that a particular viscosity can be achieved with a much lower polymer loading (Hurnaus and Plank 2015), which reduces the amount of residue that must be cleaned up. In VES-based fluids containing nanoparticles, Huang et al. (2010) showed that internal breakers could be incorporated in the micellar structure and break down the pseudocrosslinked structure through progressive oxidation without interfering with the efficacy of the nanoparticles in promoting the pseudocrosslinking. Finally, it should be noted that energized fracturing fluids containing nanoparticles require little cleanup after use.

7.4 Cement Integrity

Good cement integrity is essential for safe completion and production operations, as the cement between casing and formation prevents borehole collapse, holds the casing in place, and prevents flow between subsurface formations. In a report on well control incidents (blowouts) on the outer continental shelf of the US, the Minerals Management Service stated that, between 1992 and 2006, "[t]he most significant factors included cementing problems resulting in gas migration during or after cementing of the well casing" (Izon et al. 2007). Poor cement integrity has been cited as a contributing factor in the 2010 blowout of the Macondo well (National Academy of Engineering 2011), and presents a significant risk factor for migration of gas to the surface, with corresponding environmental and safety risks (Ingraffea et al. 2014; Watson and Bachu 2009). Cement with good integrity should have low permeability, high tensile and compressive strengths, and a relatively low Young's modulus to accommodate in-situ shear deformation without failure (Santra et al. 2012). These properties may be controlled with chemical additives. The role of nanoparticles in improving cement properties has garnered interest in industry for several years and is the subject of a number of patents (*e.g.*, Roddy 2016; Roddy et al. 2009, 2013).

Cement is a composite material made from limestone and clay, with iron and alumina sometimes added. These are fused at high temperature to form clinker, which is then ground and mixed with gypsum. The clinker mainly contains the mineral phases alite (Ca_3SiO_5 or C_3S), belite (Ca_2SiO_4 or C_2S), aluminate ($Ca_3Al_2O_6$ or C_3A), and ferrite (Ca_2AlFeO_5 or C_4AF). When mixed with water, the minerals dissolve and reprecipitate an assemblage of hydration products, including amorphous CSH gel. The CSH precipitates around solid clinker particles and eventually can replace them completely, resulting in a strong, continuous structure with some open pore space. The precipitation of CSH gel from hydration of C_2S and C_3S yields $Ca(OH)_2$, which raises the pH of the aqueous phase to 12.5–13 over the first 4 days of hydration (Michaux et al. 1989; Šiler et al. 2016). The hydration reaction may proceed for several weeks (Smith 2003).

Many studies have probed the properties of cements containing nanosilica and the effect of the large surface area of nanosilica in promoting CSH formation. Jo et al. (2007) found that cements containing nanosilica had higher compressive strengths than cements containing larger fumed silica particles. They additionally observed that the nanosilica resulted in a more compact cement structure with fewer $Ca(OH)_2$ crystals (Figure 7.17), an observation also reported by Ji (2005). Pang et al. (2014) similarly found that smaller nanosilicas resulted in faster set times. Björnström et al. (2004) used Fourier transform infrared spectroscopy (FTIR) to show that colloidal silica increased the rate of CSH formation and limited water dissolution during the hydration of C_3S, resulting in a faster set time (Figure 7.18).

Some studies have focused on optimizing nanosilica types and amounts. However, the results are inconsistent and difficult to compare, probably because different types of nanosilicas can have differing surface chemistry due to variations in silanol abundance on their surfaces (Bergna 2005). Nazari and Riahi (2011) found that the strength of self-compacting concrete increased monotonically with nanosilica addition up to 4 wt% (the largest concentration they tested). Shih et al. (2006) in contrast found that nanosilica

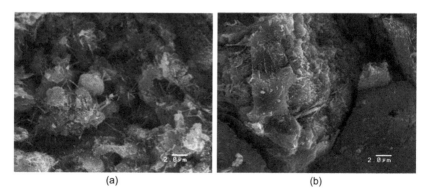

(a)   (b)

FIGURE 7.17
(a) SEM image of neat cement paste after seven days of curing. (b) SEM image of cement paste with nanosilica after seven days of curing. Note the more compact structure when nanosilica is added and the reduced number of $Ca(OH)_2$ needles. Images from Jo et al. (2007).

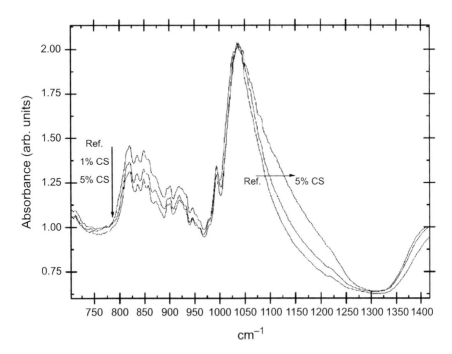

FIGURE 7.18
FTIR spectrum for cement paste amended with differing amounts of colloidal silica (CS) after curing for 4 hours. Lowering of peaks around 800–950 cm^{-1} indicates reduction in the amount of C3S present, due to increased hydrate rate in the presence of nanosilica. Broadening of peak centered around 1040 cm^{-1} is due to the increased presence of nanosilica. Figure from Björnström et al. (2004).

only increased the strength of Portland cement up to a concentration of 0.6 wt%, but decreased beyond that, which they attributed to increased nanosilica aggregation at higher concentrations. Björnström et al. (2004) and Santra et al. (2012) reported that the set acceleration effects of colloidal silica and nanosilica were enhanced with increasing silica concentration for concentrations below 5 wt%. On the other hand, Kong et al. (2013) reported that aggregated nanosilica accelerated cement hydration better than dispersed nanosilica because nanosilica adsorbed aqueous Ca and enabled faster hydration. However, they used two different types of nanosilica—precipitated nanosilica, which formed aggregates, and fumed nanosilica, which remained dispersed—so it is possible that their results were driven more by differences in surface chemistry than by aggregate size.

A small number of studies have focused on how the properties of the cement with nanosilica additives change over time. Kawashima et al. (2013) reported that nanosilica reduced the strength of cement amended with fly ash relative to control samples when tested after several weeks, which they attributed to the nanosilica consuming all the available Ca(OH)$_2$ in CSH formation, leaving none available for later-stage pozzolanic reaction with the fly ash, which would normally contribute to long-term strength. Madani et al. (2012) reported that nanosilica reduced the degree of hydration in Portland cement at ages of seven days and greater relative to neat cement, a fact that they attributed to nanosilica aggregates trapping water that could not be used in later-time hydration reactions (Figure 7.19). The role of the extremely elevated pH following the early stages of curing has similarly not been investigated, and it is not clear the degree to which nanosilica may dissolve during this time and provide aqueous hydrogen silicate or silicate ions. Björnström et al. (2004) claimed that their FTIR results did not indicate silica dissolution, but the phenomenon is not considered in other studies. A detailed and comparative study of these phenomena is lacking because data are typically not presented on surface chemistry or nanosilica aggregate size under hydration conditions.

As discussed previously, nanoparticle aggregation may be prevented by coating the particles with organic ligands. A fully hydroxylated amorphous silica surface contains 4.6 silanol groups per square nanometer (Bergna 2005), so 64% monolayer coverage of a fully hydroxylated nanosilica particle would theoretically leave 1.7 silanol groups per square nanometer exposed. Therefore, coating nanosilica with organic ligands may allow better particle dispersion while allowing access to silanol groups. An interesting additional benefit arises from the addition of organic ligands to cement. Polymer-based nanocomposites are widespread in nature in materials like oyster shells and bone (Espinosa et al. 2009). These nanocomposites are composed of plate-like mineral structures arranged around polymer chains (Okamoto 2003). The combination of mineral material with flexible polymers gives the nanocomposites flexibility and strength with limited susceptibility to flaw-induced weakness (Espinosa et al. 2009). Several recent studies have investigated polymer addition to cement; see Shahsavari and Sakhavand

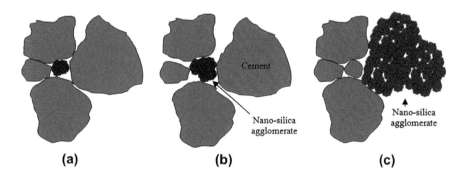

(a)   **(b)**   **(c)**

FIGURE 7.19
The importance of preventing nanosilica aggregation. (a) Small nanosilica particles can fit within the interstices of larger cement particles. (b) Aggregates of nanosilica can displace larger cement grains, increasing porosity. (c) Even larger nanosilica aggregates can create their own internal porosity and prohibit water from participating in hydration reactions. Figure from Kong et al. (2013).

(2016) for a good review. Early work by Matsuyama and Young (1999a,b,c) showed that polymer chains could be intercalated within the interlayer space of CSH. Mojumdar and Raki (2005) used FTIR to demonstrate hydrogen bonding between polyvinyl alcohol and CSH by observing a shift in the –OH bond to higher frequencies. Beaudoin et al. (2009) used ^{29}Si magic angle scanning nuclear magnetic resonance (MAS NMR) to demonstrate an increase in silicate polymerization in CSH-polymer mixtures, including PEG.

Overall, a significant number of studies have investigated the effects of nanoparticle addition on the properties of cement oilfield cement, but many details of the underlying mechanisms remain obscure and in need of detailed research.

7.5 Improved Hardware Materials

Nanoparticles and nanotechnology in general have allowed considerable advances in the development of durable, wear-resistant materials suitable for high pressure and high temperature applications (Cao 2004). As oil and gas exploration extends to more and more challenging subsurface environments, the need for high-performance drilling and completion materials has been met with nanotechnology solutions. Here we present some advances in hardware materials that improve the safety and efficiency of drilling and completion processes.

7.5.1 Drilling Hardware

Composite materials are composed of one or more dispersed materials within a matrix phase. The dispersed materials are selected to give the composite superior properties compared with the dispersed phase. An example is cobalt-cemented tungsten carbide, which is used for cutting tools; the tungsten carbide particles are dispersed in the cobalt matrix, imparting superior performance at high temperatures while resisting wear.

Nanocomposite materials are composed of nano-sized (< 100 nm) particles dispersed in the matrix phase. The small size and corresponding large surface area of the dispersed particles gives them several advantages over micron-sized dispersed particles. For instance, the Young's modulus and tensile strength of carbon nanotubes are nearly an order of magnitude larger than those of steel (Amanullah and Ramasamy 2014), imparting superior strength to carbon nanotube-based nanocomposites. Similarly, the superparamagnetic nature of magnetic nanoparticles can be used to create a soft magnetic nanocomposite, allowing easy magnetization and demagnetization (Varga 2007). The nanoparticle-amended cements discussed previously are nanocomposites that exploit the high reactive surface area of nanoparticles, imparting greater strength and faster curing compared with neat cement.

The hardware used for drilling is a good candidate for property improvement through nanocomposites. Replacing a bit or another component of the bottomhole assembly can require hours to days of downtime as the drillstring is tripped to surface and then to the bottom again. To reduce the risk of excessive downtime and the associated safety risk with repeated bit trips, studies have focused on improving the strength and durability of the drill bit itself, as well as improving the performance of other components of the bottomhole assembly. Sengupta and Kumar (2013) developed an alumina-titania nano-ceramic coating to improve the durability of drill bits. Their coating process involved plasma spraying of micron-sized alumina and titania particles. The plasma temperature was selected such that the particles partially melted, forming a structure of micron-sized grains dispersed in a matrix of nano-sized grains. The properties of the resulting nano-ceramic were not quantified explicitly, but the performance was markedly superior to conventional ceramics in cup indentation tests, hammering tests, and surface grinding tests.

Belnap et al. (2011) reported the use of elongated nanostructures such as carbon or inorganic nanotubes to improve the performance of polycrystalline diamond compact (PDC) bits at high temperature. The elongated nanostructures alleviated a key issue with PDC bits under thermal stress, which is cracking due to different coefficients of thermal expansion of the cobalt and diamond used to form the PDC composite. They showed that including elongated nanostructures in the PDC composite material resulted in better wear performance at temperatures above 750°C. The strengthening mechanism was attributed to the elongated nanostructures both mitigating the deformation effects of disparate thermal expansion coefficients and arresting crack propagation through the composite material.

As a final example, Chan et al. (2008) reported the development of a nanocomposite hardband to reduce fatigue due to drillpipe rotating inside casing. In vertical wells, wear due to drillipipe-casing contact is minimal, but in deviated and extended reach wells, the wellbore geometry can make this wear a serious concern for the casing and drillpipe. Chan et al. (2008) developed a nanocomposite coating for drillpipe and casing that reduced casing wear by a factor of nearly seven, and possessed a fracture toughness several times larger than that of conventional casing and drillpipe materials. The composition of their material was not disclosed, but the improved performance was probably due to similar mechanisms to those reported by Belnap et al. (2011) (Figure 7.20).

7.5.2 Ball Packers for Multiple Fracture Generation

One particular nanotechnology application of interest in completions is improved materials for setting balls. Setting balls are an integral component of multi-stage hydraulic fracturing. These balls are designed to sit in a ball seat, providing the hydraulic isolation necessary to stimulate only a particular section of the wellbore (Figure 7.21). A pervasive problem in hydraulic fracturing operations is the retrieval of setting balls following

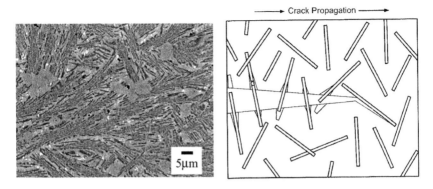

FIGURE 7.20
Left: Scanning electron microscope image of nanoparticle-amended hardband material show-ing elongated nanostructures. Figure from Chan et al. (2008). Right: Crack arresting mecha-nism in nanocomposite material. Crack propagation is inhibited by the presence of randomly aligned nanostructures. Figure from Belnap et al. (2011).

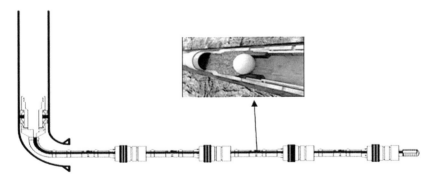

FIGURE 7.21
Setting ball use in a multi-stage hydraulic fracturing operation. Packers between each stage, indicated by yellow with black o-rings, are activated and isolated by setting balls. Inset shows a photograph of a setting ball. Figure from Xu et al. (2011).

fracturing. The balls can either be pumped to the surface or milled out. Both strategies take time and carry the risk of failure. To eliminate the need for ball removal, degradable sealing balls have been developed, allowing the balls to be dissolved through contact with completion fluid or formation fluids (Mazerov 2016; Todd et al. 2008).

Xu et al. (2011) developed a controlled electrolytic metallic (CEM) material for use in degradable setting balls. CEM was shown to provide the unique combination of high strength and in-situ degradation. The exact composition of the CEM they used was not disclosed, but its structure consisted of metal grains on the order of 150 microns dispersed in a matrix of nano-sized metallic or ceramic grains (Figure 7.22a). As with most nanocomposites, the nano-sized matrix imparted superior strength compared with the

FIGURE 7.22
(a) Scanning electron microscope of nanocomposite with metallic grains dispersed within a nanomatrix along grain boundaries. (b) Setting ball made from material in (a) prior to deployment. (c) Corroded ball after 7 days at 8000 psi and 120–160°F (49–71°C). (d) Cross-section of corroded ball in (c). Figures from Xu et al. (2011).

same material without the nano-sized matrix; the compressive strength of their materials was 3–6 times higher, and the rates of corrosion in 3% KCl brine at 200°F (93°C) were several hundred times larger than those of the conventional material. The setting balls designed by Xu et al. (2011) have been used extensively in Bakken shale wells. Laboratory analysis of one particular ball that was pumped back to surface for analysis revealed a 41% volume loss after 7 days at 8000 psi and 120–160°F (49–71°C) (Figure 7.22b,c).

7.5.3 Improved Elastomers

Elastomers are soft, viscoelastic polymers that are used for seals and flexible parts in various manufacturing applications. Examples include rubber, silicone rubber, and Viton. These materials are used in many pieces of equipment involved in the upstream oil and gas industry, including o-rings and seals. Elastomer-based components are prone to degradation at high pressure and temperature. For example, conventional o-rings made from rubber have operational limits of about 175°C and 135 MPa (19,600 psi) (Nabhani et al. 2011). As in the cases of drilling hardware and ball packers discussed previously, the addition of nanoparticles to form a nanocomposite elastomer can yield superior performance at high pressure and temperature.

Most successful applications of nanoparticles for improved elastomers have involved carbon nanotubes. The mechanism behind the property improvement is the formation of a network of interconnected nanotubes with an elastomer coating bound to their surfaces. This structure is called a cellulation structure (Endo et al. 2008; Noguchi et al. 2009) (Figure 7.23). Because the cellulation structure is continuous throughout the nanocomposite, the resulting elastomers have greater strength, but retain the flexibility of conventional elastomers (Inukai et al. 2011). The amount of carbon nanotubes required to form the cellulation structure was reported as 11% by Endo et al. (2008), but improvements in mechanical properties may be observed even at lower concentrations (Frogley et al. 2003). Similar

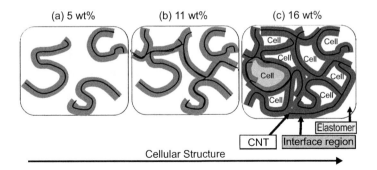

FIGURE 7.23
The cellulation process in elastomer doped with carbon nanotubes. Some elastomer adsorbs on the carbon nanotube interface (dark gray). As the weight fraction of carbon nanotubes increases from (a) to (c), eventually a load-bearing network of carbon nanotubes develops, with elastomer cells in between (light gray at (c)). Reprinted with permission from Inukai et al. (2011). Copyright 2011 American Chemical Society.

performance may be achieved through the addition of high-aspect ratio inorganic nanoparticles, such as nanoclay (Griffo and Keshavan 2009); the mechanisms of interconnected nanoparticle networks are similar to the case of carbon nanotubes.

Several novel applications of elastomer nanocomposites have been developed recently. Endo et al. (2008) made o-rings from fluorine rubber amended with multi-walled carbon nanotubes (MWNTs). They confirmed the formation of a cellulation network in the nanocomposite through transmission electron microscopy. The o-rings were tested for durability at elevated pressure and temperature and compared the results to conventional o-rings made from carbon black-filled fluorine rubber. The conventional o-rings failed after 14 hours at 260°C and 166 MPa (24,100 psi), while the nanocomposite o-rings were able to maintain a seal with no leakage for 14 hours at 260°C and 239 MPa (34,700 psi).

Griffo and Keshavan (2009) showed that nanotubes or nanoclay could be added to an elastomer to produce improved seals for roller cone bits. Slay and Ray (2010) similarly showed the efficacy of nanoparticles in improving the properties of elastomers for a variety of downhole seals. In both cases, the surfaces of the nanoparticles or nanotubes must allow for bonds to form with the elastomer through covalent bonds, van der Waals interaction, or electrostatic interaction. The properties of these bonds may be manipulated very precisely to give desired behavior. For example, Martín et al. (2012) developed a self-healing elastomer nanocomposite by amending thiol-functionalized silicone oil with silver nanoparticles. The self-healing mechanism was achieved by exploiting the fact that thiolate ligands constantly exchange at steady state on silver or gold surfaces (Figure 7.24). Further research should be able to exploit the nature of nanomaterial surfaces to yield more novel elastomer nanocomposites.

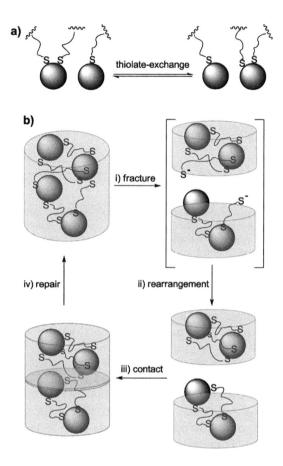

FIGURE 7.24
Self-healing mechanism in silicone elastomer doped with silver nanoparticles. (a) Thiolate ligands are constantly exchanging on the surface of the silver nanoparticles at steady state. (b) When the initially intact elastomer nanocomposite is fractured, the fracture heals by ligands bridging the fracture and bringing the faces back together. Figure from Martín et al. (2012).

7.6 Smart Coatings

Recent advances in material science have allowed the development of coatings that respond to external stimuli by changing their physical or chemical composition. Such coatings are referred to as smart or intelligent coatings. Baghdachi (2009) provides a good review of smart coating technology. Smart coatings can respond to physical stimuli (light, temperature, electromagnetic fields, acoustic waves, pH, ionic strength, solubility, etc.) or chemical stimuli (acid-base, photochemical, electrochemical, redox, and biochemical reactions). A good example of a smart coating involving nanoparticles is the use of silver nanoparticles as an antimicrobial coating. The nanoparticles remain inert until they are exposed to metabolic ion exchange reactions in a bacterial cell, at which point the silver ions take part in the ion exchange, replacing metabolically necessary ions in the cell and eventually resulting in cell death.

Many smart coating applications in the oil and gas industry are aimed at corrosion inhibition. Nanoparticles may be used to release corrosion inhibitors in response to stimuli generated by the corrosion process. Feng and Cheng (2017) developed nanocapsules of benzotriazole (BTA), a corrosion inhibitor. The nanocapsules can be dispersed in epoxy paint and break upon changes in pH. To form the nanocapsules, 70 nm silica nanoparticles were coated with a positively charged polyelectrolyte (poly-[diallyldimethylammonium], PDDAC) and then a negatively charged polyelectrolyte (sodium poly-[styrene sulfonate], SPSS). The coated nanoparticles were then added to aqueous BTA solution at low pH such that BTA existed as a cationic species, thus binding to the negatively charged nanoparticle surfaces. The nanocapsules were then dispersed in epoxy paint and coated on a steel surface. Release of BTA was controlled by pH. As corrosion proceeds at the steel surface, the pH gradually increases; as pH locally increases above 8, the BTA switches from cationic to anionic, and desorbs from the nanoparticle surface. Laboratory tests showed that the released BTA remained on the steel surface for at least 30 days (Figure 7.25).

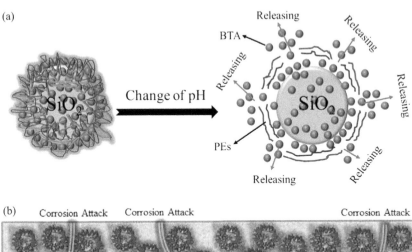

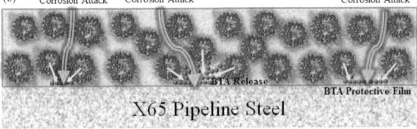

FIGURE 7.25
(a) pH-mediated release of BTA corrosion inhibitor from silica nanoparticle. (b) Schematic of self-healing mechanism in epoxy coating. Figure from Feng and Cheng (2017).

Another smart coating application is the use of nanoparticles in self-healing coatings. The process works in a similar fashion to that described above: nanoparticles with chemical agents sorbed to their surfaces, or nanocontainers with chemical agents encapsulated inside, are dispersed in the coating. The chemical agents are then released in response to an external stimulus, at which point they work to heal any defects in the coating (Samadzadeh et al. 2010). Zheludkevich et al. (2007) demonstrated that the same nanoencapsulation process reported by Feng and Cheng (2017) could heal defects in the surface coating in addition to releasing corrosion inhibitor. In their work, the polyelectrolyte coatings on the nanoparticles were observed to deteriorate as pH increased locally in response to corrosion (cf. Figure 7.25a); however, as some BTA was released, the pH decreased again, and the polyelectrolyte coating strengthened, stopping the release of the BTA. This self-healing behavior has not been tested in oilfield applications but should be the subject of further research.

Nomenclature

A	Cross-sectional area
f	Friction factor
G^*	Complex shear modulus
G'	Storage modulus
G''	Loss modulus
I	Consistency index
K	Diffusivity equation constant
k	Permeability
l	Sample length
N_{Re}	Reynolds number
n	Rheology exponent
P	Fluid pressure
P_0	Initial pore pressure
P_m	Pressure in upstream reservoir
r_1	Outer radius of annulus
r_2	Inner radius of annulus
t	Time
V	Volume of downstream reservoir
$\bar{v}$	Average flow velocity in annulus
x	Position along sample
β	Fluid compressibility
β_r	Rock compressibility
β_s	Rock matrix compressibility

$\dot{\gamma}$	Shear rate
δ	Phase lag
ε	Instantaneous strain
μ	Dynamic viscosity
μ_B	Bingham viscosity
ρ	Fluid density
σ	Instantaneous stress
σ_0	Maximum stress amplitude
τ	Shear stress
τ_0	Yield stress
φ	Interconnected porosity
ψ_n	Argument of diffusivity function

References

Abdo, J., and Haneef, M. D. (2013) Clay nanoparticles modified drilling fluids for drilling of deep hydrocarbon wells. *Appl. Clay Sci.*, 86, 76–82.

Aderibigbe, A. A., and Lane, R. H. (2013) Rock/Fluid Chemistry Impacts on Shale Fracture Behavior. Society of Petroleum Engineers, SPE International Symposium on Oilfield Chemistry, April 8–10, The Woodlands, TX, USA. https://doi.org/10.2118/164102-MS

Aftab, A., Ismail, A. R., Khokhar, S., and Ibupoto, Z. H. (2016) Novel zinc oxide nanoparticles deposited acrylamide composite used for enhancing the performance of water-based drilling fluids at elevated temperature conditions. *J. Petrol. Sci. Eng.*, 146, 1142–1157.

Alberty, M. W., and McLean, M. R. (2004) A Physical Model for Stress Cages. Society of Petroleum Engineers, SPE Annual Technical Conference and Exhibition, September 26–29, Houston, TX, USA. https://doi.org/10.2118/90493-MS

Al-Muntasheri, G. A. (2014) A critical review of hydraulic-fracturing fluids for moderate- to ultralow-permeability formations over the last decade. *SPE Production & Operations*, 29(4), 243–260.

Al-Muntasheri, G. A., Liang, F., and Hull, K. L. (2017) Nanoparticle-enhanced hydraulic-fracturing fluids: A review. *SPE Prod. Oper.*, 32(2), 186–195.

Amanullah, M., and Ramasamy, J. (2014) Nanotechnology Can Overcome the Critical Issues of Extremely Challenging Drilling and Production Environments. Society of Petroleum Engineers, Abu Dhabi International Petroleum Exhibition and Conference, November 10–13, Abu Dhabi, UAE.

American Petroleum Institute. (2016) *ASTM ANSI/API RP 13B1 4TH ED (E1) (R 2016) Recommended Practice for Field Testing Water-Based Drilling Fluids; Fourth edition; Reaffirmed March 2016; ISO 10414-1:2008.* American Petroleum Institute, Washington, DC, USA.

American Petroleum Institute. (2018) *ASTM ANSI/API RP 13B2 5TH ED (E2) Recommended Practice for Field Testing Oil-Based Drilling Fluids.* American Petroleum Institute, Washington, DC, USA.

Aston, M. S., Alberty, M. W., Duncum, S. D., Bruton, J. R., Friedheim, J. E., and Sanders, M. W. (2007) A New Treatment for Wellbore Strengthening in Shale. Society of Petroleum Engineers. SPE Annual Technical Conference and Exhibition, November 11–14, Anaheim, CA, USA. https://doi.org/10.2118/110713-MS

Baghdachi, J. (2009) Smart Coatings. In *Smart Coatings II*. ACS Symposium Series, *1002, 3–24*.

Barati, R. (2015) Application of nanoparticles as fluid loss control additives for hydraulic fracturing of tight and ultra-tight hydrocarbon-bearing formations. *J. Nat. Gas Sci. Eng.*, 27, 1321–1327.

Barati, R., and Liang, J.-T. (2014) A review of fracturing fluid systems used for hydraulic fracturing of oil and gas wells. *J. Appl. Polym. Sci.*, 131(16), 40735.

Barry, M. M., Jung, Y., Lee, J.-K., Phuoc, T. X., and Chyu, M. K. (2015) Fluid filtration and rheological properties of nanoparticle additive and intercalated clay hybrid bentonite drilling fluids. *J. Petrol. Sci. Eng.*, 127, 338–346.

Beaudoin, J. J., Raki, L., and Alizadeh, R. (2009) A 29Si MAS NMR study of modified C–S–H nanostructures. *Cement Concrete Comp.*, 31(8), 585–590.

Becker, T. E., Azar, J. J., and Okrajni, S. S. (1991) Correlations of mud rheological properties with cuttings-transport performance in directional drilling. *SPE Drill. Eng.*, 6(1), 16–24.

Belnap, J. D., Zhan, G., Sheng, X., Zhang, Y., Keshavan, M. K., Pratt, H., and Shen, Y. (2011) *US7862634B2*. United States. Retrieved from https://patents.google.com/patent/US7862634B2/en

Bergna, H. E. (2005) Colloid Chemistry of Silica: An Overview. In *Colloidal Silica: Fundamentals and Applications* (Vol. 131). CRC Press, Boca Raton, FL. pp. 9–36.

Bingham, E. C. (1922) *Fluidity and Plasticity*. New York, NY: McGraw-Hill.

Bird, R. B., Stewart, W. E., and Lightfoot, E. N. (2007) *Transport Phenomena*, 2nd edn. John Wiley & Sons, New York, NY, USA.

Björnström, J., Martinelli, A., Matic, A., Börjesson, L., and Panas, I. (2004) Accelerating effects of colloidal nano-silica for beneficial calcium–silicate–hydrate formation in cement. *Chem. Phys. Lett.*, 392(1), 242–248.

Borisov, A. S., Husein, M., and Hareland, G. (2015) A field application of nanoparticle-based invert emulsion drilling fluids. *J. Nanopart. Res.*, 17(8), 340.

Boul, P. J., Reddy, B. R., Zhang, J., and Thaemlitz, C. (2017) Functionalized nanosilicas as shale inhibitors in water-based drilling fluids. *SPE Drill. Completion*, 32(2), 121–130.

Brace, W. F., Walsh, J. B., and Frangos, W. T. (1968) Permeability of granite under high pressure. *J. Geophys. Res.*, 73(6), 2225–2236.

Brown, G. E., Henrich, V. E., Casey, W. H., Clark, D. L., Eggleston, C., Felmy, A., …Zachara, J. M. (1999) Metal oxide surfaces and their interactions with aqueous solutions and microbial organisms. *Chem. Rev.*, 99(1), 77–174.

Cai, J., Chenevert, M. E., Sharma, M. M., and Friedheim, J. E. (2012) Decreasing water invasion into Atoka shale using nonmodified silica nanoparticles. *SPE Drill. Completion*, 27(1), 103–112.

Cao, G. (2004) *Nanostructures & Nanomaterials: Synthesis, Properties & Applications*. Imperial College Press, London, UK.

Carslaw, H. S., and Jaeger, J. C. (1959) *Conduction of Heat in Solids* (2nd ed.). Oxford University Press, Oxford, UK.

Chan, A., Hannahs, D., Jellison, M. J., Breitsameter, M., Branagan, D., Stone, H., and Jeffers, G. (2008) Evolution of Drilling Programs and Complex Well Profiles Drive Development of Fourth-Generation Hardband Technology. Presented at the IADC/SPE Drilling Conference, Society of Petroleum Engineers. https://doi.org/10.2118/112740-MS

Contreras, O., Hareland, G., Husein, M., Nygaard, R., and Alsaba, M. (2014) Wellbore Strengthening in Sandstones by Means of Nanoparticle-Based Drilling Fluids. Presented at the SPE Deepwater Drilling and Completions Conference, Society of Petroleum Engineers. https://doi.org/10.2118/170263-MS

Cordier, F., and Grzesiek, S. (2002) Temperature-dependence of protein hydrogen bond properties as studied by high-resolution NMR. *J. Mol. Biol.*, 317(5), 739–752.

Crews, J. B., and Gomaa, A. M. (2012) Nanoparticle Associated Surfactant Micellar Fluids: An Alternative to Crosslinked Polymer Systems. Presented at the SPE International Oilfield Nanotechnology Conference and Exhibition, Society of Petroleum Engineers. https://doi.org/10.2118/157055-MS

Do, B. P. H., Nguyen, B. D., Nguyen, H. D., and Nguyen, P. T. (2013) Synthesis of magnetic composite nanoparticles enveloped in copolymers specified for scale inhibition application. *Adv. Nat. Sci.-Nanosci.*, 4(4), 045016.

Dupriest, F. E. (2005) Fracture Closure Stress (FCS) and Lost Returns Practices. *Presented at the SPE/IADC Drilling Conference*, Society of Petroleum Engineers. https://doi.org/10.2118/92192-MS

Dyke, O. W. V. (1949) *US2477219A*. United States. Retrieved from https://patents.google.com/patent/US2477219A/en

Emrani, A. S., and Nasr-El-Din, H. A. (2017) Stabilizing CO_2 foam by use of nanoparticles. *SPE Journal*, 22(2), 494–504.

Endo, M., Noguchi, T., Ito, M., Takeuchi, K., Hayashi, T., Kim, Y. A., … Dresselhaus, M. S. (2008) Extreme-performance rubber nanocomposites for probing and excavating deep oil resources using multi-walled carbon nanotubes. *Adv. Funct. Mater.*, 18(21), 3403–3409.

Espinosa, H. D., Rim, J. E., Barthelat, F., and Buehler, M. J. (2009) Merger of structure and material in nacre and bone—Perspectives on de novo biomimetic materials. *Prog. Mater. Sci.*, 54(8), 1059–1100.

Ettefaghi, E., Ahmadi, H., Rashidi, A., Nouralishahi, A., and Mohtasebi, S. S. (2013) Preparation and thermal properties of oil-based nanofluid from multi-walled carbon nanotubes and engine oil as nano-lubricant. *Int. Commun. Heat Mass*, 46, 142–147.

Fakoya, M. F., and Shah, S. N. (2013) Rheological Properties of Surfactant-Based and Polymeric Nano-Fluids. Presented at the SPE/ICoTA Coiled Tubing and Well Intervention Conference and Exhibition, Society of Petroleum Engineers. https://doi.org/10.2118/163921-MS

Feng, Y., and Cheng, Y. F. (2017) An intelligent coating doped with inhibitor-encapsulated nanocontainers for corrosion protection of pipeline steel. *Chem. Eng. J.*, 315, 537–551.

Feng, Y., and Gray, K. E. (2017) Review of fundamental studies on lost circulation and wellbore strengthening. *J. Petrol. Sci. Eng.*, 152, 511–522.

Fink, J. (Ed.). (2015) *Petroleum Engineer's Guide to Oil Field Chemicals and Fluids* (2nd ed.). Gulf Professional Publishing, Waltham, MA, USA.

Fischer, P., and Rehage, H. (1997) Non-linear flow properties of viscoelastic surfactant solutions. *Rheol. Acta*, 36(1), 13–27.

Founargiotakis, K., Kelessidis, V. C., and Maglione, R. (2008) Laminar, transitional and turbulent flow of Herschel–Bulkley fluids in concentric annulus. *Can. J. Chem. Eng.*, 86(4), 676–683.

Frogley, M. D., Ravich, D., and Wagner, H. D. (2003) Mechanical properties of carbon nanoparticle-reinforced elastomers. *Compos. Sci. Technol.*, 63(11), 1647–1654.

Galindo-Gonzalez, C., Feinberg, J. M., Kasama, T., Gontard, L. C., Pósfai, M., Kósa, I., ...Dunin-Borkowski, R. E. (2009) Magnetic and microscopic characterization of magnetite nanoparticles adhered to clay surfaces. *Am. Mineral.*, 94(8–9), 1120–1129.

Gao, C., Miska, S. Z., Yu, M., Ozbayoglu, E. M., and Takach, N. E. (2016) Effective Enhancement of Wellbore Stability in Shales with new Families of Nanoparticles. Presented at the SPE Deepwater Drilling and Completions Conference, Society of Petroleum Engineers. https://doi.org/10.2118/180330-MS

Gilje, E., Loutskina, E., and Murphy, D. (2017) *Drilling and Debt*. (Darden Business School Working Papers No. ID 2939603). Darden School of Business, Charlottesville, VA, USA. Retrieved from https://papers.ssrn.com/abstract=2939603

Griffo, A., and Keshavan, M. K. (2009) *US20090038858A1*. United States. Retrieved from https://patents.google.com/patent/US20090038858A1/en

Grundmann, S. R., and Lord, D. L. (1983) Foam stimulation. *J. Petrol. Technol.*, 35(3), 597–602.

Gurluk, M. R., Nasr-El-Din, H. A., and Crews, J. B. (2013) Enhancing the Performance of Viscoelastic Surfactant Fluids Using Nanoparticles. Presented at the EAGE Annual Conference and Exhibition incorporating SPE Europec, Society of Petroleum Engineers. https://doi.org/10.2118/164900-MS

Helgeson, M. E., Hodgdon, T. K., Kaler, E. W., Wagner, N. J., Vethamuthu, M., and Ananthapadmanabhan, K. P. (2010) Formation and rheology of viscoelastic "double networks" in Wormlike Micelle–nanoparticle mixtures. *Langmuir*, 26(11), 8049–8060.

Herschel, W. H., and Bulkley, R. (1926) Konsistenzmessungen von Gummi-Benzollösungen. *Kolloid-Zeitschrift*, 39(4), 291–300.

Hoelscher, K. P., De Stefano, G., Riley, M., and Young, S. (2012) Application of Nanotechnology in Drilling Fluids. Presented at the SPE International Oilfield Nanotechnology Conference and Exhibition, Society of Petroleum Engineers. https://doi.org/10.2118/157031-MS

Hoxha, B. B., Oort, E. van, and Daigle, H. (2017) How Do Nanoparticles Stabilize Shale? Presented at the SPE International Conference on Oilfield Chemistry, Society of Petroleum Engineers. https://doi.org/10.2118/184574-MS

Huang, T., and Crews, J. B. (2008) Nanotechnology applications in viscoelastic surfactant stimulation fluids. *SPE Prod. Oper.*, 23(4), 512–517.

Huang, T., Crews, J. B., and Agrawal, G. (2010) Nanoparticle Pseudocrosslinked Micellar Fluids: Optimal Solution for Fluid-Loss Control with Internal Breaking. Presented at the SPE International Symposium and Exhibition on Formation Damage Control, Society of Petroleum Engineers. https://doi.org/10.2118/128067-MS

Hurnaus, T., and Plank, J. (2015) Behavior of titania nanoparticles in cross-linking hydroxypropyl guar used in hydraulic fracturing fluids for Oil recovery. *Energ. Fuel.*, 29(6), 3601–3608.

Hwang, Y., Lee, C., Choi, Y., Cheong, S., Kim, D., Lee, K., …Kim, S. H. (2011) Effect of the size and morphology of particles dispersed in nano-oil on friction performance between rotating discs. *J. Mech. Sci. Technol.*, 25(11), 2853–2857.

Ingraffea, A. R., Wells, M. T., Santoro, R. L., and Shonkoff, S. B. C. (2014) Assessment and risk analysis of casing and cement impairment in oil and gas wells in Pennsylvania, 2000–2012. *P. Natl. Acad. Sci. U.S.A.*, 201323422. https://doi.org/10.1073/pnas.1323422111

Inukai, S., Niihara, K., Noguchi, T., Ueki, H., Magario, A., Yamada, E., …Endo, M. (2011) Preparation and properties of multiwall carbon nanotubes/polystyrene-block-polybutadiene-block-polystyrene composites. *Ind. Eng. Chem. Res.*, 50(13), 8016–8022.

Ismail, A. R., Aftab, A., Ibupoto, Z. H., and Zolkifile, N. (2016) The novel approach for the enhancement of rheological properties of water-based drilling fluids by using multi-walled carbon nanotube, nanosilica and glass beads. *J. Petrol. Sci. Eng.*, 139, 264–275.

Iyoho, A. W., and Azar, J. J. (1981) An accurate slot-flow model for non-Newtonian fluid flow through eccentric annuli. *SPE J.*, 21(5), 565–572.

Izon, D., Danenberger, E. P., and Mayes, M. (2007) Absence of fatalities in blowouts encouraging in MMS study of OCS incidents 1992-2006. *Drilling Contractor*, 63(4), 84–90.

Jaffal, H. A., El Mohtar, C. S., and Gray, K. E. (2017) Modeling of filtration and mudcake buildup: An experimental investigation. *J. Nat. Gas. Sci. Eng.*, 38, 1–11.

Ji, L., Guo, Q., Friedheim, J. E., Zhang, R., Chenevert, M. E., and Sharma, M. M. (2012) Laboratory Evaluation and Analysis of Physical Shale Inhibition of an Innovative Water-Based Drilling Fluid with Nanoparticles for Drilling Unconventional Shales. Presented at the SPE Asia Pacific Oil and Gas Conference and Exhibition, Society of Petroleum Engineers. https://doi.org/10.2118/158895-MS

Ji, T. (2005) Preliminary study on the water permeability and microstructure of concrete incorporating nano-SiO2. *Cement Concrete Res.*, 35(10), 1943–1947.

Ji, X., Chen, Y., Zhao, G., Wang, X., and Liu, W. (2011) Tribological properties of $CaCO_3$ nanoparticles as an additive in lithium grease. *Tribol. Lett.*, 41(1), 113–119.

Jo, B.-W., Kim, C.-H., Tae, G., and Park, J.-B. (2007) Characteristics of cement mortar with nano-SiO2 particles. *Constr. Build. Mater.*, 21(6), 1351–1355.

Kawashima, S., Hou, P., Corr, D. J., and Shah, S. P. (2013) Modification of cement-based materials with nanoparticles. *Cement Concrete Comp.*, 36, 8–15.

Kelessidis, V. C., Maglione, R., Tsamantaki, C., and Aspirtakis, Y. (2006) Optimal determination of rheological parameters for Herschel–Bulkley drilling fluids and impact on pressure drop, velocity profiles and penetration rates during drilling. *J. Petrol. Sci. Eng.*, 53(3), 203–224.

Kong, D., Su, Y., Du, X., Yang, Y., Wei, S., and Shah, S. P. (2013) Influence of nano-silica agglomeration on fresh properties of cement pastes. *Constr. Build. Mater.*, 43, 557–562.

Krishnan, S., Abyat, Z., and Chok, C. (2016) Characterization of Boron-Based Nanomaterial Enhanced Additive in Water-Based Drilling Fluids: A Study on Lubricity, Drag, ROP and Fluid Loss Improvement. Presented at the SPE/IADC Middle East Drilling Technology Conference and Exhibition, Society of Petroleum Engineers. https://doi.org/10.2118/178240-MS

Laik, S. (2018) *Offshore Petroleum Drilling and Production*. CRC Press, Boca Raton, FL. Retrieved from http://ebookcentral.proquest.com/lib/utxa/detail.action?docID=5257228

Li, Y., DiCarlo, D., Li, X., Zang, J., and Li, Z. (2015) An experimental study on application of nanoparticles in unconventional gas reservoir CO2 fracturing. *J. Petrol. Sci. Eng.*, 133, 238–244.

Liu, L., Fang, Z., Gu, A., and Guo, Z. (2011) Lubrication effect of the paraffin oil filled with functionalized multiwalled carbon nanotubes for bismaleimide resin. *Tribol. Lett.*, 42(1), 59–65.

Lukawski, M. Z., Anderson, B. J., Augustine, C., Capuano, L. E., Beckers, K. F., Livesay, B., and Tester, J. W. (2014) Cost analysis of oil, gas, and geothermal well drilling. *J. Petrol. Sci. Eng.*, 118, 1–14.

Ma, S., Zheng, S., Cao, D., and Guo, H. (2010) Anti-wear and friction performance of ZrO_2 nanoparticles as lubricant additive. *Particuology.*, 8(5), 468–472.

Madani, H., Bagheri, A., and Parhizkar, T. (2012) The pozzolanic reactivity of monodispersed nanosilica hydrosols and their influence on the hydration characteristics of Portland cement. *Cement Concrete Res.*, 42(12), 1563–1570.

Mao, H., Qiu, Z., Shen, Z., and Huang, W. (2015) Hydrophobic associated polymer based silica nanoparticles composite with core–shell structure as a filtrate reducer for drilling fluid at utra-high temperature. *. J. Petrol. Sci. Eng.*, 129, 1–14.

Martín, R., Rekondo, A., Echeberria, J., Cabañero, G., J. Grande, H., and Odriozola, I. (2012) Room temperature self-healing power of silicone elastomers having silver nanoparticles as crosslinkers. *Chem. Commun.*, 48(66), 8255–8257.

Matsuyama, H., and Young, J. F. (1999a) Intercalation of polymers in calcium silicate hydrate: A new synthetic approach to biocomposites? *Chem. Mater.*, 11(1), 16–19.

Matsuyama, H., and Young, J. F. (1999b) Synthesis of calcium silicate hydrate/polymer complexes: Part I. Anionic and nonionic polymers. *J. Mater. Res.*, 14(8), 3379–3388.

Matsuyama, H., and Young, J. F. (1999c) Synthesis of calcium silicate hydrate/polymer complexes: Part II. Cationic polymers and complex formation with different polymers. *J. Mater. Res.*, 14(8), 3389–3396.

Maxey, J., Crews, J., and Huang, T. (2008) Nanoparticle associated surfactant micellar fluids. *AIP Conf. Proc.*, 1027(1), 857–859.

Mazerov, K. (2016, January 12) Dissolvable Tools, Plug-Less and Wave Fracturing Among Innovations Delivering More Efficient Multistage Completions. *Drilling Contractor*. Retrieved from www.drillingcontractor.org/innovation-rd-keep-pace-need-deliver-efficient-cost-effective-multistage-completions-37893

Melrose, J. C., and Lilienthal, W. B. (1951) Plastic flow properties of drilling fluids—measurement and application. *. J. Petrol. Technol.*, 3(6), 159–164.

Melrose, J. C., Savins, J. G., Foster, W. R., and Parish, E. R. (1958) A practical utilization of the theory of bingham plastic flow in stationary pipes and annuli. *Pet. Trans., AIME*, 213, 316–324.

Michaux, M., Nelson, E., and Vidick, B. (1989) Cement chemistry and additives. *Oilfield Review*, 1(1), 18–25.

Middleton, R. S., Carey, J. W., Currier, R. P., Hyman, J. D., Kang, Q., Karra, S., … Viswanathan, H. S. (2015) Shale gas and non-aqueous fracturing fluids: Opportunities and challenges for supercritical CO2. *Appl. Energ.*, 147, 500–509.

Mojumdar, S. C., and Raki, L. (2005) Characterization and Properties of Calcium Silicate Hydrate Polymer Nanocomposites. Presented at the Annual Meeting of the American Ceramics Society, Baltimore, MD, USA.

Nabhani, N., Emami, M., and Moghadam, A. B. T. (2011) Application of nanotechnology and nanomaterials in oil and gas industry. *AIP Conf. Proc.*, 1415(1), 128–131.

National Academy of Engineering. (2011) *Macondo Well Deepwater Horizon Blowout: Lessons for Improving Offshore Drilling Safety.* National Academies Press, Washington, DC, USA.

Nazari, A., and Riahi, S. (2011) The effects of SiO2 nanoparticles on physical and mechanical properties of high strength compacting concrete. *Compos. Part B-Eng.*, 42(3), 570–578.

Nettesheim, F., Liberatore, M. W., Hodgdon, T. K., Wagner, N. J., Kaler, E. W., and Vethamuthu, M. (2008) Influence of nanoparticle addition on the properties of Wormlike Micellar solutions. *Langmuir*, 24(15), 7718–7726.

Noguchi, T., Inukai, S., Uekii, H., Magario, A., and Endou, M. (2009) Mechanical properties of MWCNT/elastomer nanocomposites and the cellulation model. Presented at the SAE World Congress and Exhibition, Detroit, MI, USA. https://doi.org/10.4271/2009-01-0606

Okamoto, M. (2003) *Polymer/Layered Silicate Nanocomposites* (Vol. 14). Rapra Technology Limited, Shawbury, UK.

Okrajni, S., and Azar, J. J. (1986) The effects of mud rheology on annular hole cleaning in directional wells. *SPE Drill. Eng.*, 1(4), 297–308.

Pang, X., Boul, P. J., and Cuello Jimenez, W. (2014) Nanosilicas as accelerators in oilwell cementing at low temperatures. *SPE Drill. Completion*, 29(1), 98–105.

Peng, Y., Hu, Y., and Wang, H. (2007) Tribological behaviors of surfactant-functionalized carbon nanotubes as lubricant additive in water. *Tribol. Lett.*, 25(3), 247–253.

Ponmani, S., William, J. K. M., Samuel, R., Nagarajan, R., and Sangwai, J. S. (2014) Formation and characterization of thermal and electrical properties of CuO and ZnO nanofluids in xanthan gum. *Colloid. Surface. A*, 443, 37–43.

Prodanović, M., and Johnston, K. P. (2017) *Development of Nanoparticle-Stabilized Foams to Improve Performance of Water-Less Hydraulic Fracturing* (Final Technical Report No. DE-FE0013723) (p. 27). US Department of Energy, Morgantown, WV, USA.

Qajar, A., Xue, Z., Worthen, A. J., Johnston, K. P., Huh, C., Bryant, S. L., and Prodanović, M. (2016) Modeling fracture propagation and cleanup for dry nanoparticle-stabilized-foam fracturing fluids. *J. Petrol. Sci. Eng.*, 146, 210–221.

Ribeiro, L., and Sharma, M. (2013) Fluid Selection for Energized Fracture Treatments. Presented at the SPE Hydraulic Fracturing Technology Conference. https://doi.org/10.2118/163867-MS

Roddy, C. W. (2016) *US9512346B2*. United States. Retrieved from https://patents.google.com/patent/US9512346B2/en?oq=US9512346B2

Roddy, C. W., Chatterji, J., and Cromwell, R. (2009) US7559369B2. United States. Retrieved from https://patents.google.com/patent/US7559369B2/en?oq=US7559369B2

Roddy, C. W., Covington, R. L., Chatterji, J., and Brenneis, D. C. (2013) *US8586512B2*. United States. Retrieved from https://patents.google.com/patent/US8586512B2/en?oq=US8586512B2

Salih, A. H., and Bilgesu, H. (2017) Investigation of Rheological and Filtration Properties of Water-Based Drilling Fluids Using Various Anionic Nanoparticles. Presented at the SPE Western Regional Meeting. https://doi.org/10.2118/185638-MS

Samadzadeh, M., Boura, S. H., Peikari, M., Kasiriha, S. M., and Ashrafi, A. (2010) A review on self-healing coatings based on micro/nanocapsules. *Prog. Org. Coat.*, 68(3), 159–164.

Santra, A. K., Boul, P., and Pang, X. (2012) Influence of Nanomaterials in Oilwell Cement Hydration and Mechanical Properties. Presented at the SPE International Oilfield Nanotechnology Conference and Exhibition . https://doi.org/10.2118/156937-MS

Sengupta, S., and Kumar, A. (2013) Nano-Ceramic Coatings—A Means for Enhancing Bit Life and Reducing Drill String Trips. Presented at the International Petroleum Technology Conference, International Petroleum Technology Conference. https://doi.org/10.2523/IPTC-16474-MS

Sensoy, T., Chenevert, M. E., and Sharma, M. M. (2009) Minimizing Water Invasion in Shales Using Nanoparticles. Presented at the SPE Annual Technical Conference and Exhibition. https://doi.org/10.2118/124429-MS

Shahsavari, R., and Sakhavand, N. (2016) Hybrid cementitious materials: Nanoscale modeling and characterization. In *Innovative Developments of Advanced Multifunctional Nanocomposites in Civil and Structural Engineering* (Vol. 62). Woodhead Publishing, Duxford, UK.

Sharma, M. M., Zhang, R., Chenevert, M. E., Ji, L., Guo, Q., and Friedheim, J. (2012) A New Family of Nanoparticle Based Drilling Fluids. Presented at the SPE Annual Technical Conference and Exhibition. https://doi.org/10.2118/160045-MS

Shih, J.-Y., Chang, T.-P., and Hsiao, T.-C. (2006) Effect of nanosilica on characterization of Portland cement composite. *Mat. Sci. Eng. A-Struct.*, 424(1), 266–274.

Šiler, P., Kolářová, I., Sehnal, T., Másilko, J., and Opravil, T. (2016) The determination of the influence of pH value of curing conditions on Portland Cement hydration. *Procedia Engineer.*, 151, 10–17.

Slay, J. B., and Ray, T. W. (2010) *US7696275B2.* United States. Retrieved from https://patents.google.com/patent/US7696275B2/en

Smith, D. K. (2003) *Cementing.* Society of Petroleum Engineers, Richardson, TX.

Taha, N. M., and Lee, S. (2015) Nano Graphene Application Improving Drilling Fluids Performance. Presented at the International Petroleum Technology Conference, Doha, Qatar. https://doi.org/10.2523/IPTC-18539-MS

Todd, B. L., Starr, P. M., Swor, L. C., Schwendemann, K. L., and Jr, T. M. (2008) US7353879B2. United States. Retrieved from https://patents.google.com/patent/US7353879B2/en

Trimmer, D. A. (1981) Design criteria for laboratory measurements of low permeability rocks. *Geophysical Research Letters*, 8(9), 973–975.

van Oort, E. (1994) A Novel Technique for the Investigation of Drilling Fluid Induced Borehole Instability in Shales. Presented at the Rock Mechanics in Petroleum Engineering . https://doi.org/10.2118/28064-MS

Varga, L. K. (2007) Soft magnetic nanocomposites for high-frequency and high-temperature applications. *J. Magn. Magn. Mater.*, 316(2), 442–447.

Vipulanandan, C., Mohammed, A., and Qu, Q. (2014) Characterizing the Hydraulic Fracturing Fluid Modified with Nano Silica Proppant. Presented at the AADE Fluids Technical Conference and Exhibition, Houston, TX, USA.

Vipulanandan, C., Mohammed, A., and Samuel, R. G. (2017) Smart Bentonite Drilling Muds Modified with Iron Oxide Nanoparticles and Characterized Based on the Electrical Resistivity and Rheological Properties with Varying Magnetic Field Strengths and Temperatures. Presented at the Offshore Technology Conference. https://doi.org/10.4043/27626-MS

Vryzas, Z., Kelessidis, V. C., Bowman, M. B. J., Nalbantian, L., Zaspalis, V., Mahmoud, O., and Nasr-El-Din, H. A. (2017) Smart Magnetic Drilling Fluid with In-situ Rheological Controllability Using Fe_3O_4 Nanoparticles. Presented at the SPE Middle East Oil and Gas Show and Conference. https://doi.org/10.2118/183906-MS

Vryzas, Z., Mahmoud, O., Nasr-El-Din, H. A., and Kelessidis, V. C. (2015) Development and Testing of Novel Drilling Fluids Using Fe_2O_3 and SiO_2 Nanoparticles for Enhanced Drilling operationS. Presented at the International Petroleum Technology Conference. https://doi.org/10.2523/IPTC-18381-MS

Vryzas, Z., Zaspalis, V., Nalbantian, L., Mahmoud, O., Nasr-El-Din, H. A., and Kelessidis, V. C. (2016) A Comprehensive Approach for the Development of New Magnetite Nanoparticles Giving Smart Drilling Fluids with Superior Properties for HP/HT Applications. Presented at the International Petroleum Technology Conference. https://doi.org/10.2523/18731-MS

Wahid, N., Yusof, M. A. M., and Hanafi, N. H. (2015) Optimum Nanosilica Concentration in Synthetic Based Mud (SBM) for High Pressure High Temperature Well. Presented at the SPE/IATMI Asia Pacific Oil and Gas Conference and Exhibition . https://doi.org/10.2118/176036-MS

Watson, T. L., and Bachu, S. (2009) Evaluation of the potential for gas and CO2 leakage along wellbores. *SPE Drill. Completion*, 24(1), 115–126.

William, J. K. M., Ponmani, S., Samuel, R., Nagarajan, R., and Sangwai, J. S. (2014) Effect of CuO and ZnO nanofluids in xanthan gum on thermal, electrical and high pressure rheology of water-based drilling fluids. *J. Petrol. Sci. Eng.*, 117, 15–27.

Wu, Y. Y., Tsui, W. C., and Liu, T. C. (2007) Experimental analysis of tribological properties of lubricating oils with nanoparticle additives. *Wear*, 262(7), 819–825.

Xu, Z., Agrawal, G., and Salinas, B. J. (2011) Smart Nanostructured Materials Deliver High Reliability Completion Tools for Gas Shale Fracturing. Presented at the SPE Annual Technical Conference and Exhibition. https://doi.org/10.2118/146586-MS

Zakaria, M., Husein, M. M., and Harland, G. (2012) Novel Nanoparticle-Based Drilling Fluid with Improved Characteristics. Presented at the SPE International Oilfield Nanotechnology Conference and Exhibition. https://doi.org/10.2118/156992-MS

Zheludkevich, M. L., Shchukin, D. G., Yasakau, K. A., Möhwald, H., and Ferreira, M. G. S. (2007) Anticorrosion coatings with self-healing effect based on nanocontainers impregnated with corrosion inhibitor. *Chem. Mater.*, 19(3), 402–411.

8

Production Operations and Flow Assurance

8.1 Introduction

Production operations and flow assurance—the practice of producing oil and gas from the subsurface and maintaining economic production rates safely and efficiently—are of paramount importance to the oil and gas industry. A wide variety of challenges exist in this area, including changes to the reservoir as a result of drilling and production, corrosion and precipitation of solids inside pipes, and properly treating produced fluids before sending to sales or disposal.

As we have seen throughout this book, technological developments involving nanoparticles to overcome these challenges take advantage of the unique properties of nanoparticles. In this chapter, we will discuss nanoparticle applications in which the nanoparticles supplement or take the place of chemicals, and applications in which the superparamagnetic properties of iron oxide nanoparticles are used for targeted heating or collection with a permanent magnet. In all cases, nanoparticles provide an alternative to the use of more hazardous chemicals and provide new functionality that cannot be achieved with other materials.

8.2 Emulsion and Colloidal Particulate Removal from Produced Fluids

Produced fluids in the oilfield are typically in the form of emulsions. Separating and handling these fluids is important both for disposal of unwanted products, like produced water, and for the handling and sale of crude oil. Emulsion separation technologies currently in use include centrifugation, gravity flotation, and various membrane and chemical treatments (Fakhru'l-Razi et al. 2009; Kokal 2005). However, the presence of micron-sized or smaller droplets in these emulsions presents a challenge because these techniques are not effective at handling such small droplets.

During the oil production process, water is brought to the surface along with oil. The produced water is the largest byproduct of production and may exceed 20 times the volume of produced oil. The treatment of produced water for re-injection or disposal is generally one of the largest oil-field operating expenses. Conventionally, both oil and water are brought to the surface and then separated at the surface. One major component of the produced water treatment is the removal of the highly stable dispersed oil. The US Environmental Protection Agency (USEPA) requires zero oil discharge for all produced water during onshore production. For offshore production, USEPA limits the discharge of oil and grease in produced water to a daily maximum of 72 mg/L and a 30-day average of 48 mg/L (40 CFR §435, 2016). However, about 60% of offshore platforms in the Gulf of Mexico are believed not to be able to reach USEPA discharge requirements. The Convention for the Protection of Marine Environment of the North-East Atlantic (OSPAR) further limits the sea discharge for dispersed oil to 40 mg/L (OSPAR 2007).

A related issue is the removal of water droplets from produced oil. The presence of water in crude oil can lead to problems during production, transport, and refining due to the effect that water droplets can have on physical properties of the liquid phase (*e.g.*, Liu et al. 2003). Much of this behavior stems from the fact that surface-active agents in the crude oil preferentially accumulate at the oil-water interface, forming a very strong emulsion that is difficult to break (Kokal 2005; McLean and Kilpatrick 1997).

Finally, with the advent of polymer flooding operations for enhanced oil recovery (EOR), the issue of treating and disposing of produced water containing residual EOR polymer has arisen. As with produced water from primary and secondary recovery operations, this produced water is often an emulsion containing small oil droplets (Duan et al. 2014), but the presence of the polymer makes separation especially difficult due to the extra emulsion stabilization afforded by the polymer (Zhang et al. 2010b).

In this section, we review several nanoparticle-assisted technologies for separating oil, water, and EOR polymer in production operations.

8.2.1 Magnetic Nanoparticle-Based Removal of Oil Droplets from Produced Water

Magnetic nanoparticles (MNPs) represent an attractive means of removing micron-size oil droplets from water. Their large surface area means that a significant amount of surface functionalization may be applied to promote attraction to oil droplets and their magnetic properties allow them to be separated using high-gradient magnetic separation (HGMS) technologies.

Ferromagnetic materials have multi-domain structures which are the small regions magnetized in a uniform direction, i.e., the magnetic moment is aligned to the same direction. Total magnetization is the vector sum of magnetization vectors for all of the individual domains. Superparamagnetic nanoparticles are particles of Fe_3O_4 (magnetite) and $\gamma\text{-}Fe_2O_3$ (maghemite) with diameters smaller

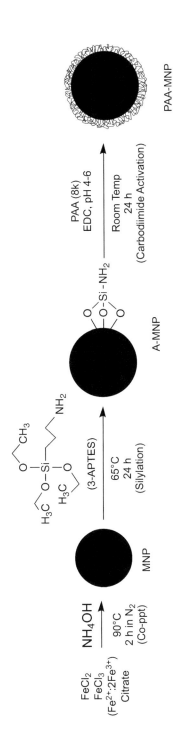

FIGURE 8.1
Summary of magnetic nanoparticle synthesis and surface coating for oil droplet removal. MNP = magnetic nanoparticle; A-MNP = amine-coated MNP; PAA-MNP = PAA-coated MNP. Figure after Ko et al. (2017).

than 100 nm. Their superparamagnetism comes from their small size, resulting in a single magnetic domain in each particle (see Figure 8.1).

One of the major advantages in using superparamagnetic nanoparticles lies in their ability to be manipulated using an external magnetic field (Rosensweig 1985). This phenomenon can be used to remove magnetic nanoparticles from solution once they have adsorbed a desired contaminant. The magnetic force on a nanoparticle can be several orders of magnitude larger than the gravitational force, allowing much faster separation from water than gravitational physical separation processes (Ko et al. 2017; Prigiobbe et al. 2015). This type of HGMS technology has been commonly applied in the mining and other industries for many years to separate magnetic minerals, for example, Fe, Co, and Ni (Yavuz et al. 2009). In HGMS, a strong magnetic gradient (up to 10^4 T/m) is produced either with a large permanent magnet or a magnetized column containing disordered material like steel wool (Moeser et al. 2002; Yavuz et al. 2009). The magnetic gradient exerts a force on magnetic materials and can move them in desired directions. HGMS has recently been applied in more complicated separations in various fields, including the separations of cells and proteins in biomedicine, as well as radionuclides, heavy metals, and non-polar organic contaminants from water in environmental engineering (Ambashta and Sillanpää 2010; Moeser et al. 2002).

The challenge in using superparamagnetic nanoparticles for removing oil droplets from produced water lies in designing a suitable surface coating that promotes attraction of the nanoparticles to the oil droplets and aggregation of the oil droplets for faster magnetic separation. Crude oil droplets typically have negative surface charges at neutral to alkaline pH values (Buckley et al. 1989; Dubey and Doe 1993), so simple electrostatic attraction can be promoted if the nanoparticles can be given a positive charge. Ko et al. (2014, 2017) showed that this could be accomplished effectively using a surface coating of (3-aminopropyl)triethoxysilane (APTES), with a terminal amine group giving the surface coating a positive charge (Figure 8.1). They were able to show that electrostatic attraction was the dominant attractive mechanism between nanoparticles and crude oil droplets by comparing results with superparamagnetic nanoparticles coated with polyacrylic acid (PAA), which gave the nanoparticles a negative charge. The APTES-coated nanoparticles were observed to quickly attach to the oil droplets, while the PAA-coated nanoparticles did not attach. In addition to being a simple attraction mechanism, the electrostatic interaction between the APTES-coated nanoparticles and oil droplets yielded an effectively electrically neutral oil droplet-nanoparticle aggregate, which was then attracted to other such aggregates through van der Waals forces (Figure 8.2). The formation of larger aggregates is then beneficial for HGMS, since the velocity of particle separation increases as the aggregate size increases (Ko et al. 2014; Prigiobbe et al. 2015).

Ko et al. (2017) showed that amine-functionalized superparamagnetic nanoparticles are effective at removing micron-size droplets of oil from water. Because of the aggregation behavior of oil droplets with attached

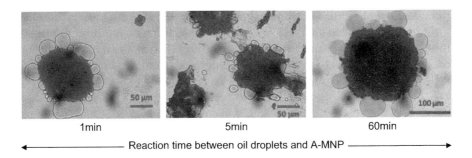

1min   5min   60min

◄────────── Reaction time between oil droplets and A-MNP ──────────►

FIGURE 8.2
Optical microscope images of oil droplets with attached magnetic nanoparticle aggregates after different mixing times. Figure from Ko et al. (2017).

nanoparticles, separation of the oil took approximately 5 minutes, as opposed to the several hours required to separate the nanoparticles themselves from aqueous dispersion. Greater than 99% of the oil could be removed using nanoparticle concentrations as low as 0.438 g/L.

8.2.2 Removal of Water Droplets from Produced Oil

Some production operations additionally require removal of water from produced oil. These water-in-oil emulsions can be particularly difficult to demulsify because asphaltenes and other surface-active compounds can make the liquid-liquid interfaces very strong (Harbottle et al. 2015). Peng et al. (2012) used ethyl-cellulose-grafted superparamagnetic nanoparticles to remove water droplets from bitumen. Similar to Ko et al. (2017), they mixed the nanoparticles with the water-in-bitumen emulsion and then removed the nanoparticle-coated water droplets with an external magnet (Figure 8.3). Aggregation of

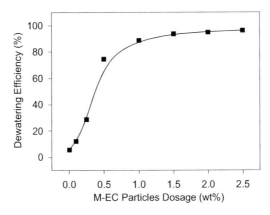

FIGURE 8.3
Amount of water removed from diluted bitumen by magnetite particles coated with ethyl cellulose (M-EC) for different particle dosing. Reprinted with permission from Peng et al. (2012). Copyright 2012 American Chemical Society.

the water droplets allowed removal of more than 80% of the initial water in 2 minutes, which is considerably faster than the 30 minutes required to remove the same amount of water by gravity separation. The nanoparticles were successfully regenerated by washing with chloroform and reused ten times with little performance degradation.

8.2.3 Removal of Remnant EOR Polymer from Produced Water

Remnant polymer from EOR activities presents a problem for produced water treatment and handling as the polymer stabilizes small droplets of oil that may be in the produced water and is itself a contaminant that must be removed prior to disposal (Duan et al. 2014; Zhang et al. 2010b). Motivated by the success of superparamagnetic nanoparticles in removing oil droplets from produced water, Ko et al. (2016) used amine-functionalized superparmagnetic nanoparticles to remove partially hydrolyzed polyacrylamide (HPAM) from water through the same aggregation and HGMS process described by Ko et al. (2017); 100% of the HPAM was removed on first application of the nanoparticles. The nanoparticles were then separated from the collected HPAM by pH adjustment above the point of zero charge, which changed the charge on the nanoparticles from positive to negative. The detached nanoparticles were then collected with a magnet and the pH of the dispersion was readjusted to the initial value. Careful control of pH is essential in this application due to the strong dependence of electrostatic charge on pH.

8.3 Inhibition of Scale/Organic Particulate Deposition

During production, deposition of inorganic and organic scale in downhole and surface equipment can represent a significant economic and safety risk. Inorganic scale originates from substances dissolved in formation water, which may exist at in-situ temperatures above 200°C and pressures above 150 MPa. Fluids injected for enhanced oil recovery, including brine, steam, supercritical CO_2, or surfactants can further promote and complicate scale deposition (Kan and Tomson 2012). Inorganic scale may include barium sulfate (barite), calcium sulfates, celestite, halite, silica, various sulfides, siderite, dolomite, and calcite (Kan and Tomson 2012). Scale may contain radioactive materials, such as ^{40}K, ^{226}Ra, and ^{228}Ra, which can readily substitute for other cations in sulfates (Fakhru'l-Razi et al. 2009). Organic particulates may be present in produced water as well and preventing their deposition on the surfaces of pipes and flowlines is important for maintaining safe and reliable production (Fakhru'l-Razi et al. 2009).

Several different technologies have been developed to address scale issues using nanoparticles. Scale-inhibiting chemicals may be used to coat or otherwise bind to nanoparticles. Phosphonates have been used for many years as scale inhibitors in oil and gas production. Tomson and coworkers at Rice University have developed a method for controlled release of phosphonates by precipitating metal-phosphonate nanoparticles using Ca, Zn, Al, or other metals (Yan et al. 2014; Zhang et al. 2010a 2011a,b). These nanoparticles can then be injected into the near-wellbore region, where they attach to mineral surfaces and gradually release small dosages of phosphonate inhibitor during production (Figure 8.4). In a similar application, Do et al. (2013) coated magnetic iron oxide nanoparticles with maleic acid-co-2-acrylamido-2-methylpropane sulfonic acid (AMPS), which is also a widely used polymeric scale inhibitor (Shi et al. 2012; Zhang et al. 2012). Rather than binding to mineral surface in the subsurface, these coated nanoparticles inhibit scale formation while dispersed in formation brine and may be detected in and removed from produced water due to the use of a magnetic core.

Another technology developed to enable controlled release of scale-inhibiting chemicals is the use of nanoparticle-stabilized double emulsions. Laugel et al. (1998) demonstrated the effectiveness of double emulsions in controlling release of hydrocortisone in topical application, with the intermediate layer of the double emulsion acting as a membrane regulating the rate of drug release. Cochet et al. (2013) proposed a similar mechanism for the controlled release of scale inhibitor in the subsurface. Nanoparticle-stabilized double

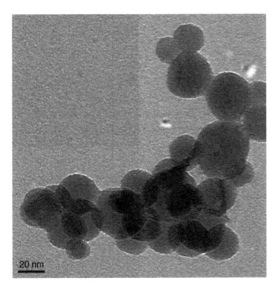

FIGURE 8.4
Transmission electron microscope image of boehmite (γ-AlO(OH)) nanoparticles with crosslinked sulfonated polycarboxylic acid for slow release of phosphonate inhibitor. Figure from Yan et al. (2014).

emulsions with the precise size and interfacial property control necessary for these applications can be generated with a microcapillary device (*e.g.*, Lee and Weitz 2008; Utada et al. 2005), but more research and development is necessary before this technology can be implemented at the field scale.

A final application of nanoparticles in scale prevention is using the nanoparticles themselves as preferential scale deposition sites. Betancur et al. (2016) and Franco et al. (2013) showed that silica nanoparticles can preferentially adsorb asphaltenes, preventing their deposition downhole and in flowlines. The high surface area and potential for surface functionalization make this application particularly attractive and is an area of ongoing research.

8.4 Flow Assurance for Oil/Gas Pipelines and Surface Facilities

Flow assurance is a production subdiscipline in which engineers strive to maintain "a reliable, manageable, and profitable flow of fluids from the reservoir to the sales point" (Denney 2002). With the advent of deepwater exploration and production, particularly in the Gulf of Mexico, subsea infrastructure design has become increasingly focused on preventing flow assurance problems and allowing easy access for intervention if necessary (Forsdyke 1997). Typical deepwater development in the Gulf of Mexico involves placing tens of kilometers of pipeline in water with temperatures of roughly 4°C and traversing changes in water depth of 300–400 m (*e.g.*, Gudimetla et al. 2006). The changes in pressure and temperature experienced by fluids as they flow through these pipelines can include precipitation of hydrates, wax, and asphaltenes, since multiple hydrocarbon phases along with water are typically present, with associated complex phase behavior (Forsdyke 1997). If any of these substances accumulate in the pipeline, flow can be inhibited or even stop completely, posing safety, environmental, and economic risks.

Flow assurance techniques generally involve a combination of chemical, mechanical, and thermal methods to prevent buildup of undesirable substances in pipelines or to remove any substances that do form. Chemical techniques typically include injection of various inhibitors to prevent the formation and accumulation of undesirable substances. These inhibitors can include various glycols or alcohols to prevent hydrate formation by changing the thermodynamic conditions for hydrate formation (Sloan et al. 2011); polymers and other organic compounds to prevent wax deposition by interfering with wax crystallization or by changing the wax rheology (Pedersen and Rønningsen 2003); and various organic compounds that either change the thermodynamics of asphaltene precipitation or maintain precipitated particles in suspension without deposition on pipeline walls (de Boer et al. 1995).

Mechanical techniques include pigging and the use of through-flowline tools, which both work to scrape deposits off the pipeline wall (Forsdyke 1997). Finally, thermal methods include pipeline burial, the use of insulated pipelines, and targeted heating (Berti 2004; Guo et al. 2005; Nysveen et al. 2005).

All these techniques have associated benefits and drawbacks. Chemical methods do not generally require production to be stopped, but the chemical additives have associated cost (Kondapi and Moe 2013), and may be environmentally hazardous if released to the ocean (Hill et al. 2002). In addition, detailed laboratory tests are required in most cases to select the mixture of inhibitors best suited for the reservoir fluids since inhibitor efficiency depends strongly on fluid composition (Borden 2013). Mechanical methods typically require production to be stopped for a period of hours to days and include the risk that pigs or tools may become stuck in the pipeline (*e.g.*, Fung et al. 2006). Pipeline burial is logistically complicated in deepwater environments, and the pipeline is more susceptible to damage due to subsurface stresses and strains and difficult to service (Nielsen et al. 1990; Palmer et al. 2003). Pipeline insulation and heating are much simpler but are still in the development stage for application in deepwater settings (Borden 2013; Kondapi and Moe 2013). There is not currently any one-size-fits-all approach to flow assurance and every location requires a field-specific flow assurance strategy that may evolve over the life of the field.

Some combination of pipeline heating and insulation provides the simplest, most broadly applicable solution, and has been widely applied in shallow water settings. However, many of the materials and equipment used in shallow water have not been tested in deepwater environments, and much research and development is ongoing to extend their application (Borden 2013). Existing technologies for pipeline heating include direct electrical heating (DEH), in which the steel pipe is heated by passing an electrical current through it (Nysveen et al. 2005), and electrically trace heated pipe-in-pipe (ETH-PiP), in which a wound heating element provides the heat source inside an insulated layer (Cam et al. 2011). Field applications of these techniques have taken place in the North Sea, the Gulf of Mexico, and offshore Brazil (Kondapi and Moe 2013; Nysveen et al. 2005; Solano et al. 2004). The largest drawbacks of these techniques are (a) power consumption and high voltages required, (b) efficiency, (c) component reliability, and (d) space requirements for topsides facilities. The power networks necessary for operation of heated pipelines require many large generators, which must be able to supply voltages of several thousand volts over distances of tens of km and require grounding infrastructure every few km (Nysveen et al. 2005). Current and thermal leaks can decrease efficiency and cause early failure of various components (Chakkalakal et al. 2014; Minami et al. 1999). Most installations require significant topsides facilities for power generation and transmission (Louvet et al. 2016). Research is ongoing to improve reliability, but this remains a power-intensive technique.

8.4.1 Hyperthermia Heating Re-Visited

Magnetic heating, called magnetic hyperthermia, is applied to selectively destroy cancerous tissues by heating the targeted area coated with magnetic particles (Pankhurst et al. 2003; Wang et al. 2010). Superparamagnetic nanoparticles can generate intense, localized heat when exposed to a magnetic field because of the energy released by the magnetization of each particle attempting to return to equilibrium. This phenomenon has been exploited in medicine for a number of years for targeted treatment of tumors (*e.g.*, Seegenschmiedt et al. 1995). The heating is described by the specific absorption rate (SAR), which is defined as c(dT/dt) where dT/dt is the change in temperature with time and *c* is the specific heat capacity (Chou 1990; Jordan et al. 1993). SAR may be calculated for a material with imbedded superparamagnetic nanoparticles as

$$\text{SAR} = \pi\mu_0^2 \frac{\varphi M_s^2 V}{k_B T} H_0^2 v \frac{2\pi v \tau}{1 + (2\pi v \tau)^2} \tag{8.1}$$

where

μ_0 is the magnetic permeability of free space
M_s is the specific magnetization of the superparamagnetic nanoparticles
V is the volume of an individual nanoparticle
H_0 is the magnetic field intensity
v is the frequency of the magnetic field
τ is the relaxation time (time constant for the return of the particle's magnetization to equilibrium)
k_B is Boltzmann's constant
T is absolute temperature (Laurent et al. 2008; Rosensweig 2002).

Equation (8.1) yields a maximum when $\tau = 1/(2\pi v)$ (Rosensweig 2002). In medicine, magnetic field frequencies on the order of hundreds of kHz are usually used, since higher frequencies tend to create hot spots because of local variations in tissue properties (*e.g.*, Andrä and Nowak 2007; Johnson et al. 1993). For a frequency of 500 kHz, the maximum SAR should be obtained with superparmagnetic nanoparticles with τ values of around 0.3 μs. Relaxation times of this order of magnitude indicate that Néel relaxation is the dominant mechanism, in which only the magnetization vector is perturbed (contrast with Brownian relaxation, in which the entire particle rotates and relaxes viscously [Brown 1963]). The Néel relaxation time τ_N for a superparamagnetic nanoparticle of radius r is given by

$$\tau_N = \tau_0 \exp\left(\frac{4\pi K r^3}{3 k_B T}\right) \tag{8.2}$$

where $\tau_0 = 10^{-9}$ s and K is the volumetric magnetic anisotropy of the nanoparticle (Hergt et al. 2004). At room temperature (298 K) and a magnetic field frequency of 500 kHz, assuming $K = 8000$ J/m³ (Rovers et al. 2009) and $k_B = 1.38 \times 10^{-23}$ m² · kg/s² · K, the maximum value of SAR corresponding to $\tau_N = 0.3$ μs noted above is given when $r = 9$ nm. SAR values of up to 400 W/g have been reported in the literature for superparamagnetic nanoparticles of this size dispersed as ferrofluid (Hergt et al. 2004).

8.4.2 MNP-Based Paint for Localized Heating

Davidson et al. (2012), Mehta et al. (2014), and Mehta (2015) showed through a combination of laboratory experiments and numerical simulations that superparamagnetic nanoparticles can be embedded in an epoxy resin paint to generate localized heating. Experiments were performed using two types of Fe_3O_4 nanoparticles purchased from Ferrotec Corporation: EMG605, which had a hydrophilic surface coating, and EMG1400, which had a hydrophobic surface coating. The nanoparticles were shown to be superparamagnetic due to lack of hysteresis in the Langevin curves (Figure 8.5) and had mean particle radii based on transmission electron microscope images of 5.3 ± 1.2 nm (EMG605) and 5.6 ± 1.6 nm (EMG1400).

Various heating experiments were performed using EMG605 and EMG1400 dispersed in epoxy paint to form a nanopaint. The different nanopaints were used to coat the inside of PVC tubes with inner diameters of 2 cm, and the temperature at the center of the tube was monitored as the nanopaint was exposed to different magnetic field strengths inside a coiled magnetic field generator operating at frequencies of 450 and 630 kHz. The hydrophilic

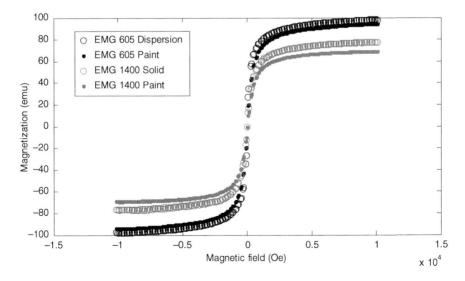

FIGURE 8.5
Langevin curves for EMG605 and EMG1400 nanoparticles in dispersion and embedded in epoxy paint. Figure from Mehta et al. (2014).

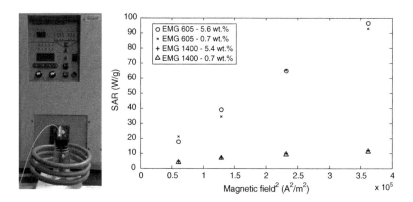

FIGURE 8.6
Left: Nanopaint-coated tube inside coiled magnetic field generator. Right: *SAR* versus magnetic field strength for different nanoparticles embedded in epoxy paint at different concentrations. Figures from Mehta (2015).

EMG605 particles provided a higher value of *SAR* for a given magnetic field strength than the hydrophobic EMG1400 particles, as high as 94.7 W/m at a magnetic field strength of 590 A/m. Interestingly, there was little to no dependence of *SAR* on nanoparticle concentration in the epoxy over the tested range of 0.7–5.6 wt% (Figure 8.6). Nanoparticle concentration did influence the observed temperature increase, however. The largest temperature increase after 2 minutes, 42.2°C, occurred for nanopaint made with 5.6 wt% of EMG605 in a magnetic field of 472 A/m. Similar results were obtained for static heating of water in the tube, although a greater nanoparticle concentration in the nanopaint was required to achieve the same temperature increment due to the larger heat capacity of water. A temperature increase of 19.4°C from room temperature was observed after 3 minutes for nanopaint containing 14.69 wt% EMG605 at a magnetic field strength of 594.7 A/m. Heating experiments performed with flowing water inside a nanopaint-coated PVC tube showed that limiting heat dissipation within the fluid being heated limits the *SAR* of the particles; higher values of *SAR* were observed when water was flowing relative to the static experiments (Figure 8.7).

The efficiency of nanopaint for removing a wax deposit was tested in the laboratory by coating a polycarbonate coupon 36 cm long by 2.5 cm wide with nanopaint containing 2.6 wt% EMG605, and then depositing paraffin wax on top of the nanopaint. After the wax cooled, the coupon was inserted inside a coiled magnetic field generator and the temperature was monitored at the wax-nanopaint interface as a magnetic field was applied. At a magnetic field strength of 725 A/m, it took 9–11 minutes to detach the wax from the coupon by melting a thin layer right at the wax-nanopaint interface. This amount of time may presumably be optimized by altering the nanoparticle loading in the nanopaint, and the experiment served mainly as a proof of concept of the method.

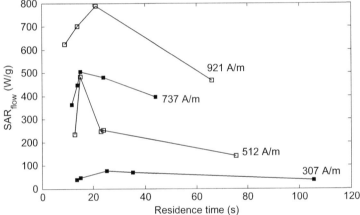

FIGURE 8.7
Top: Nanopaint-coated tube inside coiled magnetic field generator for measuring heating of fluid flowing through a pipe. Other experimental equipment is labeled. Bottom: *SAR* as a function of residence time for water flowing through the pipe. Figures from Mehta (2015).

To evaluate the feasibility of applying this technique in a pipeline, Mehta (2015) numerically modeled the heat generated by a moving magnetic field source inside the pipeline—a so-called "intelligent pig"—and the length of pipeline that might be treated by this method (Figure 8.8). The crude oil inside the pipeline was assumed to be at 25°C and the surrounding seawater was assumed to be at 5°C. A pipe-in-pipe geometry was used with a layer of insulation sandwiched between two concentric steel pipes (Varón et al. 2012). A layer of nanopaint 0.2 mm thick coated the inner wall of the pipeline. A multi-turn coil design was assumed for the intelligent pig with the axis of the coils parallel to the axis of the pipeline. The operating frequency of the coils was held constant at 500 kHz and the current was varied from 50 to 400 A. The concentration of nanoparticles in the nanopaint varied from 2 to 14

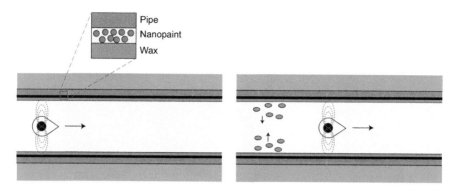

FIGURE 8.8
Intelligent pig operation inside a pipeline. Left: The intelligent pig enters the pipeline. Magnetic field lines are red, wax is brown, nanopaint is black, steel is gray. Water outside the pipeline is blue. Right: As the intelligent pig passes by, the nanopaint generates heat in response to the magnetic field and dislodges wax. Figures from Mehta (2015).

wt%, and the heat generation parameters were based on the experimental results for EMG600 nanoparticles. The results showed that, for a sufficiently high current—say, 400 A—the temperature on the inner wall of the pipeline reached the melting point of paraffin wax (50°C) in as little as 5 s even with a nanoparticle concentration as low as 2 wt%. The length of pipeline that could be treated by an intelligent pig running on a standard 147 Wh laptop computer battery was calculated to be as long as 27 km with an operating current of 300 A and 14 wt% nanoparticles in the paint (Figure 8.9).

Future work on this subject should address the optimal surface coating on the nanoparticles for best heat transfer to the paint, the best paint material for optimal heat transfer to the fluid, and a detailed hydrodynamic design of the intelligent pig.

8.5 MNP-Based Brine Hardness Reduction

Wang et al. (2014, 2017) reported a similar application for removing cations from produced water. Produced water often has salinities exceeding four times that of seawater and includes dissolved calcium, barium, sodium, strontium, and radium (Thiel and Lienhard 2014). Treating and disposing of this water remains a significant challenge, particularly in areas where shale oil and gas are being developed (*e.g.*, Lutz et al. 2013). Because of the attractive properties discussed previously, iron oxide nanoparticles have been investigated for some time as agents for removal of these cations from water (*e.g.*, Liu et al. 2013; Takafuji et al. 2004; Zhang et al. 2013). Particularly good removal efficiency has been reported in the literature using nanoparticles

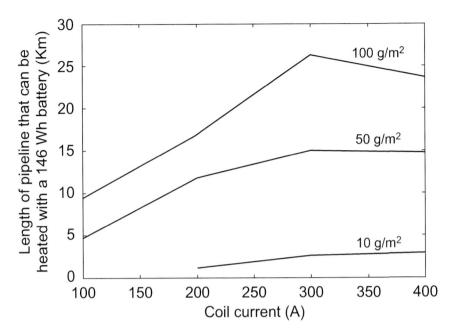

FIGURE 8.9

Length of pipeline that can be treated for wax removal using a standard laptop battery for different intelligent pig coil currents and nanoparticle loadings in the nanopaint. Figure from Mehta (2015).

coated with organic compounds that can form metal complexes or chelates. PAA is known as an effective remover of heavy metals (*e.g.*, Morales and Rivas 2015; Moulay and Bensacia 2016; Sezgin and Balkaya 2016), so Wang et al. (2014, 2017) investigated the adsorption capacity of Ca^{2+} on PAA-coated iron oxide nanoparticles under different salinity conditions and the efficiency of magnetic separation and regeneration of nanoparticles.

Wang et al. (2014, 2017) found that PAA-coated iron oxide nanoparticles had sorption capacities for Ca^{2+} as high as 57.3 mg/g at pH of 7 in a 400 mg/L Ca^{2+} solution. Increasing salinity resulted in diminished sorption capacity: in 1 wt% NaCl brine, the sorption capacity decreased roughly by a factor of 5. Decreasing the pH of the solution also reduced the sorption capacity (Figure 8.10). This is probably because higher salinity and/or lower pH screen the negatively charged surface of the coated nanoparticles, causing the nanoparticles to aggregate and reducing the exposed surface area per unit mass. Using a copolymer of PAA and poly 2-acrylamido-2-methyl-1-propanesulfonic acid (PAMPS) to coat the nanoparticle surfaces may alleviate this issue. PAMPS can help prevent contraction of the polymer chain and retain steric stabilization. Note, however, that some amount of aggregation of the nanoparticles once cations are adsorbed is advantageous for more efficient magnetic separation.

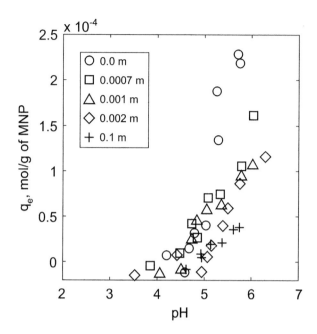

FIGURE 8.10
Adsorption capacity for Ca^{2+} on PAA-coated magnetic nanoparticles at different salt concentrations and pH levels. Figure from Wang et al. (2017).

The regeneration capability for PAA-coated iron oxide nanoparticles was tested by Wang et al. (2014). They adjusted the pH to 3.5 using acetic acid to cause the Ca^{2+} to desorb and then readjusted the pH to neutral and separated the nanoparticles. They found that, in pure water, the sorption capacity of the regenerated nanoparticles was roughly 91% of the original value, but that in 10 wt% API brine (8 wt% NaCl and 2 wt% $CaCl_2$) the sorption capacity was only 38% of the original value. This behavior could probably be improved by a copolymer surface coating.

8.6 Improved Conformance Control Using Temperature Responsive Gel

Conformance refers to the uniformity of the front of an immiscible fluid being injected into a reservoir. This is an issue of particular concern during waterflooding or CO_2 flooding as poor conformance leads to lower oil production and higher production of the injected fluids due to channeling through higher-permeability layers or viscous fingering. As a result, a tremendous number of methods have been developed over the years to ensure good conformance (Enick et al. 2012). Many of these methods include the use of polymer gels, which may be formulated to gel after an extended period of time during which they may be injected deep into the reservoir. Most conventional gels will preferentially flow through the higher-permeability layers, thus plugging them once gelled and promoting flow through lower-permeability layers. However, it is nearly impossible to prevent some gel from plugging some lower-permeability layers as well. In this section we will review an application developed involving nanoparticles to promote gel plugging only in high-permeability layers.

8.6.1 Hyperthermia Heating of Temperature Responsive Gels

Panthi et al. (2015) developed a technique for targeted heating and gelling of temperature-responsive gels using superparamagnetic nanoparticles. The principle of magnetic heating is the same as that discussed previously in Section 8.4.1. In their technique, gelling polymer, crosslinker, and iron oxide nanoparticles are injected in aqueous solution into the formation. The mixture moves preferentially into the higher-permeability layers, giving a greater concentration of polymer and nanoparticles in those layers. Next, a sonde may be lowered into the well to generate a magnetic field in the near-wellbore region to cause nanoparticle heating and gel the polymer. Since a greater concentration of nanoparticles exists in the higher-permeability layers, the gel will form in those layers first. After the magnetic field is turned off and the sonde removed, the well is flowed back to remove the ungelled polymer and nanoparticles from the lower-permeability layers.

8.6.2 Application Feasibility Experiments

Panthi et al. (2015) performed laboratory experiments to test the efficacy of iron oxide nanoparticles in heating and gelling polymers, and gelling behavior in a sandpack. They selected several different temperature-responsive polymers for their experiments: curdlan, a colloidal polysaccharide that can be dispersed in water; methyl cellulose (MC) and hydroxypropyl methyl cellulose (HPMC), water-soluble cellulose ethers; and hydrolyzed polyacrylamide (HPAM) with polyethylenimine (PEI) crosslinker, a widely used combination in conformance control applications. They tested magnetic heating using the same apparatus as that used by Mehta (2015) and a variety of iron oxide nanoparticles including EMG 605 and EMG 700 from Ferrotec and in-house synthesized PAA- and APTES-coated nanoparticles.

Gelling experiments showed that all the agents tested were able to form gels when magnetic heating was applied (Figure 8.11, top). Curdlan, MC, and HPMC formed gels more easily in the presence of salt (8 wt% NaCl + 2 wt% $CaCl_2$), while HPAM-PEI gelling was slower when salt was added. In fact, MC and HPMC formed gels at 40°C in the presence of salt, which may in fact be too low a temperature for subsurface applications. More refining of the technique would be necessary for such an application. As low as 0.28 wt% nanoparticles was needed to form gels of curdlan, MC, and HPMC after

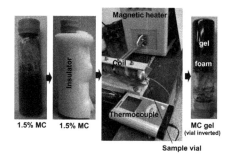

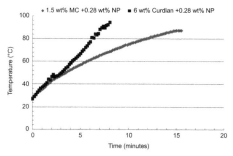

FIGURE 8.11

Top: Experimental setup for gel formation. Results for methyl cellulose gel are shown. Bottom: Temperature profile for gelling experiments with methyl cellulose and curdlan with different amounts of nanoparticles. Figures from Panthi et al. (2015). Copyright 2015, Society of Petroleum Engineers. Reproduced with permission of SPE. Further reproduction prohibited without permission.

about 10 minutes at a magnetic field frequency of 725 kHz and 21 A current (Figure 8.11, bottom), and very little difference among the different types of nanoparticles was observed in gelling efficiency.

Panthi et al. (2015) also tested nanoparticle heating-based gelling in a sandpack. They mixed 1.5 wt% MC, 2 wt% NaCl, and 0.28 wt% iron oxide nanoparticles and injected 1.6 pore volumes through a 12″ long sandpack with porosity of 27.9% and permeability of 7.93 darcies. They then heated the sandpack at 70°C for 40 minutes with the nanoparticles and then continued injecting another 0.3 pore volume. They observed that the pressure drop across the sandpack increased from 58 to 340 psi after the heating step, indicating gel formation in the pore space.

8.7 Fines Migration Control

Formation fines, which are particles smaller than 37 μm in diameter, are present in many sandstone reservoirs, particularly unconsolidated sandstones in deepwater settings (Muecke 1979). Fines can become mobilized during

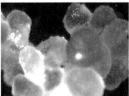

FIGURE 8.12
Left: 20/40 mesh sand coated with nanoparticles viewed at 60× magnification. Right: Formation fines attached to the nanoparticle-coated sand, also at 60× magnification. Figures from Huang et al. (2008b).

production operations through a number of triggering processes including changes in salinity, wettability, pH, flow rate, and temperature (Yuan et al. 2016), and the migration of fines can reduce reservoir permeability and adversely affect injectivity and productivity (Sarkar and Sharma 1990). Fines migration can be particularly prevalent in low salinity waterflooding as the change in water salinity causes a large change in the surface charges of colloidal particles (Lager et al. 2008). Whether this is a bad thing or not is a subject of ongoing debate, as some researchers have argued that the migrating fines can provide better sweep efficiency by preferentially clogging higher-permeability pathways and may even be able to entrain some oil in the particulate flow; see Assef et al. (2014) for a good discussion of this. However, in general, the more or less uncontrolled migration of solid particles through the reservoir is generally viewed as something to be avoided.

Because of their small size and large surface areas, nanoparticles have been investigated as a possible means to mitigate fines migration because they may carry a large electrical charge per unit volume. Many mineral surfaces, and particularly those in unconsolidated sandstone reservoirs in which fines migration is a concern, are negatively charged. In silicate minerals, this negative charge arises from the silanol groups that are present on the mineral surfaces. This causes a net repulsion between grains, and if there are not enough other forces acting on the grains to retain them in place, they can become entrained in flow. If some of the silanol groups are masked, however, the interparticle repulsion can be reduced. Ahmadi et al. (2013), Assef et al. (2014), and Ogolo et al. (2012) demonstrated that this may be accomplished by the sorption of cationic nanoparticles such as magnesia or alumina on silica grain surfaces.

Researchers at BakerHughes in the mid-2000s used this phenomenon to their advantage to develop a technique for capturing migrating fines in a proppant pack (Huang et al. 2008a,b). In this application, a nanoparticle slurry presumably containing cationic nanoparticles is mixed with proppant during a hydraulic fracturing or frac pack treatment. The nanoparticles attach to the proppant grain surfaces and act as scavengers in the proppant pack, capturing fines that migrate through (Figure 8.12). Some field applications have shown the promise of this technique to improve oil production and completely eliminate sand production from deepwater wells (Belcher et al.

2010; Huang et al. 2010). Further work is necessary to refine the technique, for instance, to develop surface coatings for nanoparticles that can optimize behavior and allow application in a wider range of environmental conditions.

References

Ahmadi, M., Habibi, A., Pourafshary, P., and Ayatollahi, S. (2013) Zeta-potential investigation and experimental study of nanoparticles deposited on rock surface to reduce fines migration. *SPE J.*, 18(3), 534–544.

Ambashta, R. D., and Sillanpää, M. (2010) Water purification using magnetic assistance: A review. *J. Hazard. Mater.*, 180(1), 38–49.

Andrä, W., and Nowak, H. (Eds.) (2007) *Magnetism in Medicine: A Handbook*, 2nd edn. Wiley, Mannheim, Germany.

Assef, Y., Arab, D., and Pourafshary, P. (2014) Application of nanofluid to control fines migration to improve the performance of low salinity water flooding and alkaline flooding. *J. Petrol. Sci. Eng.*, 124, 331–340.

Belcher, C. K., Seth, K., Hollier, R., and Paternostro, B. P. (2010) Maximizing production life with the use of nanotechnology to prevent fines migration. Society of Petroleum Engineers Presented at the International Oil and Gas Conference and Exhibition in China, Beijing, China.

Berti, E. (2004, February 1) Syntactic polypropylene coating solution provides thermal insulation for Bonga risers—Offshore. *Offshore Magazine*, 64(2).

Betancur, S., Carmona, J. C., Nassar, N. N., Franco, C. A., and Cortés, F. B. (2016) Role of particle size and surface acidity of silica gel nanoparticles in inhibition of formation damage by Asphaltene in Oil Reservoirs. *Ind. Eng. Chem. Res.*, 55(21), 6122–6132.

Borden, K. (2013) The challenges of processing and transporting heavy crude. *Oil and Gas Facilities*, 2(5), 22–26.

Brown, W. F. (1963) Thermal fluctuations of a single-domain particle. *Phys. Rev.*, 130(5), 1677–1686.

Buckley, J. S., Takamura, K., and Morrow, N. R. (1989) Influence of electrical surface charges on the wetting properties of crude oils. *SPE Reservoir Engineering*, 4(3), 332–340.

Cam, J.-F., Fisher, R., Delaporte, D., and Hall, S. J. (2011) Industrialization of electrically trace heated pipe-in-pipe (ETH-PiP). Society of Petroleum Engineers. Presented at the Offshore Europe, Aberdeen, UK.

CFR §435 (2016).

Chakkalakal, F., Hamill, M., and Beres, J. (2014) Building the world's longest heated pipeline a technology application review. 2014 IEEE Petroleum and Chemical Industry Technical Conference, San Francisco, CA, USA. pp. 481–489.

Chou, C.-K. (1990) Use of heating rate and specific absorption rate in the hyperthermia clinic. *Int. J. Hyperther.*, 6(2), 367–370.

Cochet, T., Hughes, T., Kefi, S., Lafitte, V., Tan, K., Tustin, G., and Wang, S. (2013) US8393395B2. Retrieved from https://patents.google.com/patent/US8393395B2/en (accessed 28/12/2018).

Davidson, A., Huh, C., and Bryant, S. L. (2012) Focused magnetic heating utilizing superparamagnetic nanoparticles for improved oil production applications. Society of Petroleum Engineers. Presented at the SPE International Oilfield Nanotechnology Conference and Exhibition, Noordwijk, the Netherlands.

de Boer, R. B., Leerlooyer, K., Eigner, M. R. P., and van Bergen, A. R. D. (1995) Screening of crude oils for asphalt precipitation: Theory, practice, and the selection of inhibitors. *SPE Prod. Facil.*, 10(1), 55–61.

Denney, D. (2002) Flow assurance: A π3 discipline. *J. Petrol. Technol.*, 54(9), 91–92.

Do, B. P. H., Nguyen, B. D., Nguyen, H. D., and Nguyen, P. T. (2013) Synthesis of magnetic composite nanoparticles enveloped in copolymers specified for scale inhibition application. *Adv. Nat. Sci.-Nanosci.*, 4(4), 045016.

Duan, M., Ma, Y., Fang, S., Shi, P., Zhang, J., and Jing, B. (2014) Treatment of wastewater produced from polymer flooding using polyoxyalkylated polyethyleneimine. *Sep. Purif. Technol.*, 133, 160–167.

Dubey, S. T., and Doe, P. H. (1993) Base number and wetting properties of crude oils. *SPE Reservoir. Eng.*, 8(3), 195–200.

Enick, R. M., Olsen, D. K., Ammer, J. R., and Schuller, W. (2012) Mobility and conformance control for CO2 EOR via thickeners, foams, and gels—A literature review of 40 Years of research and pilot tests. Society of Petroleum Engineers. Presented at the SPE Improved Oil Recovery Symposium, Tulsa, OK, USA.

Fakhru'l-Razi, A., Pendashteh, A., Abdullah, L. C., Biak, D. R. A., Madaeni, S. S., and Abidin, Z. Z. (2009) Review of technologies for oil and gas produced water treatment. *J. Hazard. Mater.*, 170(2), 530–551.

Forsdyke, I. N. (1997) Flow assurance in multiphase environments. Society of Petroleum Engineers. Presented at the International Symposium on Oilfield Chemistry, Houston, TX, USA.

Franco, C. A., Nassar, N. N., Ruiz, M. A., Pereira-Almao, P., and Cortés, F. B. (2013) Nanoparticles for inhibition of asphaltenes damage: Adsorption study and displacement test on Porous media. *Energ. Fuel.*, 27(6), 2899–2907.

Fung, G., Backhaus, W. P., McDaniel, S., and Erdogmus, M. (2006) To pig or not to pig: The Marlin experience with stuck pig. Offshore Technology Conference. Presented at the Offshore Technology Conference, Houston, TX, USA.

Gudimetla, R., Carroll, A., Havre, K., Christiansen, C., and Canon, J. (2006) Gulf of Mexico field of the future: Subsea flow assurance. Offshore Technology Conference. Presented at the Offshore Technology Conference, Houston, TX, USA.

Guo, B., Song, S., Chacko, J., and Ghalambor, A. (2005) *Offshore Pipelines.* Gulf Professional Publishing, Burlington, MA, USA.

Harbottle, D., Liang, C., El-Thaher, N., Liu, Q., Masliyah, J., and Xu, Z. (2015) Particle-stabilized emulsions in heavy oil processing. In *Particle-stabilized Emulsions and Colloids: Formation and Applications.* Royal Society of Chemistry, Cambridge, UK. pp. 283–316.

Hergt, R., Hiergeist, R., Zeisberger, M., Glöckl, G., Weitschies, W., Ramirez, L. P., …Kaiser, W. A. (2004) Enhancement of AC-losses of magnetic nanoparticles for heating applications. *J. Magn. Magn. Mater.*, 280(2), 358–368.

Hill, D. G., Dismuke, K., Shepherd, W., Romijn, H., Witt, I., Wennberg, K. E., … Perez, D. (2002) Reducing risk of oilfield chemicals to marine environments—Development practices, achievements and benefits. Society of Petroleum

Engineers. Presented at the SPE International Conference on Health, Safety and Environment in Oil and Gas Exploration and Production, Kuala Lumpur, Malaysia.

Huang, T., Crews, J. B., and Willingham, J. R. (2008a) Nanoparticles for formation fines fixation and improving performance of surfactant structure fluids. Presented at the International Petroleum Technology Conference, Kuala Lumpur, Malaysia.

Huang, T., Crews, J. B., and Willingham, J. R. (2008b) Using nanoparticle technology to control fine migration. Society of Petroleum Engineers. Presented at the SPE Annual Technical Conference and Exhibition, Denver, CO, USA.

Huang, T., Evans, B. A., Crews, J. B., and Belcher, C. K. (2010) Field case study on formation fines control with nanoparticles in offshore applications. Society of Petroleum Engineers. Presented at the SPE Annual Technical Conference and Exhibition, Florence, Italy.

Johnson, R. H., Robinson, M. P., Preece, A. W., Green, J. L., Pothecary, N. M., and Railton, C. J. (1993) Effect of frequency and conductivity on field penetration of electromagnetic hyperthermia applicators. *Phys. Med. Biol.*, 38(8), 1023–1034.

Jordan, A., Wust, P., Fählin, H., John, W., Hinz, A., and Felix, R. (1993) Inductive heating of ferrimagnetic particles and magnetic fluids: Physical evaluation of their potential for hyperthermia. *Int. J. Hypertherm.*, 9(1), 51–68.

Kan, A., and Tomson, M. (2012) Scale prediction for oil and gas production. *SPE J.*, 17(2), 362–378.

Ko, S., Kim, E. S., Park, S., Daigle, H., Milner, T. E., Huh, C., ...Geremia, G. A. (2017) Amine functionalized magnetic nanoparticles for removal of oil droplets from produced water and accelerated magnetic separation. *J. Nanopart. Res.*, 19(4), 132.

Ko, S., Lee, H., and Huh, C. (2016) Efficient Removal of EOR polymer from produced water using magnetic nanoparticles and regeneration/re-use of spent particles. Society of Petroleum Engineers. Presented at the SPE Improved Oil Recovery Conference, Tulsa, OK.

Ko, S., Prigiobbe, V., Huh, C., Bryant, S. L., Bennetzen, M. V., and Mogensen, K. (2014) Accelerated oil droplet separation from produced water using magnetic nanoparticles. Society of Petroleum Engineers. Presented at the SPE Annual Technical Conference and Exhibition, Amsterdam, the Netherlands.

Kokal, S. L. (2005) Crude oil emulsions: A state-of-the-art review. *SPE Prod. Facil.*, 20(1), 5–13.

Kondapi, P., and Moe, R. (2013) Today's top 30 flow assurance technologies: Where do they stand? Offshore Technology Conference. Presented at the Offshore Technology Conference, Houston, TX.

Lager, A., Webb, K. J., Black, C. J. J., Singleton, M., and Sorbie, K. S. (2008) Low salinity oil recovery—An experimental investigation. *Petrophysics*, 49(1), 28–35

Laugel, C., Baillet, A., P. Youenang Piemi, M., Marty, J. P., and Ferrier, D. (1998) Oil–water–oil multiple emulsions for prolonged delivery of hydrocortisone after topical application: comparison with simple emulsions. *Int. J. Pharm.*, 160(1), 109–117.

Laurent, S., Forge, D., Port, M., Roch, A., Robic, C., Vander Elst, L., and Muller, R. N. (2008) Magnetic iron oxide nanoparticles: Synthesis, stabilization, vectorization, physicochemical characterizations, and biological applications. *Chem. Rev.*, 108(6), 2064–2110.

Lee, D., and Weitz, D. A. (2008) Double emulsion-templated nanoparticle colloidosomes with selective permeability. *Adv. Mater.*, 20(18), 3498–3503.

Liu, G., Xu, X., and Gao, J. (2003) Study on the compatibility of asphaltic crude oil with the electric desalting demulsifiers. *Energ. Fuel.*, 17(3), 543–548.

Liu, Y., Chen, M., and Yongmei, H. (2013) Study on the adsorption of Cu(II) by EDTA functionalized Fe3O4 magnetic nano-particles. *Chem. Eng. J.*, 218, 46–54.

Louvet, E., Giraudbit, S., Seguin, B., and Sathananthan, R. (2016) Active heated pipe technologies for field development optimisation. Offshore Technology Conference. Presented at the Offshore Technology Conference Asia, Kuala Lumpur, Malaysia.

Lutz, B. D., Lewis, A. N., and Doyle, M. W. (2013) Generation, transport, and disposal of wastewater associated with Marcellus Shale gas development. *Water Resour. Res.*, 49(2), 647–656.

McLean, J. D., and Kilpatrick, P. K. (1997) Effects of asphaltene solvency on stability of water-in-crude-oil emulsions. *J. Colloid. Interf. Sci.*, 189(2), 242–253.

Mehta, P. (2015) *Application of Superparamagnetic Nanoparticle-based Heating for Non-abrasive Removal of Wax Deposits from Subsea Oil Pipelines* (MS). University of Texas at Austin, Austin, TX, USA.

Mehta, P., Huh, C., and Bryant, S. L. (2014) Evaluation of superparamagnetic nanoparticle-based heating for flow assurance in subsea flowlines. International Petroleum Technology Conference. Presented at the International Petroleum Technology Conference, Kuala Lumpur, Malaysia.

Minami, K., Kurban, A. P. A., Khalil, C. N., and Kuchpil, C. (1999) Ensuring flow and production in deepwater environments. Offshore Technology Conference. *Presented at the Offshore Technology Conference*, Houston, TX, USA.

Moeser, G. D., Roach, K. A., Green, W. H., Laibinis, P. E., and Hatton, T. A. (2002) Water-based magnetic fluids as extractants for synthetic organic compounds. *Ind. Eng. Chem. Res.*, 41(19), 4739–4749.

Morales, D. V., and Rivas, B. L. (2015) Poly(2-acrylamidoglycolic acid-co-2-acrylamide-2-methyl-1-propane sulfonic acid) and poly(2-acrylamidoglycolic acid-co-4-styrene sodium sulfonate): synthesis, characterization, and properties for use in the removal of Cd(II), Hg(II), Zn(II), and Pb(II). *Polym. Bull.*, 72(2), 339–352.

Moulay, S., and Bensacia, N. (2016) Removal of heavy metals by homolytically functionalized poly(acrylic acid) with hydroquinone. *Int. J. Ind. Chem.*, 7(4), 369–389.

Muecke, T. W. (1979) Formation fines and factors controlling their movement in porous media. *J. Petrol. Technol.*, 31(2), 144–150.

Nielsen, N.-J. R., Lyngberg, B., and Pedersen, P. T. (1990) Upheaval buckling failures of insulated buried pipelines: A case story. Offshore Technology Conference. Presented at the Offshore Technology Conference, Houston, TX, USA.

Nysveen, A., Kulbotten, H., Lervi, J. K., Bomes, A. H., and Hoyer-Hansen, M. (2005) Direct electrical heating of subsea pipelines—technology development and operating experience. In *IEEE Trans. Ind. Applic.* (Vol. 43). Denver, CO, USA. pp. 177–187.

Ogolo, N. C., Olafuyi, O. A., and Onyekonwu, M. (2012) Effect of nanoparticles on migrating fines in formations. Society of Petroleum Engineers. Presented at the SPE International Oilfield Nanotechnology Conference and Exhibition, Noordwijk, the Netherlands.

OSPAR. (2007) Convention for the protection of the marine environment of the northeast Atlantic.

Palmer, A. C., White, D. J., Baumgard, A. J., Bolton, M. D., Barefoot, A. J., Finch, M., …Baldry, J. a. S. (2003) Uplift resistance of buried submarine pipelines: comparison between centrifuge modelling and full-scale tests. *Géotechnique*, 53(10), 877–883.

Pankhurst, Q. A., Connolly, J., Jones, S. K., and Dobson, J. (2003) Applications of magnetic nanoparticles in biomedicine. *J. Phys. D Appl. Phys.*, 36(13), R167.

Panthi, K., Mohanty, K. K., and Huh, C. (2015) Precision control of gel formation using superparamagnetic nanoparticle-based heating. Society of Petroleum Engineers. Presented at the SPE Annual Technical Conference and Exhibition, Houston, TX, USA.

Pedersen, K. S., and Rønningsen, H. P. (2003) Influence of wax inhibitors on wax appearance temperature, pour point, and viscosity of waxy crude oils. *Energ. Fuel.*, 17(2), 321–328.

Peng, J., Liu, Q., Xu, Z., and Masliyah, J. (2012) Novel magnetic demulsifier for water removal from diluted bitumen emulsion. *Energ. Fuel.*, 26(5), 2705–2710.

Prigiobbe, V., Ko, S., Huh, C., and Bryant, S. L. (2015) Measuring and modeling the magnetic settling of superparamagnetic nanoparticle dispersions. *J. Colloid. Interf. Sci.*, 447, 58–67.

Rosensweig, R. E. (1985) *Ferrohydrodynamics*. Cambridge University Press, Cambridge, UK.

Rosensweig, R. E. (2002) Heating magnetic fluid with alternating magnetic field. *J. Magn. Magn. Mater.*, 252, 370–374.

Rovers, S. A., van der Poel, L. A. M., Dietz, C. H. J. T., Noijen, J. J., Hoogenboom, R., Kemmere, M. F., ... Keurentjes, J. T. F. (2009) Characterization and magnetic heating of commercial superparamagnetic iron oxide nanoparticles. *J. Phys. Chem.*, 113(33), 14638–14643.

Sarkar, A. K., and Sharma, M. M. (1990) Fines migration in two-phase flow. *J. Petrol. Technol.*, 42(5), 646–652.

Seegenschmiedt, M. H., Fessenden, P., and Vernon, C. C. (1995) *Thermoradiotherapy and Thermochemotherapy: Biology, Physiology, Physics*. Springer Science & Business Media, Berlin.

Sezgin, N., and Balkaya, N. (2016) Adsorption of heavy metals from industrial wastewater by using polyacrylic acid hydrogel. *Desalin. Water. Treat.*, 57(6), 2466–2480.

Shi, W.-Y., Ding, C., Yan, J.-L., Han, X.-Y., Lv, Z.-M., Lei, W., ...Wang, F.-Y. (2012) Molecular dynamics simulation for interaction of PESA and acrylic copolymers with calcite crystal surfaces. *Desalination*, 291, 8–14.

Sloan, E. D., Koh, C., and Sum, A. K. (2011) *Natural Gas Hydrates in Flow Assurance*. Gulf Professional Publishing, Burlington, MA, USA.

Solano, R. F., de Azevedo, F. B., Vaz, M. A., and de Oliveira Cardoso, C. (2004) Design and installation of buried heated pipelines at the Capixaba North terminal offshore Brazil (Vol. 3). Presented at the ASME 23rd International Conference on Offshore Mechanics and Arctic Engineering, Vancouver, Canada. pp. 77–86.

Takafuji, M., Ide, S., Ihara, H., and Xu, Z. (2004) Preparation of poly(1-vinylimidazole)-grafted magnetic nanoparticles and their application for removal of metal ions. *Chem. Mater.*, 16(10), 1977–1983.

Thiel, G. P., and Lienhard, J. H. (2014) Treating produced water from hydraulic fracturing: Composition effects on scale formation and desalination system selection. *Desalination*, 346, 54–69.

Utada, A. S., Lorenceau, E., Link, D. R., Kaplan, P. D., Stone, H. A., and Weitz, D. A. (2005) Monodisperse double emulsions generated from a microcapillary device. *Science*, 308(5721), 537–541.

Varón, L. a. B., OrlandeH. R. B., and Vianna, F. L. V. (2012) Estimation of the convective heat transfer coefficient in pipelines with the Markov chain Monte Carlo method. In *Blucher Mechanical Engineering Proceedings* (Vol. 1). pp. 2014–2025.

Wang, Q., Prigiobbe, V., Huh, C., and Bryant, S. L. (2017) Alkaline earth element adsorption onto PAA-coated magnetic nanoparticles. *Energies*, 10(2), 223.

Wang, Q., Prigiobbe, V., Huh, C., Bryant, S. L., Mogensen, K., and Bennetzen, M. V. (2014) Removal of divalent cations from brine using selective adsorption onto magnetic nanoparticles. International Petroleum Technology Conference. Presented at the International Petroleum Technology Conference, Kuala Lumpur, Malaysia.

Wang, X., Tang, J., and Shi, L. (2010) Induction heating of magnetic fluids for hyperthermia treatment. *IEEE. T. Magn.*, 46(4), 1043–1051.

Yan, C., Kan, A. T., Wang, W., Yan, F., Wang, L., and Tomson, M. B. (2014) Sorption study of γ-AlO(OH) nanoparticle-crosslinked polymeric scale inhibitors and their improved squeeze performance in porous media. *SPE J.*, 19(4), 687–694.

Yavuz, C. T., Prakash, A., Mayo, J. T., and Colvin, V. L. (2009) Magnetic separations: From steel plants to biotechnology. *Chem. Eng. Sci.*, 64(10), 2510–2521.

Yuan, B., Moghanloo, R. G., and Zheng, D. (2016) Analytical evaluation of nanoparticle application to mitigate fines migration in porous media. *SPE J.*, 21(6), 2317–2332.

Zhang, J., Zhai, S., Li, S., Xiao, Z., Song, Y., An, Q., and Tian, G. (2013) Pb(II) removal of $Fe_3O_4@SiO_2$-NH_2 core–shell nanomaterials prepared via a controllable sol–gel process. *Chem. Eng. J.*, 215–216, 461–471.

Zhang, P., Fan, C., Lu, H., Kan, A. T., and Tomson, M. B. (2011a) Synthesis of crystalline-phase silica-based calcium phosphonate nanomaterials and their transport in carbonate and sandstone porous media. *Ind. Eng. Chem. Res.*, 50(4), 1819–1830.

Zhang, P., Kan, A. T., Fan, C., Work, S. N., Yu, J., Lu, H., …Tomson, M. (2011b) Silica-templated synthesis of novel zinc-DTPMP nanomaterials: Their transport in carbonate and sandstone media during scale inhibition. *SPE J.*, 16(03), 662–671.

Zhang, P., Shen, D., Fan, C., Kan, A., and Tomson, M. (2010a) Surfactant-assisted synthesis of metal-phosphonate inhibitor nanoparticles and transport in porous media. *SPE J.*, 15(3), 610–617.

Zhang, Q., Ren, H., Wang, W., Zhang, J., and Zhang, H. (2012) Molecular simulation of oligomer inhibitors for calcite scale. *Particuology*, 10(3), 266–275.

Zhang, Y., Gao, B., Lu, L., Yue, Q., Wang, Q., and Jia, Y. (2010b) Treatment of produced water from polymer flooding in oil production by the combined method of hydrolysis acidification-dynamic membrane bioreactor–coagulation process. *J. Petrol. Sci. Eng.*, 74(1), 14–19.

9

Reservoir Sensing

9.1 Introduction

Mainly spurred by the Advanced Energy Consortium organized by the University of Texas at Austin (Barron et al. 2010), active research efforts have been made in recent years on the use of nanoparticles to obtain more accurate information on reservoir rock and fluid in-situ properties and on the dynamics of resident and injected fluids. Broadly, three different approaches were taken: (1) addition of a nanoparticle dispersion in a portion of the injection fluid bank, for waterflooding or EOR, and the detection of their location and movement remotely; (2) addition of a nanoparticle dispersion in the injected fluid which then detects certain properties of the reservoir rock and/ or fluids, and carries the information with the nanoparticles (*e.g.*, Agenet et al. 2011). When the particles are produced, they can be "interrogated" to retrieve the data; and (3) deployment of the nanoparticle-based sensing device downhole, either in the wellbore or at the near-wellbore zone of the reservoir formation. In the following section, the first approach of the remote reservoir imaging will be described. The second approach, known as either "intelligent tracer" or "nanoreporter," will be described in Section 9.3. The third approach of downhole monitoring is briefly described in Section 9.4.

9.2 Enhanced Reservoir Imaging

While the remote sensing of rock and fluids properties deep in the reservoir with the injection of a "contrast" agent is a very attractive idea, a major difficulty is that both the excitation signal for probing (such as a magnetic field oscillation) and the measured response signal generally attenuate very quickly in the reservoir rock (Al-Ali et al. 2009), and picking up any meaningful response signal is quite difficult, unless the distance between the excitation signal generator and the receiver is quite close. In this section, the current development status of this concept will be described.

9.2.1 Remote Detection of Magnetic Nanoparticle Bank Employing EM Crosswell Tomography

When the superparamagnetic nanoparticles (SNPs) are subjected to the external magnetic field, they magnetize and generate tiny induced magnetic field around them, as described in Chapter 5. Therefore, when a bank of SNP dispersion is injected into a reservoir and is then subjected to an external magnetic field oscillation, the nanoparticle ensemble generates a "halo" of induced field which can be remotely detected at an observation well, as schematically shown in Figure 9.1. For the oscillatory magnetic field generation and the measurements of the induced field response, the existing technology of the crosswell electromagnetic (EM) tomography can be utilized (Al-Ali et al. 2009; Al-Shehri et al. 2013). We briefly introduced electromagnetic tomography in Chapter 5. While the EM tomography is currently employed to identify a reservoir zone which has an electrical conductivity contrast, *e.g.*, due to the injection of water which has a salinity lower than the resident salinity, the aim of the proposed use of SNP is to create a zone of magnetic susceptibility contrast to delineate its location and movement. Once the induction responses are measured at the observation well, a set of Maxwell equation solutions will be obtained with the assumed locations/configurations of the nanoparticle dispersion zone; and an inversion algorithm is employed to identify the exact location/configuration. Rahmani et al. (2015a) and Hu et al. (2016), employing different Maxwell equation solution algorithms, carried out simulation studies to evaluate the field application feasibility of the technique described above. Rahmani et al.'s (2015a) work will be described in some detail below for a better understanding of the remote sensing idea of utilizing the SNP ensemble and its magnetization.

9.2.1.1 *Fundamental Physics for Conductive vs. Magnetic Slug*

Before the simulation study is described in some detail, the key physical principles for the remote detection of the injected fluid bank, which shows the "contrast" of its electrical conductivity, or magnetic susceptibility, against the background reservoir medium will be first briefly described. Figure 9.1 shows a schematic of the secondary magnetic field generation for a small bank of (a) electrically conductive nanoparticle dispersion and (b) SNP dispersion. For the conductive dispersion bank, the transmitter signal induces eddy currents that flow in the formation in horizontal planes. These currents generate secondary magnetic fields that are proportional to the contrast between the nanoparticle bank and the background conductivities. The receiver array in the offset well can thus measure the vertical component of this secondary magnetic field (Figure 9.1a). The physics of the secondary field generation by a bank of the SNP dispersion is notably different. With the external magnetic field

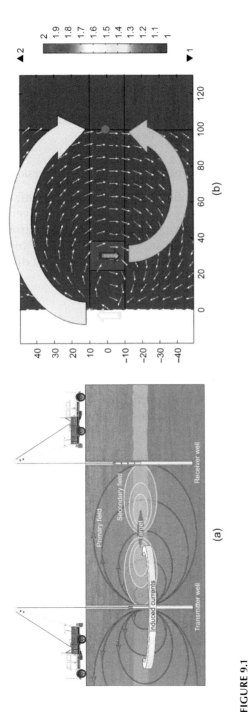

FIGURE 9.1

Schematic of secondary magnetic field generation for (a) a bank of electrically conductive particle dispersion, and (b) a bank of SNP dispersion (red box in 2D cross-section) for 100 m separation between source (small arrow left side) and receiver (red dot at $x = 100$, $y = 0$). The large green arrow is the primary field and the smaller arrow is the secondary field induced by the bank of the injected "contrast" dispersion. From Rahmani et al. (2015a). Copyright 2015, Society of Petroleum Engineers. Reproduced with permission of SPE. Further reproduction prohibited without permission.

application, the SNP bank (the red rectangle in Figure 9.1b) produces a secondary magnetic field that can be also measured at the location of the receiver (the red dot in Figure 9.1b). The magnetic response in this case is independent of the electrical current (Rahmani et al. 2015a) and hence the SNP bank can be thought of as an induced magnetic dipole, which is in phase with the primary magnetic field. The secondary (induced) magnetic dipole is proportional to the target magnetic susceptibility times the value of the primary magnetic field and is dependent on distance from source with r^3 geometrical spreading.

9.2.1.2 Effective Magnetic Susceptibility of the SNP Dispersion

The first step in assessing the utilization of SNP as a magnetic contrast agent is the calculation of the effective magnetic susceptibility versus SNP concentration, i.e., how much magnetic contrast to the background reservoir medium can be generated with the injection of SNPs. For this purpose, the Langevin expression for the magnetic susceptibility, χ, of a dilute dispersion of SNPs can be used (Rosensweig 1997):

$$\chi = \frac{\pi}{18}\phi\mu_0 \frac{M_d^2 d^3}{k_B T} \tag{9.1}$$

where
- M_d is the saturation magnetization of the material at the absolute temperature T
- ϕ is the volume fraction of the particles
- d is the particle diameter (assuming spherical particles)
- μ_0 is the absolute magnetic permeability of vacuum equal to $4\pi \times 10^{-7}$ Wb/(A·m)
- k_B is the Boltzmann constant

The magnetic permeability, μ, of the ferrofluid (i.e., aqueous dispersion of SNPs) varies linearly with susceptibility: $\mu = 1 + \chi$. Please refer to Chapter 5 for definitions and for details on the origin of this equation (especially Section 5.2.3).

The next step is to calculate the effective magnetic permeability (μ_{eff}) of the ferrofluid-saturated rock using effective medium theory, which accounts for the contributions of the rock matrix and of the ferrofluid and the residual oil occupying the rock porosity. There is a choice in the approaches to effective medium theory, resulting in minor differences in the μ_{eff} value. Among various methods of calculating μ_{eff}, the implicit Bruggeman equation is commonly employed (Sihvola 1999):

$$\left(1-(1-S_{\text{or}})\beta\right)\frac{\mu_e - \mu_{\text{eff}}}{\mu_e + 2\mu_{\text{eff}}} + (1-S_{\text{or}})\beta\frac{\mu_e - \mu_{\text{eff}}}{\mu_i + 2\mu_{\text{eff}}} = 0 \tag{9.2}$$

where

μ_i is the magnetic permeability of the ferrofluid calculated using Equation (9.1)

μ_e is the magnetic permeability of the rock matrix

β is the porosity of the formation

S_{or} is the residual oil saturation within the rock.

$\beta(1 - S_{or})$ is thus the volume fraction occupied by ferrofluid within the rock.

9.2.1.3 Feasibility Simulation Model

For their feasibility simulation, Rahmani et al. (2015a) employed a simple reservoir model which is a homogeneous formation; the injector well is aligned along the axis of symmetry of the 2D axisymmetric model, identified with a dash-dot black vertical line on the left in Figure 9.2. The SNP bank propagates in a single-phase piston-like displacement away from the injector toward the observer well located 100 m away. A bank of ferrofluid is injected increasing the relative magnetic permeability of the formation by a factor of two (to $\mu_{eff} = 2$ from 1). The reservoir layer in which we inject the fluids is 20 m thick. The injection bank size is 20 m in height and 31 m in radius at the end of the ferrofluid injection, amounting to 9.61% of the pore volume of the reservoir. The magnetic source (identified with a gray dot at left boundary in

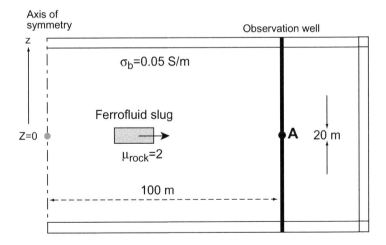

FIGURE 9.2

Schematic of the simple axisymmetric reservoir model. A bank of an aqueous dispersion of SNPs (shown as a shaded rectangle) is injected at the central well (the left boundary, with the center of the injection layer shown as a gray dot). The magnetic field oscillation is generated at the injection well with the generator (a point source) moving up and down the well. The magnetization response measurement well is shown as a heavy black vertical line. From Rahmani et al. (2015a). Copyright 2015, Society of Petroleum Engineers. Reproduced with permission of SPE. Further reproduction prohibited without permission.

Figure 9.2) is a point magnetic dipole oriented upwards with a unit magnetic moment and initially located at the center of the reservoir layer ($z = 0$). The casing of both injection and observer wells is assumed to be non-conductive and non-magnetic.

9.2.1.4 A Brief Summary of the Simulation Study

Changing the excitation source's vertical position, relative to that of the injection zone at the well, could drastically change the magnetization response at the measurement well. Figure 9.3 shows the magnetic tomography map (% $|H_z|$ anomaly; see Equation 4 of Rahmani et al. [2015a]) for the SNP bank shown in Figure 9.2, at different radial distances away from the injection source, at the frequency of 10 Hz. In the tomography maps, the magnetic anomaly, represented by the color scale with dark red color signifying best detectability, is shown with changing the source position (horizontal axis) and the receiver position (vertical axis). The figure suggests that at the early stages of the injection, the magnetic response dramatically changes with the transmitter position. The anomaly undergoes an abrupt transition (from −42% to +20% in the earliest stage) when the transmitter moves across the boundary of the dispersion bank layer (i.e., $z = -20$ m and $z = 20$ m). As the SNP bank progresses toward the receiver, i.e., observer, well (subsequent panels), however, the sensitivity to the source position decreases. The final panel shows that the magnetic anomaly can be clearly detected at the receiver position. The measurements exhibit a high level of vertical sensitivity when the bank is either close to the source or to the receivers.

The simulation study showed that the ferrofluid magnetic response is significant at low frequencies whereas an electrically conductive bank is hardly detectable. This could enable field application because most wells are cased with steel which suppresses the signal at high frequencies. Moreover, the simulations indicate that the sensitivity of magnetic measurements at the early stages of the flood is significantly higher for the ferrofluid slug, both at low and at high (500 Hz) frequencies. The ferrofluid magnetic response, although high when close to the source and receivers, drops off quickly toward the inter-well region. Moving the transmitter and receivers relative to the slug also provides a noticeable increase in the vertical sensitivity of the magnetic measurements. When the slug is close to the injection point, moving the transmitter provides high resolution measurements and moving the receivers has much less impact on the measurements. As the slug approaches the observer well, the measurements exhibit higher sensitivity to receiver position and less sensitivity to the transmitter location. It is therefore deemed that with use of low-frequency magnetic oscillation (<100 Hz), if the magnetic susceptibility contrast is of the order of ~0.1, meaningful measurements of induction responses are possible. Built upon the promising feasibility simulation outcome, Rahmani et al. (2015b) extended their simulation investigation to see if the SNP bank injection can be employed to enhance the

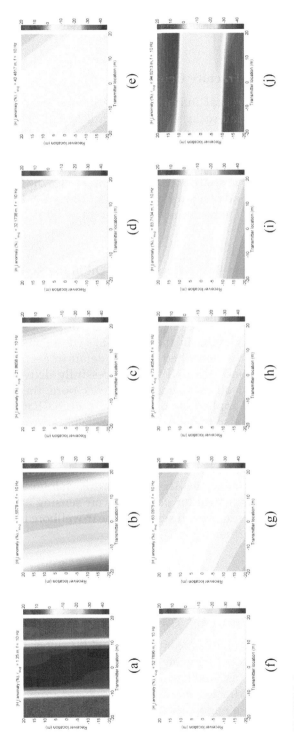

FIGURE 9.3
Magnetic tomography map (% |H_z| anomaly) for the SNP bank shown in Figure 9.2, for different radial-distance positions of the advancing bank, from the injection source. From (a) to (j): 1.3, 11.6, 21.9, 32.2, 42.5, 52.8, 63.1 83.7, and 94 m from the injection well. The color scale identifies the magnetic anomaly, the vertical axis corresponds to the receivers' vertical location, and the horizontal axis corresponds to the magnetic oscillation source's vertical position. From Rahmani et al. (2015a). Copyright 2015, Society of Petroleum Engineers. Reproduced with permission of SPE. Further reproduction prohibited without permission.

characterization of the reservoir heterogeneity, which again showed a positive application potential.

9.2.2 EM-Based Hydraulic Fracture Detection

Employing a physical/theoretical interpretation technique similar to the above (Hu et al. 2017), a small-scale, shallow-depth field test has recently been successfully carried out to remotely image proppants in hydraulic fracture networks (Ahmadian et al. 2018; LaBrecque et al. 2016). Since the test employed coke breeze and steel shots as the "contrast" agents, rather than the SNPs as used in the above study and by Al-Shehri et al. (2013), it is outside the scope of this book and is merely cited here for reference. Interested readers are also referred to the recent related publications by Palisch et al. (2016, 2017).

9.2.3 Magnetic Nanoparticle-Based Detection of Oil in the Reservoir

Another potential application of SNP for remote detection of the presence of oil in subsurface formations has been explored (Prodanovic et al. 2010; Ryoo et al. 2012), which relies on the targeted delivery of SNPs to the oil/water interfaces of the oil ganglia in the reservoir. (The related surface coating optimization effort is described in some detail in Section 2.7.2.) In the proposed method, once the SNPs are successfully delivered and attached to the interface of oil meniscus in the rock pore, the external magnetic field oscillation is applied. As described earlier in Chapter 5, the external field gradient will also make the SNPs oscillate, causing the oil/water interface to fluctuate and generate a tiny density wave. Such interfacial fluctuation and the resulting acoustic wave could be detected and analyzed (Oh et al. 2006; Ryoo et al. 2012). Due to the quick attenuation of the source signal as discussed at the beginning of this chapter, this technique is proven not to be practical for deep reservoir sensing at present. However, it has potential for improved sensing of near-wellbore formation properties and warrants further study.

9.3 Nanoparticles as Intelligent Tracers

In the above, use of nanoparticles as "contrast agents" (i.e., whose presence in the reservoir can be detected remotely) was described; and in this subsection, use of nanoparticles as "intelligent tracers" (i.e., from which the reservoir information can be extracted *after* they are produced from reservoir) is described. Different uses of nanoparticles as intelligent tracers have been reported, as described below.

9.3.1 Nanoreporter

In one category of "intelligent tracers," the surface coating on the nanoparticle is designed to change its nature in a specific manner, when the particles contact a certain target fluid component or a mineral component in the reservoir, or some other reservoir properties. When the nanoparticles are produced with oil and/or water, the changes in their surface coating is tracked in relation to the nanoparticle effluent concentration and time, from which the desired reservoir property can be deduced (thus, known as "nanoreporter"). For example, Berlin et al. (2011) developed a nanoreporter which can detect the presence of oil in the reservoir. They developed carbon clusters (HCC) with <40 nm length and 1 nm width, with a hydrophobic core and hydrophilic surface prepared from single-walled carbon nanotubes by a harsh oxidation procedure. Covalently attaching polyethylene glycol (PEG) to the carboxylic acid groups on the surface of HCC renders the nanoparticle's dispersion stability in water. Hydrophobic compounds ("oil marker") introduced into PEG-HCCs are sequestered on the hydrophobic domains of the HCC core, as schematically shown in Figure 9.4. Such design of the nanoparticle allows not only its long-distance transport through the reservoir rock carrying the hydrophobic compounds, but also the selective release of the "oil marker" when they contact the resident oil. In a similar nanoparticle design, oxidized carbon black (OCB) was used as the core whose surface was functionalized with polyvinyl alcohol (PVA). As the "oil marker" hydrophobic compound, 2,2′,5,5′-tetrachlorobiphenyl was employed. To evaluate the transportability of the nanoparticles, corefloods with sandstone and limestone rock cores were carried out, which showed acceptable results (see below).

The above nanoparticles, however, suffered from thermal stability problems, such as being prone to agglomeration in seawater at a high temperature (70°C), and accordingly showed poor transportability in rock. To remedy the problem, Hwang et al. (2012) used a carboxyl group-functionalized carbon black (fCB) as the nanoparticle core, which is less oxidized than the above OCB core. The subsequently PVA-functionalized fCB exhibited an improved dispersion stability and transported better in the rock cores. As the "oil marker" compound, triheptylamine was also employed. For both studies (Berlin et al. 2011; Hwang et al. 2012), corefloods with rock core in which oil is present demonstrated the retention of the "oil marker" chemical, indicating that the "nanoreporter" performed its design function.

9.3.2 Development and Field Testing of a Fluorescent Nanoparticle Tracer

Saudi Aramco has an active research effort to illuminate the reservoir using nanoparticles, taking two different approaches: the first is to use nanoparticles ("A-dots") as a nanoreporter which can be optically detected from the produced fluids (Kanj et al. 2008, 2009). They reported a successful field

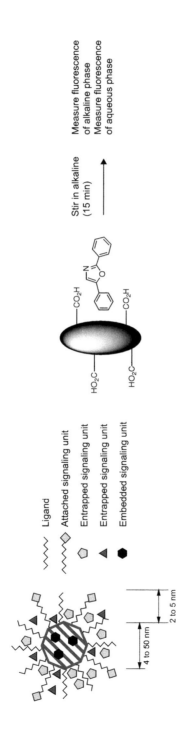

FIGURE 9.4

"Nanoreporter" nanoparticle which carries a hydrophobic "oil marker" chemical in it, for hydrocarbon detection in the oil reservoir rock. Adapted from information in Berlin et al. (2011) and Hwang et al. (2012).

testing of the method (Kanj et al. 2011). The second approach is to use SNPs for the mapping of the reservoir fluid bank, utilizing the fact that electromagnetic wave's speed slows down when it passes through magnetic media (Al-Shehri et al. 2013). In view of the fact that the fluorescent nanoparticle tracer developed by Aramco has already been field-tested, their work is described here in some detail.

A-dot is a carbon-based fluorescent nanoparticle whose average diameter is ~8 nm, and which was made to withstand the harsh reservoir conditions of the Ghawar field in Saudi Arabia. It thus maintains its dispersion stability at temperatures exceeding 100°C and salinities exceeding 120,000 ppm in total dissolved solids and in the presence of carbonate rocks (limestone, dolostone). The fluorescent light emission of A-dot is virtually independent of the excitation wavelength in the 400–500 nm range, making it a good emitter detectable at low concentrations (in the range of 50–100 ppm). In order to test the transportability of the nanoparticles in the Ghawar reservoir rock (average porosity of 20.3% and brine permeability of 9.9 md), corefloods were carried out by injecting a 20% pore volume bank of 10 ppm A-dot dispersion. Figure 9.5 shows the effluent concentration (dashed curve) and cumulative recovery (solid curve) of A-dots from a coreflood, revealing an excellent transportability with >93% recovery of the injected nanoparticles. Kanj et al. (2011) also reported their field test operation and results. The A-dot injection and production were carried out in a huff-and-puff mode, at the same well which was originally a producer with 99% water cut. Figure 9.6 shows the effluent concentration (solid curve) and cumulative recovery (dotted curve) of A-dots from the field test, with ~82% recovery of the injected nanoparticles, but with a significant delay in the effluent profile. The field test therefore demonstrated that a nanoparticle tracer can be made to maintain its dispersion stability and transportability, in-situ in the reservoir even under very harsh conditions.

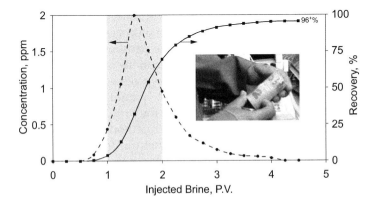

FIGURE 9.5
A-dot concentration in the effluent (dashed curve) and cumulative recovery (solid curve) from a coreflood with a composite carbonate rock core. The gray area represents the injected bank size. From Kanj et al. (2011).

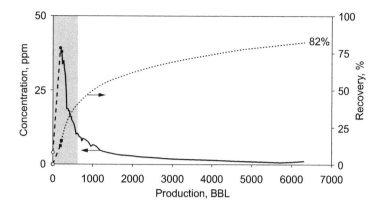

FIGURE 9.6
A-dot concentration as a function of the water production from the huff-n-puff field test well (solid curve). The dotted curve is the cumulative recovery and the gray area represents the injected bank size. From Kanj et al. (2011).

9.3.3 Magnetic Nanoparticle Tracers

Creation of multiple hydraulic fractures along a horizontal well is a widely and actively practiced technique to stimulate the production of oil and/or gas from shale and other low-permeability reservoirs. As the production begins from the horizontal well, oil, gas, and/or water from all fractures are mixed together and produced at the well head. If a quantitative resolution of each fluid from each fracture can be made and the well-performing and poorly performing fractures can be identified, it will greatly help to make decisions on any remedial actions and to predict more accurately the production performance and the ultimate hydrocarbon recovery from the well. Also, when an improved oil recovery (IOR) process is implemented at an oil reservoir, the ability to assess the process performance at an early stage of its operation can greatly help the optimal management of the process. Because the IOR processes are usually implemented as multiple-pattern floods with many injectors and producers, it is difficult to distinguish the source of the oil produced at a production well, i.e., what portion of the produced oil is mobilized by the IOR fluid injected at which injection well. A quantitative identification of the origin of the produced oil will greatly help the process optimization.

Huh et al. (2015) earlier proposed a novel method that allows such quantitative resolution of production from each fracture, or from each flood pattern, by analyzing non-invasively and continuously the different paramagnetic nanoparticle tracers that had been added to the fracturing fluids or to the injection water for IOR. Even though the feasibility of the proposed technique has been demonstrated only in the laboratory, in view of its potentially significant benefit, and also because it is a very effective utilization of the unique magnetization response of SNPs, the technique will be described in some detail here. Below, the basic principle of detecting the presence of SNPs will be first explained, followed by description of the technique for composition analysis for a mixture of SNPs with differing magnetic properties.

9.3.3.1 Measurement of Non-Linear Magnetization
Response to Imposed Magnetic Oscillation

In employing the paramagnetic nanoparticles as a tracer, we utilize their two unique properties. First, by applying a steady magnetic field gradient to a dilute dispersion of nanoparticles, we can force the nanoparticles to a desired direction, so that they can be concentrated in a small volume, thereby improving their detectability. The collection technique is a well-established technology known as the high-gradient magnetic separation (HGMS) method (Gerber and Birss 1983; Moeser et al. 2004). Second, when a paramagnetic nanoparticle dispersion is subjected to a varying magnetic field strength (H), a unique magnetization response (M) results, which is known as the Langevin equation (see Chapter 5 and Figure 5.13b):

$$\frac{M}{M_s} = \coth(\alpha) - \frac{1}{\alpha} \equiv L(\alpha) \tag{9.3}$$

where
$M_s = \phi M_d$ is the saturation magnetization of the dispersion with $\phi =$ volume fraction of nanoparticles
$M_d =$ bulk magnetization of the nanoparticle solid.

In Equation (9.3),

$$\alpha \equiv \frac{\pi \mu_0 M_d d^3 H}{6kT} \tag{9.4}$$

where
$\mu_0 =$ vacuum permeability
$d =$ nanoparticle diameter
$T =$ absolute temperature.

As can be seen from the Langevin relation given above, the magnetization response depends on the nanoparticle size and the bulk magnetization of the metal oxide that forms the nanoparticle core. Recall that for superparamagnetic nanoparticles, the Langevin curve goes through the coordinate origin (since for $H = 0$, $M = 0$).

A common method of measuring the nanoparticle concentration (or in general, the amount of magnetic material) is to measure the magnetic susceptibility, as defined by Equation (9.1) above. In particular, the value at the coordinate origin (i.e., the slope of the curve there) is typically reported to represent the magnetic characteristics of the material. It can be measured with a susceptibility meter, which is usually done at a fixed frequency.

Based on the effective magnetic susceptibility versus concentration calibration curve (developed for the particular nanoparticle and fluid combination, as described in Section 9.1.1), the nanoparticle concentration can be non-invasively and instantly determined from the susceptibility measurements (Biederer et al. 2009). Resolving the relative fractions of the nanoparticles from a mixture is, however, difficult with the above method. A novel technique is described below.

As introduced in Chapter 5, the magnetic particle imaging (MPI) technique of distinguishing different nanoparticles magnetically is based on the fact that the magnetization curve is dependent on the size of nanoparticle, d, as shown above in Equation (9.4). This is also shown in Figure 9.7a (Chantrell et al. 1978); and Figure 9.7b shows how these different sizes respond to magnetization (such as in MPI application). The temporal change in the particle magnetization $M(t)$ induces a voltage in a receive coil. Different metal alloy oxides, which have different Langevin curve characteristics, can also be employed to expand the range of choice of different paramagnetic nanoparticles as tracers.

9.3.3.2 Composition Analysis from Paramagnetic Nanoparticle Mixture in Fluid

Because the concentrations of nanoparticles as tracers will be very small, the induced voltages from nanoparticle of each size are linearly additive, so that the total voltage signal detected by the receiving coil is the weighted sum of individual signals generated by nanoparticle of each size. The induced voltage signal is periodic with base frequency, f_0. Due to the non-linearity of the magnetization curve, the induced voltage signal contains the excitation frequency f_0 as well as harmonics (i.e., integer multiples) of f_0, as shown in Figure 9.8. The estimation of concentration of each nanoparticle component is dependent on the non-linearity of the magnetization curve at each particle size. An inversion computer program which calculates the composition of the nanoparticle mixture from the measured signal as described above has been developed and the detectability of different sized SNPs has been demonstrated in laboratory tests (Huh et al. 2015).

The schematics of the envisioned SNP tracer measurement apparatus is shown in Figure 9.9. Here, the magnetic excitation coil and the receiver coil are put around a section of the oil-carrying pipeline which is magnetically transparent. As the SNPs that had been earlier injected at the injection wells together with the EOR injection fluids are produced with oil, their magnetization responses can be measured and their concentrations could be deduced from the above technique. Since the concentrations of SNPs would be extremely low, to improve their detection resolution, they may need be magnetically concentrated (Gerber and Birss 1983; Moeser et al. 2004), as briefly mentioned above.

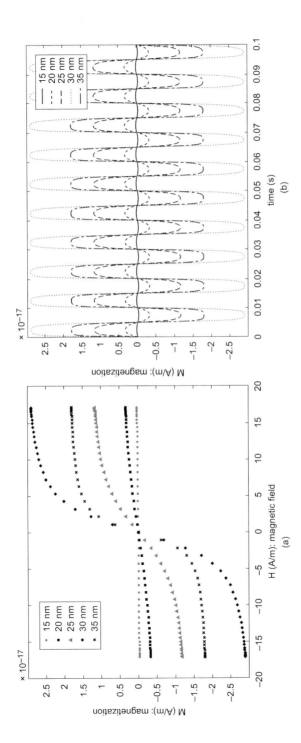

FIGURE 9.7

(a) Characteristic magnetization curves of SNPs of different sizes. (b) Magnetization responses of the SNPs of different sizes, with application of a sinusoidal magnetic field oscillation. From Huh et al. (2015).

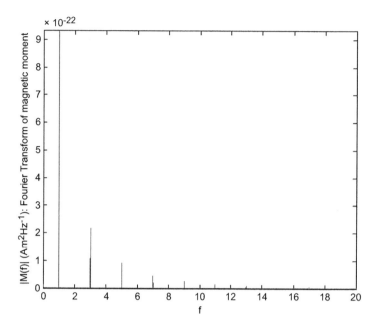

FIGURE 9.8
Fourier spectrum of total induced voltage signal. From Huh et al. (2015).

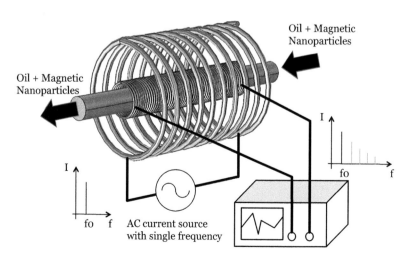

FIGURE 9.9
Schematics of the envisioned SNP tracer measurement apparatus. From Huh et al. (2015).

9.4 Downhole Monitoring

Utilizing the unique ability of some nanoscale materials that can detect minute amounts of chemicals, or accurately measure nanoscale changes in mechanical and other physical properties, active research efforts are being made to develop nanosensors (Bogue 2004, 2008; Shelley 2008). The key advantage of nanosensors is the very high surface area that the nanomaterials such as CNT offer for detection of extremely minute property changes. While the majority of the developments are for innovative biosensors for medical and security purposes, gas and chemical sensors are beginning to reach the market, which potentially could be utilized for upstream oil applications. Barron et al. (2010) provide some examples of oilfield applications development. The biggest challenge for such applications will be how the tiny, fragile nanosensors can be protected from the harsh downhole conditions for their proper functioning.

References

Agenet, N., Moradi-Tehrani, N., and Tillement, O. (2011) Fluorescent Nanobeads: A New Generation of Easily Detectable Water Tracers (IPTC 15312), International Petroleum Technology Conference, Nov. 15–17, Bangkok, Thailand.

Ahmadian, M., LaBrecque, D., Liu, Q., Slack, W., Brigham, R., Fang, Y., Banks, K., Hu, Y., Wang, D., and Zhang, R. (2018) Demonstration of Proof of Concept of Electromagnetic Geophysical Methods for High-Resolution Illumination of Induced Fracture Networks (SPE 189858), SPE Hydraulic Fracturing Technology Conference, Jan. 23–25, The Woodlands, TX, USA.

Al-Ali, Z. A., Al-Buali, M. H., AlRuwaili, S., Ma, S. M., Marsala, A. F., Alumbaugh, D., DePavia, L., Levesque, C., Nalonnil, A., and Zhang, P. (2009) Looking deep into the reservoir. *Oilfield Rev.*, 21(2), 38–47.

Al-Shehri, A. A., Ellis, E. S., Servin, J. M. F., Kosynkin, D. V., Kanj, M. Y., and Schmidt, H. K. (2013) Illuminating the Reservoir: Magnetic Nanomappers (SPE 164461), SPE Middle East Oil and Gas Show and Conference, Mar. 10–13, Manama, Bahrain.

Barron, A. R., Tour, J. M., Busnaina, A. A., Jung, Y. J., Somu, S., Kanj, M. Y., Potter, D., Resasco, D., and Ullo, J. (2010) Big things in small packages. *Oilfield Rev.*, 22(3), 38–49.

Berlin, J. M., Yu, J., Lu, W., Walsh, E. E., Zhang, L., Zhang, P., Chen, W., Kan, A. T., Wong, M. S., and Tomson, M. B. (2011) Engineered nanoparticles for hydrocarbon detection in oil-field rocks. *Energy & Environ. Sci.*, 4(2), 505–509.

Biederer, S., Knopp, T., Sattel, T. F., Lüdtke-Buzug, K., Gleich, B., Weizenecker, J., Borgert J., and Buzug, T. M. (2009) Magnetization response spectroscopy of superparamagnetic nanoparticles for magnetic particle imaging. *J. Phys. Appl. Phys.*, 42(20), 205007.

Bogue, R. (2008) Nanosensors: A review of recent progress, *Sensor Rev.*, 28(1), 12–17.

Bogue, R. W. (2004) Nanotechnology: What are the prospects for sensors? *Sensor Rev.*, 24(3), 253–260.

Chantrell, R. W., Popplewell, J., and Charles, S. W. (1978) Measurements of particle size distribution parameters in ferrofluids. *IEEE Trans. Magn.*, MAG-14, 975–977.

Gerber, R., and Birss, R. R. (1983) *High Gradient Magnetic Separation*. Research Studies Press, John Wiley & Sons, New York.

Gleich, B. and Weizenecker, J. (2005) Tomographic imaging using the nonlinear response of magnetic particles. *Nature*, 435, 1214–1217.

Hu, Y., Yu, Z., Zhang, W., Sun, Q., and Liu, Q. H. (2016) Multiphysics coupling of dynamic fluid flow and electromagnetic fields for subsurface sensing. *IEEE J. Multiscale Multiphys. Comput. Tech.*, 1, 14–25.

Hu, Y., Fang, Y., LaBrecque, D., Ahmadian, M., and Liu, Q. H. (2017) Reconstruction of high-contrast proppant in hydraulic fractures with Galvanic measurements. *IEEE Trans. Geosci. Remote Sensing*, 56(4), 2066–2073.

Huh, C., Milner, T. E., Nizamidin, N., Wang, B., and Pope, G. A. (2015) Hydrophobic paramagnetic nanoparticles as Intelligent Crude Oil Tracers. *US Patent:* US 20,150,376,493, December 31, 2015.

Hwang, C.-C., Wang, L., Lu, W., Ruan, G., Kini, G. C., Xiang, C., Samuel, E. L., Shi, W., Kan, A. T., and Wong, M. S. (2012) Highly stable carbon nanoparticles designed for downhole hydrocarbon detection. *Energy & Environ. Sci.*, 5(8), 8304–8309.

Kanj, M. Y., Funk, J. J., and Afaleg, N. I. (2008) Towards In-Situ Reservoir Nano-Agents, NSTI Nanotechnology Conference, Boston, MA, USA.

Kanj, M. Y., Funk, J. J., and Al-Yousif, Z. (2009) Nanofluid Coreflood Experiments in the ARAB-D (SPE 126161), SPE Saudi Arabia Section Technical Symposium, May 9–11, Al-Khobar, Saudi Arabia.

Kanj, M. Y., Rashid, M., and Giannelis, E. (2011) Industry First Field Trial of Reservoir Nanoagents (SPE 142592), SPE Middle East Oil and Gas Show and Conference, Sept. 25–28, Manama, Bahrain.

LaBrecque, D., Brigham, R., Denison, J., Murdoch, L., Slack, W., Liu, Q. H., Fang, Y., Dai, J., Hu, Y., and Yu, Z. (2016) Remote Imaging of Proppants in Hydraulic Fracture Networks Using Electromagnetic Methods: Results of Small-Scale Field Experiments (SPE 179170), SPE Hydraulic Fracturing Technology Conference, Feb. 9–11, The Woodlands, TX, USA.

Moeser, G. D., Roach, K. A., Green, W. H., Hatton, T. A., and Laibinis, P. E. (2004) High-gradient magnetic separation of coated magnetic nanoparticles. *AIChE J.*, 50(11), 2835–2848.

Oh, J., Feldman, M. D., Kim, J., Condit, C., Emelianov, S., and Milner, T. E. (2006) Detection of magnetic nanoparticles in tissue using magneto-motive ultrasound, *Nanotechnology*, 17(16), 4183.

Palisch, T., Al-Tailji, W., Bartel, L., Cannan, C., Czapski, M., and Lynch, K. (2016) Recent Advancements in Far-Field Proppant Detection (SPE 179161) SPE Hydraulic Fracturing Technology Conference, Feb. 9–11, The Woodlands, TX, USA.

Palisch, T., Al-Tailji, W., Bartel, L., Cannan, C., Zhang, J., Czapski, M., and Lynch, K. (2017) Far-field Proppant Detection Using Electromagnetic Methods (SPE 184880) SPE Hydraulic Fracturing Technology Conference, Jan. 4–28, The Woodlands, TX, USA.

Prodanovic, M., Ryoo, S., Rahmani, A. R., Kuranov, R. V., Kotsmar, C., Milner, T. E., Johnston, K. P., Bryant, S. L., and Huh, C. (2010) Effects of Magnetic Field on the Motion of Multiphase Fluids Containing Paramagnetic Nanoparticles in Porous Media (SPE 129850), SPE Improved Oil Recovery Symposium, Tulsa, OK, USA.

Rahmani, A. R., Bryant, S., Huh, C., Athey, A.E., Ahmadian, M., Chen, J., and Wilt, M. (2015a) Crosswell magnetic sensing of superparamagnetic nanoparticles for subsurface applications. *Soc. Petrol. Eng. J.*, 1067–1082.

Rahmani, A. R., Bryant, S. L., Huh, C., Ahmadian, M., Zhang, W., and Liu, Q. H. (2015b) Characterizing Reservoir Heterogeneities Using Magnetic Nanoparticles (SPE 173195), SPE Reservoir Simulation Symposium, Feb. 23–25, Houston, TX, USA.

Rosensweig, R. E. (1997) *Ferrohydrodynamics.* Dover Publications, Minneola, NY.

Ryoo, S., Rahmani, A. R., Yoon, K. Y., Prodanović, M., Kotsmar, C., Milner, T. E., Johnston, K. P., Bryant, S. L., and Huh, C. (2012) Theoretical and experimental investigation of the motion of multiphase fluids containing paramagnetic nanoparticles in porous media. *J. Petrol. Sci. Eng.*, 81, 129–144.

Shelley, S. (2008) Nanosensors: Evolution, not revolution … yet. *Chem. Eng. Progr.*, 104, 8–12.

Sihvola, A. H. (1999) *Electromagnetic Mixing Formulas and Applications.* Institution of Electrical Engineers, London, UK.

10

Enhanced Oil Recovery: Foams and Emulsions

10.1 Introduction

Probably the most important uses of nanoparticles for enhanced oil recovery (EOR) are (1) to stabilize CO_2 foams which are then employed as improved mobility control agents for CO_2 EOR processes; and (2) to stabilize light oil-in-water emulsions which serve as an improved mobility control agent for heavy oil recovery processes, as well as to lower the heavy oil viscosity when the emulsion's internal phase contacts the resident heavy oil (Bennetzen and Mogensen 2014). These two applications are the main topics of discussion for this chapter. As was described in Section 4.2, it is well known that nanoparticles act like surfactants at the gas/liquid or liquid/liquid interface, thereby stabilizing foams and emulsions. The similarities and differences between nanoparticles and surfactants are well described by Binks (2002).

When and where a low-cost and steady CO_2 source is available, CO_2 flooding is probably the most reliable EOR process so far available, and large-scale field implementations have been successfully and economically made in the Permian Basin of West Texas and other areas. Extensive literature on CO_2 EOR processes is available and excellent monographs (Jarrell et al. 2002; Lake et al. 2014; Sheng 2011) on CO_2 EOR processes are available. One critical weakness of the CO_2 EOR process is the low viscosity and low density of CO_2. In order to maintain reservoir pressure, CO_2 is usually injected into a reservoir as a supercritical fluid, at which condition (*e.g.*, at the critical-point state of 1069.4 psi and 87.8°F) its density is ~0.8 gr/mL and its viscosity is ~0.03 cp. With such very low viscosity, displacing oil with CO_2 generally creates a severe adverse mobility condition, as well as a severe tendency for gravity segregation. To circumvent this critical weakness, extensive research has been carried out to develop CO_2 foam as an improved mobility control agent, as the apparent viscosity of foam can be orders of magnitudes higher than the CO_2 viscosity. Again, excellent reviews on the surfactant-stabilized CO_2 foams and their use to improve the volumetric sweep of reservoir during the CO_2 EOR processes are available, *e.g.*, by Rossen (1996).

While foams and emulsions stabilized by colloidal solid particles have been widely used for an immense variety of industrial and consumer applications (Binks 2002, 2017; Binks and Horozov 2006), their use for enhanced oil recovery purposes has been limited until recently. This is because the colloidal solids cannot be transported long distance within oil reservoirs, *e.g.*, from injection well to production well, generally being filtered almost immediately upon injection into the reservoir rock. Now, because nanoparticles are two orders of magnitude smaller than colloids and thus can migrate through the pore throats in sedimentary rocks (Roberts et al. 2012), their use opens up new possibilities for EOR. Nanoparticles being solids, foams/emulsions stabilized with nanoparticles can withstand the high-temperature and/or high-salinity reservoir conditions for extended periods, thereby overcoming some key limitations of foams/emulsions stabilized by surfactants. This can substantially expand the range of reservoirs to which EOR can be applied. Finally, nanoparticles can carry additional functionalities such as superparamagnetism (Melle et al. 2005), which offers potential to manipulate the behavior of the foams/emulsions with the application of an external magnetic field.

10.2 CO_2 Foams for Improved Mobility Control

A great challenge for the foam application for EOR is maintaining its long-term stability in the reservoir, especially for the high-salinity and/or high-temperature reservoirs where the commonly used surfactants tend to either degrade or precipitate. As described in Chapter 4, the nanoparticle-stabilized foams generally have a more robust stability than the surfactant-stabilized foams; and they are more tolerant of the high-salinity and high-temperature conditions (Adkins et al. 2007; Dickson et al. 2004). Prompted by the above benefits, an active research effort has been made as described in detail below. Because use of nanoparticle-stabilized CO_2 foams for EOR application is a new research area, it requires multiple development steps starting from the optimal surface coating design for nanoparticles, to the injection of an aqueous nanoparticle dispersion into the reservoir at which CO_2 EOR operation is either already ongoing or is planned, so that nanoparticle-stabilized CO_2 foam is generated in the reservoir. Therefore, the multi-scale design of nanoparticle-stabilized CO_2 foam process is first described, step-by-step, in Sections 10.2.1 through 10.2.4.

10.2.1 Multi-Scale Design of CO_2 Foam Process

Figure 10.1 schematically shows the design strategy for developing a nanoparticle-stabilized CO_2 foam field trial, which builds in complexity from the initial, laboratory benchtop characterization of nanoparticles up to field-scale modeling. The first step of nanoparticle stability characterization,

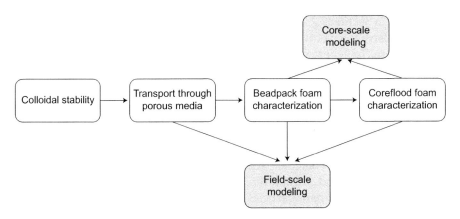

FIGURE 10.1
Design strategy flow diagram for nanoparticle-stabilized CO_2-in-water foam field trial. After Worthen et al. (2015). Copyright 2015, Society of Petroleum Engineers. Reproduced with permission of SPE. Further reproduction prohibited without permission.

usually in the reservoir-condition brine at reservoir temperature, allows rapid screening of various nanoparticle candidates, where unstable particles are excluded from further testing. The proper design of nanoparticle coatings is key to passing this step, as will be described in more detail in the next sub-section. In the second step (*nanoparticle transport through porous media*), single-phase flow experiments are carried out to ensure that the nanoparticles themselves can be transported through reservoir rock with minimal retention. Next (*beadpack foam characterization*), foam experiments are initially performed in small beadpacks, which can be quickly carried out at different process conditions. In the next step (*coreflood foam characterization*), corefloods are carried out to demonstrate feasibility of foam generation and transport in reservoir rock with the reservoir-condition fluids (without and with oil) at reservoir temperature and pressure. Results from beadpacks and corefloods are then modeled at the core-scale to evaluate modeling techniques (*core-scale modeling*). Finally, the field-scale modeling is performed to evaluate whether to proceed with the development of a field trial, and to design the pilot test (*field-scale modeling*). Successfully designing a nanoparticle-stabilized foam process thus requires the above multi-scale investigation. The first two steps of *colloidal stability characterization* and *nanoparticle transport through porous media* will be discussed in the following sub-sections 10.2.1.1 and 10.2.1.2. The subsequent tasks will be discussed in sub-sections 10.2.2 to 10.2.4.

10.2.1.1 Stability of Nanoparticles in Brine

For the nanoparticles to serve as an effective foam stabilizer that has to survive for a long period of time flowing through a long distance in the reservoir, they have to stay as individual particles without being aggregated. As described in Section 2.2, the surface coating should be able to generate a sufficient repulsive force between nanoparticles in the reservoir brine at reservoir temperature. To generate the nanoparticle-stabilized foams, the particle concentration needs to be above a certain value to maintain foam stability, especially in a high salinity reservoir condition (Espinoza et al. 2010). However, when the nanoparticles are concentrated, the enhanced collisions between nanoparticles could potentially result in greater aggregation (Holthoff et al. 1996). The aggregates could then be more likely trapped at rock pore throats, thus being lost and unable to serve as the foam stabilizer as intended. Aggregation of nanoparticles (and colloidal particles) normally occurs when the energy barrier in electrostatic and entropic repulsive interaction between particles is overcome. In the case of nanoparticles, where transport behavior is dictated by Brownian motions, the initial particle concentration is a key factor governing the aggregation kinetics (Elimelech and Omelia 1990; Chen and Elimelech 2006). For oil reservoir applications, the reservoir brine usually has a high salinity so that the electrostatic repulsion between particles is virtually eliminated (Vandesteeg et al. 1992). To prevent the instability of nanoparticles in the aggregation-inducing, high-salinity environment, the surface treatment of nanoparticles should provide steric repulsion between particles keeping them stable along their long travel distance, thereby ensuring the longevity of nanoparticle-stabilized foam in the reservoir (Nguyen et al. 2014).

So far with only a few exceptions, silica nanoparticles have been employed for CO_2 foam process design. To investigate the silica nanoparticle's stability, i.e., under what conditions and how rapidly the aggregation of a candidate nanoparticle occurs, the nanoparticle dispersion that is either obtained from a vendor or synthesized in-house needs to be characterized. The initial size distribution of the nanoparticles is first measured by dynamic light scattering (DLS) and transmission electron microscopy (TEM). The stock suspension is then diluted in the reservoir brine at different proportions to obtain the desired nanoparticle concentration for the study of the aggregation. The samples are usually prepared at two different pH levels (*e.g.*, 3 and 8.5). The low pH value is close to the pH of CO_2-saturated water. The higher pH condition is that of a typical reservoir, without CO_2. Figure 3.6 from Kim et al. (2015) shows measurements of the zeta potential of the nanoparticles, which is a good indicator of the particle's electrostatic repulsive force. As pH is reduced, the magnitude of zeta potential decreases and the point of zero charge is slightly below pH 2. This result demonstrates that the inter-particle electrostatic interaction would be less pronounced at low pH. In API brine (i.e., 8 wt% NaCl and 2 wt% $CaCl_2$), the nanoparticles showed no significant surface charge (~0 mV) in the pH range tested. This result supports the negligibility of surface charge at high-salinity conditions due to the suppression of the electrical double layer and/or the complexation with cations in the dispersion (Jaisi et al. 2008).

The nanoparticle characterization is followed by the *aggregation test*, in which the nanoparticle dispersion sample is kept in a stationary condition, and the particle aggregate size change over the test period is monitored at regular time intervals by measuring it with DLS. Sampling is usually continued until the aggregate sizes is over 1000 nm. Figure 10.2(a) from Kim et al. (2015) shows measurements over time of the hydrodynamic diameter at pH 3. Regardless of brine composition and nanoparticle concentration, the aggregation was insignificant over 250 hrs. Only a few samples showed a tiny increase (*e.g.*, 20 nm to 30 nm) in the particle size when the nanoparticle concentration and the calcium content were increased. The role of calcium is expected to be less pronounced at low pH due to the reduced magnitude of the surface charge hindering the interaction between calcium ions and the nanoparticle surface. Though the overall surface charge of the nanoparticles was nearly neutralized in API brine, calcium binding might occur due to the surface charge heterogeneity (Bouyer et al. 2001). On the other hand, Figure 10.2(b) shows the aggregation behavior was notably different at pH 8.5, exhibiting a drastic increase in the aggregate size within 100 hrs at conditions identical to the pH 3 experiments.

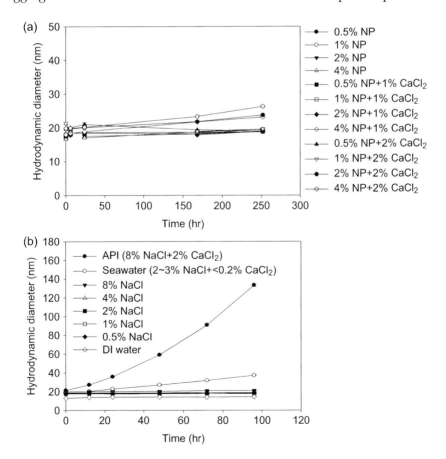

FIGURE 10.2
(a) Effect of calcium cation and nanoparticle concentrations on aggregation at pH 3 and 8% NaCl brine. (b) Effect of salinity on aggregation at pH 8.5. From Kim et al. (2015).

However, aggregation was negligible in the absence of calcium ions at pH 3 and 8.5, revealing the role of calcium in the aggregation—the bridging of particles via calcium binding onto the surface functional groups (Chen and Elimelech 2006). The initial nanoparticle concentration was another key parameter controlling the aggregation process. Considering that Brownian motion is a dominant mechanism controlling nanoparticle transport, more nanoparticles in the same volume space would lead to more collisions, resulting in faster aggregation. In cases with higher than 1% silica nanoparticle concentration in API brine, the aggregation continued until the suspension turned into a gel. The gel formation suggests the aggregation among the smaller aggregates might be possible through the calcium bridging.

10.2.1.2 Transport of Nanoparticle Dispersion in Porous Media

Before the nanoparticle transport test is carried out in reservoir rock samples, which in many cases either are not available or should be employed for only limited testing, sand columns which are packed with well-characterized sand grains can be employed, especially when the candidate reservoir's main lithology is sandstone. Also, if understanding the detailed mechanism of nanoparticle deposition on rock surface is desired, the quartz crystal microbalance with dissipation (QCM-D) can be employed to measure the deposition of nanoparticles on a model silica surface, as briefly described below.

- *Sandpack Test*: The schematic test setup is shown in Figure 10.3(a). The nanoparticle dispersion is flowed through a glass cylindrical column packed with well-characterized sands (*e.g.*, an Ottawa sand with mean diameter of 350 μm). The nanoparticle suspension is injected from the accumulator to the column by pumping at controlled flow rate to yield a desired nanoparticle residence time in the column. The influent and effluent samples are then diluted in 3 wt% HNO_3 (trace metal grade) and analyzed using inductively coupled plasma-optical emission spectrometry (ICP-OES). Kim et al. (2015) carried out a series of the column tests with variable mean aggregate diameter at a fixed Darcy velocity (71 m/day). Here, the nanoparticles were pre-aggregated prior to injection, and aggregation during the transport in porous media was assumed negligible as the residence time of particles in the column was very short (order of minutes). Figure 10.3(b) shows the overall retention of the nanoparticle aggregates was insignificant when the average aggregate size was smaller than 235 μm. However, 24~28% of the injected mass of aggregates larger than 1 μm was retained.

Figure 10.3(c) shows another set of the column tests performed (Kim et al. 2015) at different Darcy velocities (7.1, 14.2, and 71 m/day). Without pre-aggregation (Exp. #1, 2, and 7), the nanoparticles were rarely retained, showing an almost identical result to the high Darcy velocity test. This result suggests that the individual nanoparticles were sterically repulsive enough not to be deposited onto the surface of the sand grains. However, once the nanoparticles

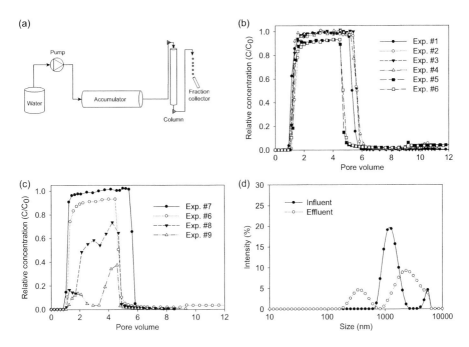

FIGURE 10.3
(a) Schematic view of transport test setup. (b) Effect of nanoparticle (or aggregate) size on transport of nanoparticles (in API brine at pH 8.5). (c) Effect of flow rate on transport of nanoparticle aggregates (in API brine at pH 8.5). (d) Aggregate size distribution of influent and effluent (samples from Exp. #5 – high velocity; 71 m/day). From Kim et al. (2015).

were injected after pre-aggregation under high-salinity conditions (which induced the aggregation), the retention was increased up to 88% as Darcy velocity decreased from 71 m/day to 7.1 m/day (Exp. #9 of Figure 10.3(c)). This high retention is considered mainly due to the pore straining of the aggregates. The change in aggregate size distribution between influent and effluent (Figure 10.3(d) for Exp. #5) indicates the possible breakdown of the aggregates by the applied hydrodynamic force, affecting the retention of the aggregates. Moreover, the fluctuations in the retention profiles at the low velocity tests imply the repeated straining and the breakdown process across the column (Legg et al. 2014). Overall, the transport of the aggregates revealed irregular effluent profile due to the complicated physical pore straining and the hydrodynamic force that breaks the loosely aggregated particles.

- *Coreflood Test with Reservoir Rock Sample*: Typically, a 1-inch diameter by 12-inch long section of reservoir rock (*e.g.*, Boise sandstone) is sealed in an epoxy column and saturated with water overnight under vacuum. The core is then flushed with the background brine at least 30 pore volumes before the test is started. To establish the steady-state condition, the nanoparticle dispersion is usually injected at least for 4 pore volumes and followed by the particle-free background brine. The influent and effluent samples can again be diluted in 3 wt% HNO_3 and analyzed using inductively coupled plasma-optical emission spectrometry. Understanding the retention of nanoparticles in porous media is important to predict their transport at the field scale and minimal retention of nanoparticles is desirable. As described by Worthen et al. (2015), the surface treated nanoparticles impose sufficient repulsion to overcome the attraction onto the sandstone at the high salinity condition. The transport data obtained from these experiments is considered in reservoir-scale modeling below to estimate foam behavior.

- *QCM-D Deposition Test*: One technique that can quantitatively provide the detailed mode of deposition of nanoparticles on solid surface is the quartz crystal microbalance with dissipation (QCM-D) apparatus. QCM-D utilizes the slight change in frequency when an electrical pulse of prescribed frequency is applied to a well-defined model solid surface to measure a minute change in its weight with deposition of chemicals or solid particles. When a nanoparticle dispersion is slowly aggregating, a sample volume of it is flown on top of a model silica surface and the nanoparticle deposition versus time can be measured. Kim et al. (2015) measured the deposition of the silica nanoparticle aggregates flowing above a plane silica surface with three different aggregate sizes (70, 460, and 1200 nm) in API

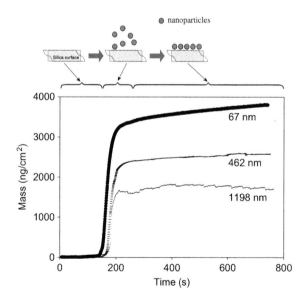

FIGURE 10.4

Variations of adsorbed mass of silica nanoparticle aggregates onto silica surface during the deposition test (2% silica nanoparticles in API brine). The illustration at the top of this figure shows the principle of QCM-D technology regarding how the adsorbed mass changes as nanoparticles deposit onto silica surface. The lag phase at the initial stage represents the time to reach the silica surface from the sample vial. From Kim et al. (2015).

brine. The deposition test was carried out immediately after measuring the aggregate sizes, assuming the sizes nearly unchanged during the test period lasting less than one hour. Figure 10.4 shows that the adsorbed mass of nanoparticles decreased with an increase of aggregate sizes consistent with the simulation results by other researchers (*e.g.*, Nelson and Ginn 2011) that the deposition of particles would decrease as particles grow in size up to ~1000 nm due to the reduced Brownian motion. It is anticipated that the size-dependent deposition trend is valid with surface-treated silica aggregates in the size range tested here. As long as the experimental conditions are maintained, a similar deposition trend is expected during the transport tests in columns filled with the silica-type packing such as sand.

10.2.2 Generation of CO_2 Foam and its Stability

Because the purpose of CO_2 foam is to decrease the mobility of the CO_2 phase, the characterization of the apparent viscosity of CO_2 foam flowing in porous media, with respect to various reservoir properties and operating parameters, is an important task for a successful design of the nanoparticle-stabilized CO_2 foam process. Before the experimental results on the foam's apparent viscosity is described, different ways to quantify the foam's apparent viscosity are first defined. Pressure drop data with and without nanoparticles are used in calculating apparent viscosity, normalized viscosity, and core mobility reduction factor as defined in the next section.

10.2.2.1 Apparent Viscosity and Core Mobility Reduction Factor

In the next two sub-sections, the results from the foam generation experiments are reported in terms of the apparent viscosity and the mobility reduction factor. The measured ΔP in the beadpack (or core) allows for calculation of the apparent viscosity of the foam (μ_{app}) using Darcy's law:

$$\mu_{app,\,beadpack} = \frac{k_{beadpack} \cdot A \cdot \Delta P}{q \cdot L} \tag{10.1}$$

where $k_{beadpack}$ is the permeability of the beadpack (or core), A is the cross-sectional area of the beadpack, q is the total volumetric flow rate, and L is the length between the ΔP cells. $k_{beadpack}$ was determined by measuring the pressure drop of water with a known viscosity and flow rate. In coreflood experiments, the core mobility reduction factor (MRF) is the ratio of the apparent viscosity of the foam at a given total flow rate to the apparent viscosity of the CO_2-water mixture at the same total flow rate (baseline):

$$MRF = \frac{\mu_{app}\ of\ foam}{\mu_{app}\ of\ baseline} = \frac{\Delta P_{Core}\ of\ foam}{\Delta P_{Core}\ of\ baseline} \tag{10.2}$$

10.2.2.2 Beadpack Foam Generation and Characterization

The CO_2-in-water (C/W) foams are typically first formed and characterized in an apparatus as schematically shown in Figure 10.5 (Worthen et al. 2013a,b). The pressure gradient caused by the flow of foam through porous media is measured with differential pressure transducers, attached to the upstream and downstream sides of the beadpacks. The beadpacks of different permeabilities can be prepared, *e.g.*, of 22 Darcy (filled with 180 μm spherical glass beads, with porosity of 0.34) or 1.2 Darcy (with 30–50 μm beads). To test the feasibility of C/W foam formation with nanoparticles under representative reservoir conditions, a series of beadpack experiments can be quickly carried out with candidate nanoparticles that are colloidally stable to identify potential systems for coreflood studies. Figure 10.6(a) from Worthen et al. (2013a) gives foam apparent viscosities in 1.2 and 22 Darcy beadpacks. The foams appear to be weakly shear thinning, where the highest viscosity of ~10 cp was achieved at an interstitial velocity of ~30 ft/day (shear rate of ~90 s^{-1}). White foams were observed in the view-cell for all conditions (see inset pictures). Interestingly, changing from 0.5 wt% to 1 wt% nanoparticles and from a 1.2 Darcy to 22 Darcy beadpack seemed to have minimal effect on the apparent viscosity. This suggests that (1) the interface was nearly saturated with nanoparticles when greater than ~0.5 wt% nanoparticles are present in the aqueous phase and (2) the foam viscosity is largely dictated by the flow conditions such as shear rate experienced by the foam.

Worthen et al. (2013a) also investigated the influence of a surfactant (Figure 10.6(b)), where a zwitterionic surfactant (LAPB) was chosen as a model surfactant. Compared with the nanoparticle-only systems (Figure 10.6(a)), added surfactant causes an increase in viscosity of 1~2 orders of magnitude, as well as demonstrating much stronger shear thinning behavior. These experiments were conducted in DI water, where the nanoparticles are not

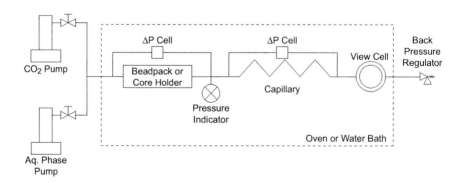

FIGURE 10.5
Schematic diagram of foam generation and characterization apparatus. After Worthen et al. (2015). Copyright 2015, Society of Petroleum Engineers. Reproduced with permission of SPE. Further reproduction prohibited without permission.

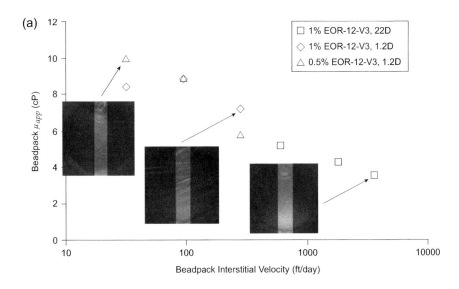

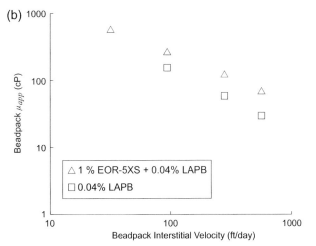

FIGURE 10.6

(a) Foam viscosity of 1% EOR-12-V3 in beadpacks, foam quality of 0.8, 9.6% TDS brine at 67°C and 3300 psia. Beadpack permeability as indicated in legend. Foam is slightly shear thinning. (b) Strong shear thinning and high viscosity of 0.04% LAPB surfactant (with (triangles) and without (squares) nanoparticles) in 1.2D beadpack, DI water, 50°C, 2800 psia, quality of 0.8. From Worthen et al. (2015). Copyright 2015, Society of Petroleum Engineers. Reproduced with permission of SPE. Further reproduction prohibited without permission.

effective foam stabilizers (Aroonsri et al. 2013), but their presence caused the foam viscosities to approximately double. This is attributed to attraction between the negatively charged nanoparticles and positively charged surfactant which causes nanoparticles to adsorb at the CO_2/water interface to increase foam stability (Worthen et al. 2013a). LAPB switches from zwitterionic to cationic under the low pH conditions of the CO_2-saturated aqueous phase and this may electrostatically interact with the nanoparticle surface (Worthen et al. 2013a).

10.2.2.3 Coreflood Foam Generation and Characterization

After the successful demonstration of foams generated in beadpacks, the foams are then tested in reservoir rock cores. The C/W foams are usually formed and characterized in an apparatus similar to that shown in Figure 10.5; but a rock core is used in a coreholder with pressure taps. For the coreflood experiments (Aroonsri et al. 2013) discussed in this and next subsections, Boise sandstone (porosity of ~ 30% and permeability of 1~3 Darcy, 1-inch diameter by 6 or 12-inch long) was used. Figure 10.7 from Aroonsri et al. (2013) shows the apparent viscosity results of foam generation experiments at three nanoparticle concentrations and compared them against the baseline apparent viscosity (circles). In all three experiments, foam was successfully generated at the shear rate of 200 s^{-1} and above, even with the nanoparticle concentration as low as 0.1 wt%. As the shear rate increased, the foam became more viscous in all three cases, indicating the importance role of energy input in forming nanoparticle-stabilized foams. However, at a given same shear rate where foams were formed, larger nanoparticle concentrations generated more viscous foam than the lower concentrations. The increased nanoparticle concentration may have increased the nanoparticle density on the interface, providing a more rigid barrier at the bubble surface. By monitoring the pressure drop in the upstream, middle, and downstream thirds of the core, it was clearly shown that foam generation started in the upstream section of the core, then propagated through the rest of the core.

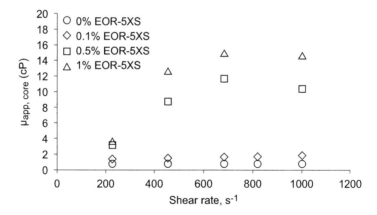

FIGURE 10.7

Foam apparent viscosity in a Boise sandstone core versus shear rate in the core. Nanoparticle concentration was varied from 0.1 to 1% in 7.2% TDS brine; dispersion was co-injected with CO_2 at 57°C and 2200 psia with a foam quality of 0.75. The black circles indicate the apparent viscosity with no added nanoparticles (i.e., with no foam present). From Aroonsri (2014).

Figure 10.8 from Aroonsri et al. (2013) shows foam apparent viscosity in Boise sandstone cores at residual oil saturation (using decane as an oil phase) versus pore volumes injected. In this series of experiments, the foam, pre-generated by co-injection through a beadpack using 1 wt% silica nanoparticle (5 nm d) dispersion in 4 wt% NaCl, was injected into the core at residual oil condition (S_{or} = 30%). Three other injection schemes were also investigated: brine-CO_2 flood (co-injected at the same phase rates as the foam experiment with no nanoparticles), liquid CO_2 only (no nanoparticles), and a 1PV CO_2 foam slug stabilized by nanoparticles. We see that nanoparticle-stabilized CO_2 foam decreased the mobility of the injected CO_2, even under the influence of residual hydrocarbon phase. At steady state, the continuous flooding of pre-generated foam yielded the apparent viscosity as high as 4.5 cp, which can be inferred as the apparent viscosity behind the foam front. Comparing between different types of injection, the continuous foam flood gave the mobility reduction factor of around 3 times over the brine/CO_2 flood baseline and as high as almost 30 times over the continuous liquid CO_2 injection. In terms of residual oil recovery, the continuous foam flood, CO_2/brine flood, and continuous CO_2 flood, all showed that residual oil saturation could be reduced from ~30% down to 6–7% (Aroonsri 2014). These results demonstrated that the ability of nanoparticle-stabilized CO_2 foam in

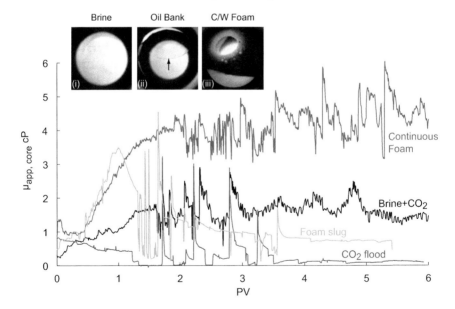

FIGURE 10.8
Apparent viscosity measured across the Boise sandstone core at residual dodecane saturation (S_{or} = 0.30) for four injection schemes: continuous foam, continuous brine with CO_2 (no nanoparticles), a 1 PV foam slug, and pure CO_2. All experiments were performed at 23°C and 2000 psia. Inset: Sequential photographs of viewcell during continuous foam injection experiment. Images show (1) the viewcell initially filled with brine; (2) the oil bank, where the black arrow indicates the oil-brine interface; and (3) the steady state foam. From Aroonsri (2014).

enhanced oil recovery process was as good as the continuous CO_2 flood or the water-alternating-CO_2 flood, but with the added benefit of conformance control. Based on view cell observations (Figure 10.8 inset), an oil bank ahead of the foam front was formed in all experiments.

10.2.2.4 Factors That Control CO_2 Foam Generation

In this sub-section, key findings on the effect of nanoparticle concentration and adsorption of surfactant on nanoparticle surface, on CO_2 foam generation, will be discussed.

- *Effects of Nanoparticle Concentration*: Espinosa et al. (2010) found a minimum of 0.05 wt% PEG-coated silica nanoparticles in deionized water was required for generation of CO_2-in-water foams at 21.1°C, 1350 psia, and a CO_2/water phase ratio (by volume) of 5:1 (83.3% foam quality) and 11:1 (91.7% foam quality) in a beadpack filled with 180 micron diameter glass beads at a shear rate of ca. 1300 s^{-1}. Smaller nanoparticle concentrations (0.025 wt% and 0.01 wt%) in the aqueous phase were unable to stabilize foams at the same conditions. CO_2/water phase ratios up to 19:1 (95% foam quality) were tested and 0.05 wt% nanoparticles were found to successfully stabilize CO_2-in-water foams. When foam was generated, it had 2 to 18 times more resistance to flow than the same fluids without nanoparticles. Foams were generated at temperatures up to 95°C. Foam generation by co-injection of the fluids requires a threshold shear rate. Worthen et al. (2013a) found that bare colloidal silica nanoparticles were unable to stabilize CO_2-in-water foams at 35°C and 2800 psia even at a large concentration of 1 wt% because the hydrophilic-CO_2philic-balance (HCB) of the nanoparticles was very large and the nanoparticles did not adsorb strongly at the interface. Thus, the PEG coating on the silica nanoparticles imparted the proper HCB for adsorption at the CO_2-water interface (Worthen et al. 2013a), but enough nanoparticles had to be present to successfully stabilize a CO_2-in-water foam.

10.2.3 Apparent Viscosity of CO_2 Foams in Porous Media

10.2.3.1 Apparent Foam Viscosity from Beadpack Experiments

Utilizing the findings from studies of nanoparticle-stabilized CO_2-in-water foams in beadpacks, this sub-section describes how the foam's apparent viscosity is affected by various parameters, such as salinity, CO_2/water ratio, and beadpack shear rate.

- *Salinity Effect on C/W Foam Formation and Viscosity*: Figure 10.9(a) from Aroonsri et al. (2013) shows the foam's apparent viscosity in beadpack, calculated from the pressure data using Equation (10.1). The CO_2-in-water (C/W) foams were generated at 50°C and 2800 psia in a 22.5 Darcy beadpack at a shear rate of 1130–2260 s^{-1}. Two different kinds of silica nanoparticles were employed (see Aroonsri et al. 2013). With no salt (TDS=0), the PEG-coated silica nanoparticles produced a weak foam with viscosity of 3 cp, but the 5-nm (EOR-5XS) and 10-nm (EOR-12) silica nanoparticles did not produce a foam. As the salinity increased, the viscosity increased for all nanoparticles tested. Significantly, the EOR-5XS and EOR-12 nanoparticles demonstrated a "no foam" to "foam" transition when the aqueous phase was changed from DI water (TDS=0) to synthetic seawater (TDS=3.5 wt%). Increasing the salinity caused the nanoparticles to behave more hydrophobically which resulted in a lower HCB and an improved contact angle (closer to 90°) at the CO_2/water/nanoparticle interface (Worthen et al. 2013a). This process is akin to "salting out" an ionic surfactant, whereby increasing salinity causes the surfactant to behave more hydrophobically. Similarly, Espinosa et al. (2010) found that increasing salinity from DI water to 2% NaCl at a 0.1 wt% nanoparticle concentration and from 2% to 4% NaCl at 0.5 wt% nanoparticle concentration caused an increase in foam viscosity. However, increasing salinity too much would likely cause the loss of colloidal stability of the nanoparticles, which would cause aggregation and prevent the nanoparticles from flowing through porous media. These results indicate that the salinity of the aqueous phase can be a key parameter for the formation of viscous C/W foams.

- *C:W Ratio Effect on C/W Foam Formation and Viscosity*: Figure 10.9(b) from Aroonsri et al. (2013) shows the foam's apparent viscosities as functions of foam quality (CO_2 fraction). The shear rate in the beadpack and in the capillary tube were 2260 s^{-1} and 1150 s^{-1}, respectively. Results for partially hydrophobic silica nanoparticles (50% SiOH) are plotted for comparison as they were generated using the same apparatus and under the same operating conditions (Worthen et al. 2013a). By varying the C:W ratio, a maximum viscosity was identified at a foam quality of 0.75 for C/W foams stabilized with all nanoparticles tested. The viscosity of the foam is expected to reach a maximum as quality is increased and then drops almost to the gas phase viscosity as the quality approaches unity, i.e., no water (Xue et al. 2016a,b). When the foam quality is too small, the bubbles are very dilute and lamellae between them are not formed, and the viscosities are not expected to be more than a few times that of the solvent. When the quality is too high, the lamellae are very thin and rupture or, in the very extremely high quality regime, the system does not have sufficient liquid to form lamellae. The critical quality (i.e., gas volume fraction) that provides the peak viscosity before its sudden collapse ranges from 0.75 to as high as 0.98 (Espinosa et al. 2010; Xue et al. 2016a,b). While no foam or very low-viscosity foam was formed at qualities above 0.95 in general, Xue et al. (2016b) was able to raise the critical quality to

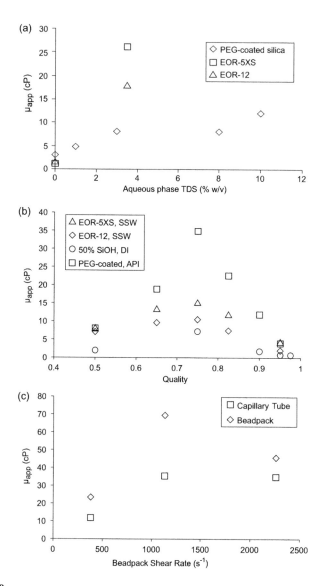

FIGURE 10.9

(a) Apparent viscosity from capillary viscometer of C/W foam (with 1 wt% PEG-coated silica at 50°C and 2800 psia; with a 9:1 volume ratio of CO_2 to water) as a function of salinity. EOR-5XS and EOR-12 were tested in DI water and synthetic seawater (SSW) with a 3:1 volume ratio of CO_2 to water. (b) Apparent viscosity from capillary viscometer of C/W foam (with 1 wt% PEG-coated silica at 50°C and 2800 psia) as a function of foam quality (CO_2 volume fraction). PEG-coated silica was tested in API brine; EOR-5XS and EOR-12 were tested in SSW. Wacker 50% SiOH particles are partially modified with dichlorodimethylsilane and were tested in DI water (see details in Worthen et al. 2013a). (c) Apparent viscosity of C/W foam (1% PEG-coated silica in API brine at 50°C and 2800 psia), measured in a capillary viscometer and in beadpack, as a function of shear rate. From Aroonsri et al. (2013). Copyright 2013, Society of Petroleum Engineers. Reproduced with permission of SPE. Further reproduction prohibited without permission.

0.98 with combined use of silica nanoparticle, polymer, and surfactant. The minimum foam quality investigated in beadpacks thus far was 0.5, where a marked reduction in foam viscosity occurred (see Figure 10.9(b) and Worthen et al. 2013a). No foam is expected at very low qualities, but the exact quality where no measurable increase in viscosity occurs has not yet been identified for nanoparticle-stabilized foams. Further studies of the optimum quality for maximum viscosity would be warranted as a function of porous media, shear rate, temperature, pressure, and nanoparticle concentration. Overall, these results indicate that the CO_2:water ratio of the foam is a key parameter which can dictate whether a viscous CO_2-in-water foam is formed.

- *Effect of Shear Rate*: Figure 10.9(c) from Aroonsri et al. (2013) shows the foam's apparent viscosities as functions of shear rate. CO_2-in-water foams were generated with 1 wt% PEG-coated silica nanoparticles in API brine at 50°C and 2800 psia in a 22.5 Darcy beadpack at a shear rate of 380–2260 s^{-1}. The apparent viscosity of the foams increased from 12 cp (measured in the capillary tube) at a beadpack shear rate of 380 s^{-1} to a steady value of ~35 cp at 1130 s^{-1} and 2260 s^{-1}. (The shear rate in the capillary tube is about half the shear rate in the beadpack.) In the beadpack, an apparent peak viscosity of 69 cp was observed at 1130 s^{-1}, which decreased to 46 cp at 2260 s^{-1}.

These results suggest a more subtle interpretation of the concept of a critical shear rate. Very weak foams may be generated at lower shear rates while more viscous foams may be generated at higher shear rates. At very low shear rates, no foam is expected to form as there is not enough shear force provided to form small bubbles (Walstra and Smulders 1998). Espinosa et al. (2010) found a "critical shear rate" of ~4000 s^{-1} in a similar beadpack when the experiments were done at 95°C and 1350 psia with 0.5 wt% 3-nm silica nanoparticles in DI water. Overall, these results indicate that the shear rate (determined by the porous media properties and the fluid flow rate) is a key parameter which may determine whether a viscous C/W foam is formed.

10.2.3.2 Apparent Foam Viscosity in Reservoir Rocks

Tables 10.1 and 10.2 list recent publications on nanoparticle-stabilized foam coreflood experiments, using silica nanoparticles and using nanoparticles made of other than silica, respectively. As can be seen from the lists, most of the experiments tested CO_2 foams stabilized with silica nanoparticles. The nature of foam's internal phase (CO_2, N_2, or air at what temperature and pressure) and of the nanoparticles employed are provided, as well as some other relevant information. In view of the familiarity of the experimental details, the studies carried out by the authors' coworkers (Aroonsri et al. 2013; Espinosa et al. 2010) will be discussed in detail in this sub-section. Works carried out by other researchers will then be described.

Aroonsri et al. (2013) carried out foam generation experiments both in unfractured and fractured cores of Boise sandstone, Berea sandstone, and

TABLE 10.1

Studies on Silica Nanoparticle-Stabilized Foams for Mobility Control (Unless otherwise specified, all are for CO_2 foams.)

References	Particle size[1] (nm); Surface Coating	Pressure (psia); Temperature (°F)	Salinity (wt%)	Porous Medium	Porosity (%); Permeability (md)
Espinosa et al. 2010	5; PEG[2]	1350–1400; 50, 75, 90, 95	2–4 (NaCl)	G[3]	
Mo et al. 2012	17;	1200–1500; 25	2 (NaCl)	S1[4]	17.4; 33
Worthen et al. 2012	5, 6, 100–200; DCDMS[5], PEG	1200–3000; 35, 50		G	
Aroonsri et al. 2013	10, 20; PEG, proprietary	2000–2800; 50	0	G, S1, S2[6], L[7]	22–29; 1800, 200, 7
Yu et al. 2013	17, 20;	1200; 20	2 (NaCl)	S1	17.4, 20.5; 33, 270
Worthen et al. 2013a	5–30; DCDMS	1200–3000; 50	0.1, 8 (NaCl)	G	34; 22500
Worthen et al. 2013b	28; CAPB[8]	19.4 MPa; 50	0–3 (NaCl)	G	34; 22500
Yu et al. 2013, 2014a,b	10, 12, 70;	1200; 25	2 (NaCl)	G	
Mo et al. 2014	17–20;	1200–2500; 25–60	2 (NaCl)	S1, L, D[10]	20.5, 18.4, 18.5; 270, 106, 295
Singh and Mohanty 2014, 2015	20; PEG with AOS[9]	100; 25	1–10 (NaCl)	S1	23.5, 22.3, 21.3; 357, 313, 383
Kim et al. 2016a,b	5, 12, 25, 80;	2200; 70	API brine[11]	P[12], S2	27.5–31.7, 42; 5500, 2600–4800
Xue et al. 2016a,b	Colloidal; LAPB with AOS	3000; 50	0–3 (KCl)	G	; 2300

(Continued)

TABLE 10.1 (CONTINUED)

Studies on Silica Nanoparticle-Stabilized Foams for Mobility Control (Unless otherwise specified, all are for CO_2 foams.)

References	Particle size[1] (nm); Surface Coating	Pressure (psia); Temperature (°F)	Salinity (wt%)	Porous Medium	Porosity (%); Permeability (md)
Singh and Mohanty 2016	20; PEG with AOS	14.7; 55, 75	1, 2, 4, 8 (NaCl)	P	30;
San et al. 2017	7–20;	1500; 25, 40, 65	1–10 (NaCl), 0.1–1 ($CaCl_2$)	S1	19.1–19.5; 192–203
Emrani et al. 2017; Emrani and Nasr-El-Din 2017	100, 140; AOS	14.7–800; 77–250	1–10 (NaCl)	S3[13]	20; 120–170

Notes: [1]Nominal or average diameter; [2]Poly(ethylene oxide); [3]Glass beadpack; [4]Berea sandstone; [5]Dichloro-dimethyl-silane; [6]Boise sandstone; [7]Indiana limestone; [8]Cocamidopropyl-betaine; [9]α-olefin sulfonate; [10]Dolomite; [11]8 wt% NaCl, 2 wt% $CaCl_2$; [12]Ottawa sandpack; [13]Buff Berea sandstone.

TABLE 10.2

Studies on Foams in Which Nanoparticles Other Than Silica Particles Are Used as a Stabilizer

References	Particle Type	Particle diameter[1] (nm); Surface Coating	Pressure (psia); Temperature (°F)	Salinity (wt%)	Porous Medium	Porosity; Permeability (md)
Singh et al. 2015	TTFA[2]	80–120; Anionic, non-ionic surfactants	14.7; 25	1 (NaCl)	S1[3]	18.9; 300
Eftekhari et al. 2015	Fly ash	100–200; AOS[3]		0.5, 1, 5 (NaCl); 0.9, 1.9, 9.5 (CaCl$_2$)	S4[4]	; 2000
Singh and Mohanty 2016	Alumina-coated silica	20; Triton CG-110, AOS	14.7; 25		S1	22, 20, 22, 18; 606, 442, 585, 125
Kalyanaraman et al. 2017	PECNP[5]	140, 442; Sulfonic N120	1300, 1800; 40	2 (KCl)	L[6]	19, 16, 17; 292, 356, 387
Emrani et al. 2017	Fe$_2$O$_3$	50; AOS	14.7–800; 77–250	1–10 (NaCl)	S3[7]	20; 120–170

Notes: [1]Nominal or average diameter; [2]Thermally treated fly ash; [3]Berea sandstone; [4]Bentheimer sandstone; [5]Polyelectrolyte complex nanoparticle; [6]Indiana limestone; [7]Buff Berea sandstone.

Indiana limestone. Foam generation in a fractured cement core was also studied. Stable foam was successfully generated in every type of core. Results of unfractured Boise sandstone corefloods will be first discussed, followed by results from the fractured cores and other types of rocks.

- **Foam Generation in Sandstone Core:** Figure 10.10 shows the foam experiment results from the coreflood with a Boise sandstone core (Case B4, below). Apparent viscosities and MRF (as defined by Equation 10.2 above) are plotted against the volume of injected fluid in pore volume unit (PV) for 5 different flow rates. At each injection rate, the pressure drop measured across the core sharply increases, but reaches a steady state, demonstrating that the nanoparticle-stabilized CO_2 foam can be transported in sandstone rock without trapping. To better understand the foam transport in rock, the foam apparent viscosity was also calculated from pressure drop measured across the capillary tube installed downstream of the core. The apparent viscosity from the capillary tube is lower than the corresponding value from the core, because the foam bubble movement is not constrained by the narrow tortuous passage in rock. With the knowledge of baseline apparent viscosity obtained from CO_2-DI water co-injection, the MRF was determined; and Figure 10.11 shows MRF from four different corefloods (Cases B1, B2, B3, B4) as a function of effective shear rate, which is calculated as

$$\gamma_{eff} = 6.0 \left[\frac{3n+1}{4n} \right]^{n/(n-1)} \left[\frac{U_w}{\sqrt{k_{aq} S_w \phi}} \right] \tag{10.3}$$

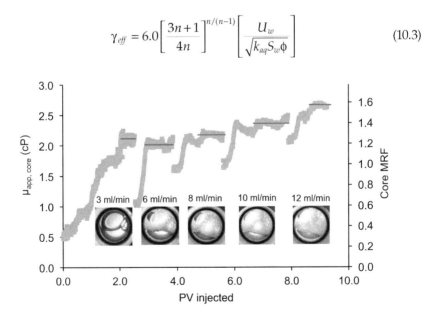

FIGURE 10.10
Apparent viscosity in Boise sandstone from coreflood B4: 1% PEG-coated silica in DI water was co-injected with CO_2 with phase ratio of 1:1 at 2000 psia and 23°C. Baseline core apparent viscosity (without nanoparticle) is 1.69 cp. From Aroonsri et al. (2013). Copyright 2013, Society of Petroleum Engineers. Reproduced with permission of SPE. Further reproduction prohibited without permission.

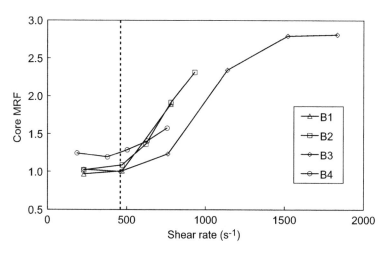

FIGURE 10.11
Core MRF versus shear rate for Boise sandstone corefloods exhibit critical shear rate for foam generation at 460 s^{-1}. All experiments were at 2000 psia and 23°C, with phase volume ratio of 1:1 (1 wt% PEG-coated silica in DI water). From Aroonsri et al. (2013). Copyright 2013, Society of Petroleum Engineers. Reproduced with permission of SPE. Further reproduction prohibited without permission.

where $C = 6.0$ is a constant; U_w is Darcy velocity; k_{aq} is permeability; and ϕ is porosity. For the corefloods B1, B2, B3, and B4, 1 wt% dispersion of 5-nm silica nanoparticles in DI water was co-injected with CO_2, at 1:1 phase volume ratio, at 2000 psia and 23°C. Permeabilities of the cores for B1, B2, B3, and B4 were 1650, 1650, 1720, and 1940 md, respectively. The apparent viscosity for the two lower flow rates are distinctly smaller than those for the three higher flow rates, indicating that at those low rates, no foam or only very weak foam was generated. Once shear rate exceeds a certain value, the core MRF sharply increases with increasing shear rate. This threshold behavior suggests an existence of critical shear rate for foam generation in Boise sandstone matrix. For these specific cases in Boise sandstone, the critical shear rate for foam generation is estimated to be 460 s^{-1}. We note, however, that very weak foam was sometimes observed below the threshold shear rate.

CO_2 flooding is sometimes implemented for EOR in reservoirs which contain natural fractures. CO_2 viscosity being so low, the injected CO_2 will zip through the fracture network bypassing all the oil in the matrix zone. The use of foam therefore becomes much more important than its application to unfractured reservoirs. To study foam generation in a fractured rock, a fracture was created in Boise sandstone core and the co-injection of 1 wt% dispersion of 5-nm silica nanoparticles and CO_2 was carried out as above. The measurement of MRF versus shear rate clearly showed that the concept of threshold shear rate for foam generation applies not only for unfractured rock but also for the fracture in the rock. An interesting aspect of these foam floods in fractured cores is that, with the foam generation in the fracture, more of the injected fluid can be *diverted* (allocated) to the matrix zone despite the fact that foam is generated in the matrix zone also. For more detailed discussion on flow allocation into fracture and matrix zone, refer to Aroonsri et al. (2013).

- *Factors Affecting Threshold Shear Rate for Corefloods*: As reported by a number of researchers, the shear rate thresholds are evident in generating nanoparticle-stabilized CO_2 foam, but their specific values vary. In beadpacks (permeability ~20 D), the threshold shear rate for foam generation with CO_2 at 1350 psia and 0.5 wt% 5-nm silica nanoparticles in DI water was less than 1400 s^{-1} at 23°C, but increased to ~2600 s^{-1} at 70°C and to ~4000 s^{-1} at 90°C (Espinosa et al. 2010). This suggests that the threshold increases as the CO_2 density decreases. Flow in a beadpack can also exhibit a significantly smaller threshold (below 400 s^{-1}) at a higher pressure (2800 psi, 50°C), again due to high CO_2 density.

The threshold shear rates change with rock type even though other experimental conditions are kept the same. The threshold shear rate tends to decrease when the rock permeability becomes lower. The shear rate threshold observed in consolidated rock is 470 s^{-1} in the high permeability Boise sandstone (1700 mD) and about 300 s^{-1} in the lower permeability Berea sandstone (200 mD). Both values are smaller than the thresholds at similar conditions in beadpacks. This suggests that smaller pore throats facilitate the creation of bubbles and thereby reduce the shear rate needed to generate foam. The two fractured Boise sandstone corefloods carried out by Aroonsri et al. (2013) have fracture aperture sizes of 104 and 65 microns and critical shear rates of 3700 s^{-1} and 5800 s^{-1}, respectively. On the other hand, the fractured cement core, with an aperture of 61 micron, exhibits a threshold shear rate smaller than 3000 s^{-1}. These thresholds are all much larger than those observed in beadpacks and sandstone cores, suggesting that smaller constrictions facilitate bubble creation.

10.2.4 Foam Transport Modeling

As with surfactant-stabilized CO_2 foams, the main objective of developing the nanoparticle-stabilized foams is so that they can be employed to remedy the severe adverse mobility problem during conventional CO_2 flooding. It is therefore necessary to model the foam flow in porous media, which is carried out in two steps: (1) Matching of laboratory coreflood results to obtain the foam transport model parameters; and (2) utilizing the matched model parameters, carrying out field-scale simulations. While extensive modeling efforts have been made on the surfactant-stabilized foam transport (see, *e.g.*, Ma et al. 2015; Rossen 2013 for comprehensive reviews), development of models to describe the dynamics of nanoparticle-stabilized foams in porous media has only just begun (Lotfollahi et al. 2017; Prigiobbe et al. 2016; Worthen et al. 2015). In matching the results from the laboratory foam transport experiments carried out with glass beadpacks, Prigiobbe et al. (2016) employed the surfactant-stabilized foam transport model available in literature to see if the surfactant-stabilized foam model could be equally employed to describe the nanoparticle-stabilized foam transport. They found that even though the model parameter values are quite different from those for the surfactant-stabilized foams, the existing foam model can adequately describe the nanoparticle-stabilized foam behavior in porous media. This suggests that the mechanisms governing the surfactant-stabilized foam flow also apply for the flow of nanoparticle-stabilized foam. For the reservoir-scale screening simulations, therefore, the existing foam transport model was also employed, albeit with the model parameter values quite different from the surfactant-stabilized foam parameters.

10.2.4.1 Modeling of Laboratory Foam Flow Experiments

For the 1-D foam flow modeling, Prigiobbe et al. (2016) considered the incompressible, immiscible two-phase flow of CO_2 and water (without oil) in a porous medium. Simplifying assumptions were made that the volume of water, nanoparticles, and a small amount of surfactant that constitute the foam lamellae is negligible and is considered to be a part of the CO_2 gas phase. Adopting the existing foam model (*e.g.*, Kovscek et al. 1995; Kam and Rossen 2003), the foam dynamics can be tracked by solving the fractional flow equation for water saturation, S_w, and the population balance for the density of the lamellae, n_f:

$$\phi \frac{\partial S_w}{\partial t} + u_t \frac{\partial f_w}{\partial x} = 0 \tag{10.4}$$

$$\phi \frac{\partial (S_g n_f)}{\partial t} + u_t \frac{\partial (f_g n_f)}{\partial x} = \phi S_g (r_g - r_c) \tag{10.5}$$

where ϕ is porosity; u_t is the total flux which is the sum of the CO_2 and water fluxes (u_w, u_g, respectively); $S_g (= 1 - S_w)$ is CO_2-phase saturation; r_g and r_c are the rates of generation and collapse of the foam lamellae, respectively; and f_w is the fractional flow of the water phase (with $f_g = 1 - f_w$). Fractional flow is related to the fluid phase mobilities by

$$f_w = \frac{k_{rw} / \mu_w}{k_{rw} / \mu_w + k_{rg} / \mu_g^f} \tag{10.6}$$

where k_{rw} and k_{rg} are the relative permeabilities of the water and gas phases, respectively, which we assume of Corey type; μ_w is the water viscosity (Pa.s); and μ_g^f is the effective gas viscosity (Hirasaki and Lawson 1985).

$$\mu_g^f = \mu_g^o + C_f \frac{n_f}{\left[u_g / (\phi S_g) \right]^{1/3}} \tag{10.7}$$

where C_f is a model parameter. The water and gas fluxes, u_w and u_g, are given by the Darcy's law for the water and gas phases.

- *Rate of generation*: In a bulk foam, a bubble forms when the shear stress exceeds the Laplace pressure of the bubble. When a foam flows through a porous medium, additional mechanisms are present, namely lamella leave-behind, gas-bubble snap-off, and lamella division (Ransohoff and Radke 1988; Rossen 1996). While the Laplace pressure-triggered foam (generally coarse) is largely controlled by the flow velocity, the latter mechanisms are responsible for the formation of a strong foam (large n_f) and are regulated by the pressure gradient (Gauglitz et al. 2002; Tanzil et al. 2002). In the application of interest, where strong foam is required, snap-off and lamella division are the most important generation phenomena. Gauglitz et al. (2002) observed that a minimum pressure gradient (∇p^{min}) exists for the transition from weak or coarse foam to strong foam, which is inversely proportional to the permeability. However, contrary to a N_2 foam, a surfactant-stabilized CO_2 foam presents a negligible ∇p^{min} and can form even at low-pressure gradients (Gauglitz et al. 2002). Moreover, experiments reported by Yu et al. (2014a,b) suggest that in the presence of nanoparticles, foam generation is still governed by pressure gradient. On the basis of these observations, we formulated the rate of generation as (Kam and Rossen 2003)

$$r_g = C_g \nabla p^m \qquad (10.8)$$

where C_g and m are two model parameters. This equation was formulated on the basis of the experiments performed by Rossen and Gauglitz (1990), where it was observed that foam generation results from the mobilization of the lamellae at a minimum pressure gradient (∇p^{min}), which depends on the initial population density in the pore space and scales inversely with the permeability of the medium, i.e., $\propto 1/k$.

- *Rate of Destruction*: In bulk, foam is destabilized due to drainage of liquid within the lamella because of capillary suction (coalescence), which occurs after film thinning to a critical level and hole formation as well as gas diffusion from small to large bubbles (Murray and Ettelaie 2004). For a foam migrating through a porous medium, experimental observations suggest that the dominant mechanism is the former (Jiménez and Radke 1961; Rossen and Lu 1997). Coalescence is a spontaneous process of thin film rupture, which is regulated by capillary pressure and foam quality. During the displacement of a bubble through the pores, P_c increases, reaching its critical value (P_c^*). Therefore, several lamellae rupture within the foam simultaneously, the foam becomes coarser, and P_c decreases. During the displacement of the coarse foam, P_c increases again. It is this process of lamellae rupture and coarse foam displacement which maintains the capillary pressure around its limiting value (Rossen 1996). In the presence of particles stabilizing a foam, the capillary pressure still controls the coalescence mechanism of a bubble and its value depends on particle properties and particle arrangement within the lamella (Kaptay 2003). Large positive value of P_c^* ensures that a thin liquid film between the bubbles of a foam can withstand a higher capillary pressure force. Considering the Leveret J-function which relates P_c to S_w, the rate of coalescence can be formulated as

$$r_c = C_c n_f \left(\frac{S_w}{S_w - S_w^*} \right)^n \tag{10.9}$$

where S_w^* is the minimum water saturation value corresponding to P_c^* for which the lamella ruptures, while C_c and n are two model parameters.

Assuming local equilibrium is reached at steady state, the rate of foam generation equals the rate of foam destruction (i.e., $r_g = r_c$) and therefore combining Equations (10.8) and (10.9), foam texture at steady state is given by

$$n_f = \frac{C_g \nabla p^m}{C_c \left(\dfrac{S_w}{S_w - S_w^*} \right)^n} \tag{10.10}$$

substituting this equation in to Equation (10.7), the effective gas viscosity writes

$$\mu_g^f = \mu_g^0 \frac{C_f C_g}{C_c} \frac{\nabla p^m}{\left(\dfrac{S_w}{S_w - S_w^*} \right)^n} \frac{1}{\left(\dfrac{u_g}{\phi S_g} \right)^{1/3}} \tag{10.11}$$

which, substituted in Equation (10.8), allows to determine the pressure gradient and to formulate the objective function, $F = (\nabla p^{mod} - \nabla p^{mes})^2$, where ∇p^{mod} and ∇p^{mes} are the calculated and measured pressured gradient values at steady state. In the first step of the parameter estimation, model parameters and a combination of them (namely, $C_f C_g / C_c$, n, m, and S_w^*) were then determined by minimizing F and their uncertainty was also calculated using the Jacobian matrix upon optimization. The measurements were divided in three groups. Two of them contained results from tests performed using the column of 1.2 D permeability, distinguished by the use of nanoparticles. A third group included all the experiments performed using the column of 22.5 D permeability. This approached was chosen on the basis of the significant difference in foam behavior due to the change in solution composition and porous medium permeability. Figure 10.12 shows the calculated average trends of μ_g^f as a function of f_g (Equation 10.11) upon this optimization.

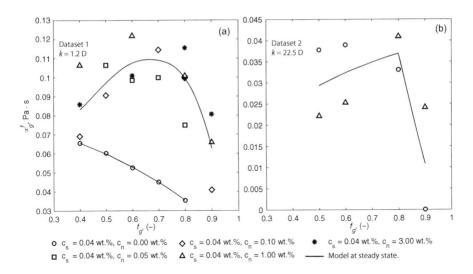

FIGURE 10.12

Effective gas viscosity (μ_g^f) as a function of foam quality (f_g) calculated upon optimization of the pressure gradient at steady state. Experiments performed using the columns of (a) 1.2 D permeability; and (b) 22.5 D permeability. From Prigiobbe et al. (2016).

Similar to the trend of the pressure gradient, μ_g^f increases with f_g, reaches a maximum, and then decreases as the foam quality approaches unity. At low foam quality, CO_2 bubbles are almost spherical, with thick lamellae between them, and the viscosities are not expected more than a few times that of the surfactant solution. As the quality of the foam increases, the concentration of the lamellae increases as well, resulting in a larger μ_g^f due to the deformation of the bubble through the pores of the medium. At large f_g, namely 0.8–0.95, either the lamellae are too thin and they therefore easily rupture, or the water saturation is too low, i.e., there is not enough liquid to support the formation of the lamellae. Consequently, μ_g^f becomes smaller approaching the value of μ_g^0. In the lower permeability column, μ_g^f is the largest when nanoparticles are used. Its maximum is 0.11 Pa.s for $f_g \sim 0.75$. This value is almost twofold of the maximum attained when only a surfactant is added, namely $\sim$0.067 Pa.s at $f_g \sim 0.4$. Moreover, the maximum moves toward larger foam quality values in the presence of nanoparticles because the nanoparticles by adsorbing at the CO_2-water interface enhance the stability of the thin films present in high lamellae density foam. In the high permeability column, no significant effect of nanoparticles was observed and the calculated maximum of μ_g^f is approximately 0.035 Pas for f_g as large as 0.8. In the second step of the parameter estimation, we completed the determination of the model parameters, inverting the pressure difference measured during the transient displacement of three selected experiments, i.e., 6, 25, and 34. The resulting match of the pressure gradient

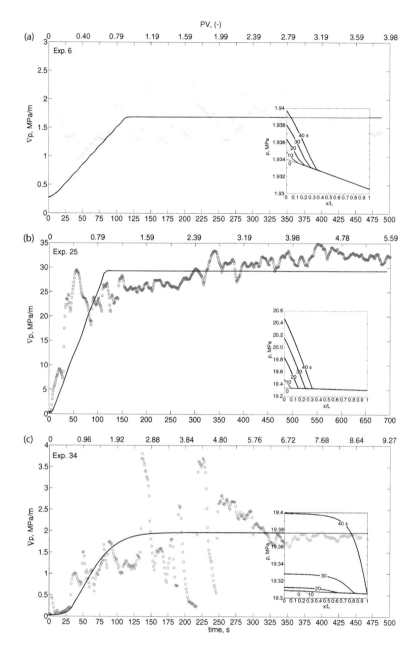

FIGURE 10.13
Pressure gradient (∇p, MPa/m) as a function of time during the transient foam displacement in: (a) exp. 6, $f_g = 0.6$; (b) exp. 25, $f_g = 0.6$; (c) exp. 34, $f_g = 0.8$. The symbols indicate: circles, measurements; solid lines, model for which the number of grid points was 20. The inserts show the pressure as a function of normalized distance (x/L) at different times. From Prigiobbe et al. (2016).

are shown in Figure 10.13. The model describes well the period during the formation of the foam, but it cannot capture the oscillations of pressure after the breakthrough. This might be ascribed to the assumption behind the model that local equilibrium is attained at steady state. However, the measurements show that this hypothesis is not entirely correct and it might be only a simplification of the overall foam behavior. While the model partially fails, it catches the earliest and the latest parts of the experiment. Large oscillations have been measured, but this is due to the strong dynamics of the foam generation and destruction phenomena. The model does not contain mathematical constitutive equations that describe such oscillations and can only predict the average behavior of the foam before and during steady state. Therefore, the match between the model and the experiments can be considered reasonably satisfactory.

10.2.4.2 Field-Scale Foam Injection Modeling

Reservoir simulations are necessary to see the effect of nanoparticle-stabilized CO_2-in-water foam in the field. An approximate simulation work to model the generation and transport of nanoparticle-stabilized CO_2-in-brine foam in multi-layered rock formations, reported in Worthen et al. (2015), is briefly described. In the field-injection feasibility simulation, the model parameters that were obtained from the above coreflood matches were utilized. Such simulation is carried out to assist with the analysis and interpretation of existing field pressure data, design of foam injection schemes, and optimization of the best injection scenario. The key idea is to determine how the nanoparticle-stabilized foam can deflect the flow from the higher permeability streaks into the lower permeability streaks (Figure 10.14(a)); Figure 10.14(c) shows the permeability variations per layer that is used in these guiding simulations; for simplicity in viewing, the results show the permeability increasing linearly with depth (between 50–500 mD). No crossflow among the layers is assumed. Figure 10.14(b) shows model predictions of the front position and pressure distribution at one time for the cases of with and without foam. Here, CO_2 is co-injected with either nanoparticle suspension or nanoparticle-free brine at a gas volume fraction of 0.75 for 10 hours at a flow rate of 500 bbl/day. For the nanoparticle-free brine, the injected fluid moves fastest through the high permeability layers. With the nanoparticle suspension, the front speed is reduced in the high permeability layers and increased in the low permeability layers. This is also associated with an increase in pressure at the injection well (>100 psi where the pressure response increase from 120 psi in the absence of foam to 250 psi when foam is generated as shown in Figure 10.14(b)). Much of this difference is due to the flow dependence of the viscous foam as found in the core studies in Figure 10.12. The high permeability zones (bottom of the reservoir) have a high enough initial flow velocity near the well to create the foam (shear rate > 400 s^{-1}). This foam has a higher apparent viscosity

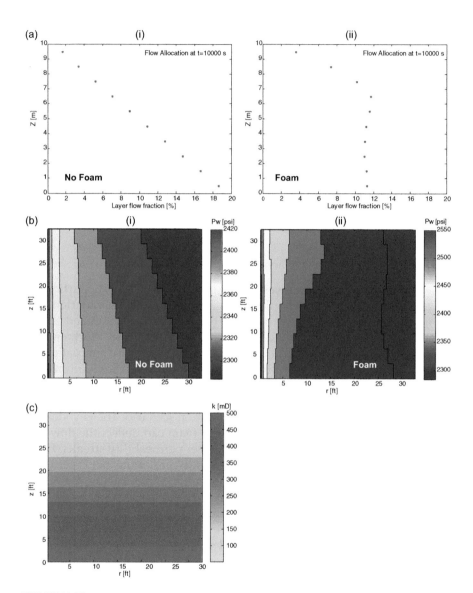

FIGURE 10.14

Simulated (a) flow allocation and (b) pressure profiles after 2.8 h co-injection of CO_2 and 9.6% TDS brine (b(i)) and 0.5 wt% EOR-5XS particles (b(ii)) at 0.75 gas fraction at a constant flow rate of 500 bbl/day. (c) The permeability field implemented in the simulations; intrinsic permeability increases linearly from 50mD to 500mD with depth and is assumed uniform along each layer in the proximity of injection well (i.e., $r < 30$ ft). From Worthen et al. (2015). Copyright 2015, Society of Petroleum Engineers. Reproduced with permission of SPE. Further reproduction prohibited without permission.

than the no foam case, slowing down the flow in these zones. The low permeability zones (top of the reservoir) do not exceed the necessary shear rate, thus maintaining a lower overall viscosity. The combination results in a better sweep and a higher pressure difference.

10.2.5 CT Scan and Micromodel Experiments for Foam Flow Mechanisms

To properly understand the mechanism(s) of foam flow in porous media, a direct observation of foam/lamella's dynamic behavior in rock pores would be most effective, and a number of researchers carried out the experiments (1) to observe macroscopically the foam flow patterns in the reservoir rock cores employing a CT scan; and (2) to observe microscopically the movement of foam bubbles and lamellae in micromodels and microfluidic devices, as described in some detail below.

10.2.5.1 CT Scan Experiments

Because there is microscopic, pore-level heterogeneity even in small rock core samples, when the low-viscosity CO_2 is injected into a core filled with water (and sometimes oil), the CO_2 rapidly fingers through the water phase, seeking high-permeability streaks. When the apparent viscosity of the CO_2 phase drastically increases with the generation of foam, the CO_2-phase flow pattern even in laboratory-scale cores will change considerably. While a direct visual observation of the CO_2 and foam behavior in cores is obviously not possible, Aminzadeh et al. (2012a,b; 2013a,b), DiCarlo et al. (2011), and Zhang et al. (2017) employed a CT scan to quantitatively measure the changes in flow pattern in the absence and presence of nanoparticle-stabilized foam in rock cores. Another objective of the above CT scan observations is to understand how the nanoparticle-stabilized foam is generated in-situ in the rock. This is because, when the foam-based mobility control is desired during a CO_2 EOR process, injecting a surface-produced (pre-made) foam into the reservoir is not practical. The high apparent viscosity of pre-made foam develops a high pressure gradient at the near-wellbore zone, causing injectivity problem. Therefore, as generally practiced with the surfactant-stabilized foam, small alternate banks of CO_2 and the nanoparticle-containing water are injected so that the alternate banks mix in the reservoir and the foam is generated in-situ from a small distance away from the wellbore. The CT scan observations can help understand the in-situ foam generation mechanism(s).

Aminzadeh et al. (2012a,b; 2013b) measured the in-situ saturations of CO_2 and water in real time by placing a rock core horizontally in a Universal Systems HD-350 modified medical scanner which scans the whole core at 1 cm intervals. CT scanning measures the local density of the materials occupying each voxel; since CO_2 is less dense than the brine or nanoparticle dispersion, the measured density can be converted to saturation by a linear interpolation between CT values of the voxel saturated with the brine (or nanoparticle dispersion) and CO_2. The resolution of the scanner was 0.30 mm in the scan plane with a scan thickness of 1 cm. The core was scanned every 10 minutes for early time during the displacement and every 20 minutes after the CO_2 had broken through. In the control experiment, the core was preloaded with a 2 wt% NaBr (doping agent) after it was vacuumed overnight. Liquid CO_2 was injected to the core at 0.02 cm/min at 24°C. The injection pressure was varied from 1350–1400 psi during the flood as the outlet pressure remained constant at 1350 psi using a back pressure regulator. In order to prepare the same core saturated with nanoparticle dispersion, since the same core was used for nanoparticle case as it was used for control case, more than 20 PV of brine was injected after the core was depressurized to the atmospheric pressure gradually. Then, the core was saturated with nanoparticle dispersion (5 wt% nanoparticle and 2 wt% NaBr) by injecting 2 PV of the nanoparticle dispersion. CO_2 was then injected to displace the nanoparticle dispersion at a flux of 0.02 cm/min.

Figures 10.15(a)–(c) show the set of saturation distributions versus space and time when CO_2 is injected to displace a 5 wt% nanoparticle dispersion. Figure 10.15(a) shows the lateral CT scans along the core after 0.26 PV of CO_2 injection. This is the same core at almost the same PV injected as the control experiment (Figure 10.16); thus, the only difference in conditions between Figures 10.15 and 10.16 is the presence of the nanoparticles in the in-situ brine. When compared with Figure 10.16(a), the preferential flow paths are wider and more smeared out. For instance, without nanoparticles (Figure 10.16(a)), three distinct CO_2 preferential flow paths are seen near the top of the core in slices 7–9. In the displacement with the nanoparticles, the preferential paths show signs of merging: the whole top half of the core has a much more uniform CO_2 saturation. This merging is evident in all slices to some degree; slices 11 and 12 also clearly show this merging. Generally, the CO_2 is much more diffuse and less distinct in the nanoparticle case than in the control case. Figure 10.15(b) shows the side view of the core at 0.26 PV of CO_2 injection into the nanoparticle dispersion and can be compared with Figure 10.16(b). Some of the same heterogeneity-dominated flow can be seen in Figure 10.15(b) with a large CO_2 region in the lower half between 15 and 20 cm, and another large CO_2 region in the upper half at 10 cm. But there are important differences with Figure 10.16(b) where the effect of the nanoparticles can be seen: (1) the leading CO_2 tongue ($x > 30$ cm) is not visible in the nanoparticle case; and (2) the entrance region ($x < 10$ cm) has a higher saturation of CO_2 in the nanoparticle case than the control case. These results match the observations of the longitudinal slices of wider fingers, leading to a slower front movement and a higher overall CO_2 saturation behind the front. Figure 10.15(c) shows the lateral CT scans (y-z plane) at the distance 7 cm from the core inlet at different times. The saturation distribution here shows that after 0.11 PV, CO_2 fingers break through from top of the core; this is a slightly later arrival time than for the control case. Once the fingers break through, they are also wider than the control case (later pore volumes).

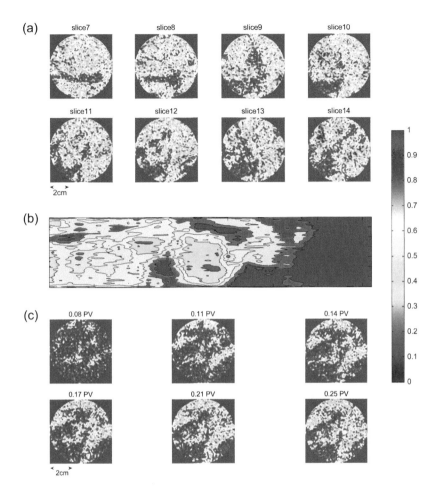

FIGURE 10.15
CT scans for liquid CO_2 injection into a Boise sandstone initially filled with 2% NaBr brine and 5 wt % silica nanoparticles. (a) Water saturation distribution along the core after 0.26 PV of CO_2 injection. Each slice is 1 cm apart longitudinally. (b) Side view of the core at the same time (0.26 PV) of CO_2 injection shows slightly higher saturation behind the front, wider finger behind the front, and less gravity over ride. (c) Saturation distribution of the 7th slice at different times. Saturation change is the result of finger growth with time. From Aminzadeh et al. (2012a). Copyright 2012, Society of Petroleum Engineers. Reproduced with permission of SPE. Further reproduction prohibited without permission.

The results of the experiments with and without nanoparticles and with and without pre-equilibration of brine with CO_2 can be summarized as follows:

1. The core used in the displacements with pre-equilibrated brine was heterogeneous, and this heterogeneity plays a large role in determining the CO_2 flow path. The presence of nanoparticles does not override the influence of heterogeneity, but their effect can still be

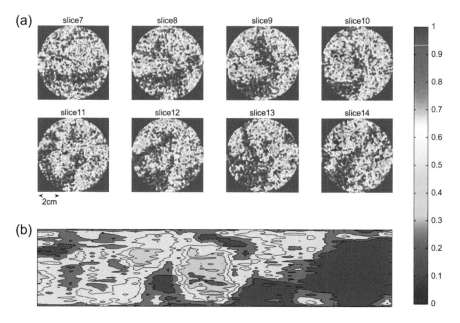

FIGURE 10.16

CT scans for the control case of liquid CO_2 injection into a Boise sandstone initially filled with 2% NaBr brine. (a) Water saturation distribution along the core after 0.25 PV of CO_2 injection. Each slice is 1 cm apart longitudinally. (b) Side view of the core at the same time (0.25 PV) of CO_2 injection, non-uniform saturation distribution and gravity segregation can be seen. From Aminzadeh et al. (2012a). Copyright 2012, Society of Petroleum Engineers. Reproduced with permission of SPE. Further reproduction prohibited without permission.

> ascertained. CO_2 displacing a nanoparticle-laden brine shows a slightly slower displacement front, a larger saturation of CO_2 near the inlet, and wider and more smeared CO_2 flow paths when compared with CO_2 displacing nanoparticle-free brine.
>
> 2. The core used in the displacements with non-equilibrated brine was relatively homogeneous. For non-equilibrated brine, CO_2 displacing a nanoparticle-laden brine shows very small differences (which may be at the noise level) when compared with CO_2 displacing pure brine. Both displacements, however, showed very little or no preferential flow.

Zhang et al. (2017) extended the above studies by Aminzadeh et al. (2012a,b; 2013a,b) and DiCarlo et al. (2011) by focusing on how the CO_2-phase trapping occurs and how the addition of nanoparticles affects the trapping behavior.

10.2.5.2 Micromodel Experiments

In order to directly observe how CO_2 foam bubbles are generated, coalesce and move in rock pore channels, various kinds of micromodels have been employed, even though the micromodels are generally idealized 2-D representation of a porous medium, and the pore dimension is still much bigger than that of real rock pores. Nevertheless, they provide valuable information on the dominant mechanisms of foam generation, stability, and transport. They also help to understand the mechanisms of mobilization of bypassed oil, when the foam phase provides an apparent viscosity much higher than the CO_2 phase. Study of the nanoparticle-stabilized foam behavior in micromodels has been carried out by Guo and Aryana (2016), Guo et al. (2017), Li et al. (2016), Nguyen et al. (2014), Sun et al. (2014), and Yekeen et al. (2017a). All of the above studies except Nguyen et al. (2014), who used nanoparticles only, investigated the synergy between nanoparticles and surfactant in stabilizing foams. Other than Li et al. (2016), whose experiments were run at 1160 psi, all experiments were carried out at low-pressure (<200 psi) conditions.

Nguyen et al. (2014) used a micromodel which was made of borosilicate glass making the model surface water-wet, and contained cylindrical pillars of 200 and 280 µm-diameter (D) generating ~120–200 µm-D pores (70 µm depth). They used 12 nm-D silica nanoparticles coated with 50% and 75% dichlorodimethylsilane coverage. Within micro-confined media, the nanoparticle-stabilized CO_2 foam stayed stable for more than 10 days; and the foam injection displaced 15% more oil than the conventional CO_2 injection. Li et al. (2016) carried out both micromodel experiments and sandpack floods to study the synergy between sodium dodecyl sulfate (SDS) and partially hydrophobic silica nanoparticles (NP) on CO_2 foam stability. They found a certain optimum ratio of SDS/NP for best foam stability (0.17 for their process condition), and also found that the destabilizing effect of crude oil for surfactant-stabilized foam is much less with the addition of nanoparticles.

Guo and Aryana (2016) employed etched glass micromodels whose flow channels are images from a sandstone cross-section, and studied the synergy between silica nanoparticle or nano-clay and a mixture of alpha-olefin sulfonate (AOS) and lauramidopropyl betaine (LAPB). Sun et al. (2014) also employed the etched glass micromodels, whose flow channels (with 40 µm depth and 50–80 µm width) are images from Shengli oilfield rock samples, to investigate the effects of silica nanoparticles on the SDS-stabilized N_2 foam stability and enhanced oil recovery potential. In their comprehensive study on the effects of silica or aluminum-oxide (Al_2O_3) nanoparticle addition on the SDS-stabilized CO_2 foam stability, Yekeen et al. (2017a,b) also used etched glass micromodels (with channel depth of 100 µm and average widths of 200, 600, and 800 µm). Foam stability decreased in presence of salts until the transition salt concentration. Beyond the transition salt concentration, foam stability generally increased with the increasing salt concentrations. The presence of nanoparticles increased the foam half-life and decreased the transition salt concentrations. The dominant mechanism of foams flow process was identified as bubble-to-multiple bubble lamellae division. The pore scale mechanisms of residual oil displacement by foam were found to be both direct displacement and emulsification of oil, which were independent of the pore geometry of the etched glass micromodels. There was lamellae detaching and collapsing during the SDS-foam flow in presence of oil. Both SiO_2-SDS and Al_2O_3-SDS foams propagated successfully in porous media in the presence of oil with almost 100% microscopic displacement efficiency.

Before leaving this section on the use of silica nanoparticles to stabilize CO_2 foam, we note that extensive research has been carried out to develop a very high-quality CO_2 foam to significantly reduce the amount of water used for hydraulic fracturing of oil wells (Qajar 2016; Xue et al. 2016a,b). The details are described in Section 7.3.2 (CO_2 foam "waterless" fracturing fluids).

10.3 CO_2 Foams Stabilized by Fly Ash and Other Nanoparticles

The fact that Pickering emulsions and foams can be generated with many different kinds of solid materials, as long as the surface wettability requirement is properly met (Binks 2002), prompted researchers to utilize materials that are either very low cost or waste products to make emulsions and foams (*e.g.*, Gonzenbach et al. 2006a,b; Luo et al. 2016). Table 10.2 lists recent literature on foams that are stabilized by nanoparticles other than silica nanoparticles. One innovative way of generating nanoparticle-stabilized CO_2 foam for EOR application is to use nanomilled fly ash as the foam stabilizer (Lee et al. 2015; Singh et al. 2015). Because fly ash is a little-to-negative value waste product from coal-burning power plants, the technique has the dual benefit of sequestering not only CO_2 but also fly ash. A similar application has recently been reported by Guo et al. (2017) and Eftekhari et al. (2015), the latter of which generated the nano-size fly ash particles from high-power sonication, instead of the nanomilling as we describe below. In this section, the process of preparing the nano-size fly ash particles by nanomilling is first described, followed by the optimized formulation of stable foams with the fly ash nanoparticles. Experiments on the apparent viscosity of fly ash-stabilized foams flowing in porous media are then described.

10.3.1 Nanomilling of Fly Ash

As described in detail in Chapter 2, virtually all nanoparticles are made using the "bottoms up" methods and the "top down" methods are rarely employed. For this reason, the most common top-down method of nanomilling is described here in some detail.

10.3.1.1 Grinding Equipment

Wet grinding is one of the top-down approaches used for production of mineral nanoparticles. Among various ultrafine grinding devices, stirred ball mills are known to be most effective for nanoparticle production owing to simplicity of operation, high size reduction rate, and relatively low energy consumption. The grinding setup and its operation employed by Lee et al. (2015) are described here.

Grinding is usually conducted in two stages using two different types of stirred ball mills. The first-stage grinding is performed in an attrition to reduce the size of the fly ash particles so that 90% passes through 10-µm mesh. The grinding chamber is filled (leaving ~0.3 void volume fraction) with zirconia balls of 3-mm diameter with density of 5.9 g/cm^3. Wet grinding is performed at 30 wt%-solid slurry to fill 80% of the mill chamber. A stirrer axis fitted with four bars is rotated at 800 rpm (stirrer tip speed: 3.55 m/s). The second-stage grinding is conducted in a beads mill, which is loaded with zirconia balls, filled to ~65% of the mill volume. Different sizes of zirconia beads (100, 300, and 500 µm in diameter) can be tested as grinding media for optimal performance. The milling chamber contained a rotor rotating at 3750 rpm (stirrer tip speed: 8.55 m/s). The mill was operated in the re-circulation mode: the product from the first-stage grinding was passed into the feed vessel and repeatedly pumped back to the milling chamber with an integrated, continuously adjustable pumping and stirring system. The mean residence time for each pass-through was 78 s. The solids concentration of the slurry was 10 wt%. The samples were taken from the outlet of the mill at regular intervals and analyzed for size distribution using laser diffraction and dynamic light scattering.

10.3.1.2 Effect of Grinding Energy on Produced Particle Size

Figure 10.17 from Lee et al. (2015) shows the variation of the median size of the fly ash particles as a function of the energy input in the two-stage grinding with a 3 mm/500 µm grinding media combination. The samples were ground and the mean sizes of the particles were measured after 3 min of sonication with the addition of a 0.1 wt% non-ionic surfactant to disperse the particles. In the first-stage grinding, the +70-mesh fly ash sample showed the fastest size reduction. This may be because the relatively weaker carbonaceous materials were concentrated in the +70-mesh fly ash sample. The slope of the curve suggests that, at first, the grinding rate significantly increased in the second-stage grinding, but it decreased as the grinding proceeded. Obviously, it became more difficult to grind particles as the particle size approached the nano-size range. In terms of the size reduction rate, there was not much difference between the three fly ash samples, as they followed the same trend (Figure 10.17). The final median size of the particles after

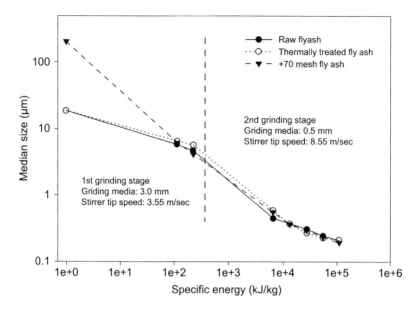

FIGURE 10.17
Variation of median size of fly ash as a function of the energy input in the two-stage grinding test. From Lee et al. (2015).

recirculating the samples 64 times was approximately 200 nm at a specific energy of ~105 kJ kg^{-1}.

10.3.1.3 Effect of Milling Bead Size in the Second-Stage Grinding

The grinding media size is one of the critical factors for determining the grinding efficiency. In general, the rate of breakage of smaller particles is higher for smaller media sizes. However, as the media size decreases, the collision force decreases, which reduces the fineness of the ground product. Therefore, there may be an optimum media size for the production of sub-micron-sized particles. Usually, fly ash contains still unburned carbon residues which provides some hydrophobic character to the fly ash. An extended thermal treatment burns off the carbon residue, and such thermally treated flay ash (TTFA) samples were also tested by Lee et al. (2015). Figure 10.18 shows the effect of the media size on grinding the TTFA samples. Contrary to the general perception that smaller media are more effective for grinding fine particles, grinding with 300 μm and 500 μm media showed higher rates of size reduction than with 100 μm media. At the later stage of grinding, the rate of size reduction decreased, approaching a size limit of approximately 200 nm. In the submicron particle size range, the viscosity of the product suspension increased owing to the strong particle–particle interactions. It seems that the 100 μm grinding media were too light to overcome the hydro-dynamic force exerted by the viscous suspension and to give the particles a

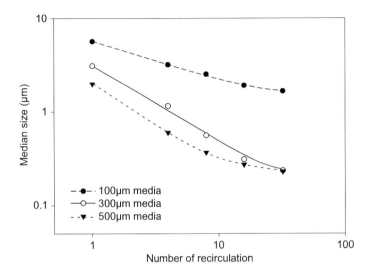

FIGURE 10.18
Median size of fly ash as a function of the recirculation number for three different sizes of grinding media. From Lee et al. (2015).

sufficient impact force for breakage. In the following sub-sections 10.3.2 to 10.3.4, the optimal formulations of fly ash-stabilized CO_2 foams and their characterization for flow in porous media, as studied by Singh et al. (2015), will be described.

10.3.2 Dispersion Stability Analysis

As described above in Section 10.2.2.2, when the nanoparticle's surface wettability is such that it cannot work as an effective foam stabilizer, a low concentration of an appropriate surfactant can be added to bring about the right HCB, seeking synergy between nanoparticle and surfactant (Binks 2017; Worthen et al. 2013b). In attempting to use nanomilled fly ash as a CO_2 foam stabilizer, this approach of adding a low concentration of surfactant can be employed to achieve the right hydrophilic-CO_2philic balance (HCB), because it is preferable to utilize the fly ash without any additional processing. In order to find a surfactant that will bring the best synergy with the TTFA nanoparticles, Singh et al. (2015) screened two cationic (designated as: C1, C2), three non-ionic (N1, N2, N3), and two anionic (A1, P1) surfactants. Just as the case with use of silica nanoparticles in Section 10.2.1.1, the dispersion stability of the TTFA nanoparticle with different surfactant addition was first investigated by mixing 0.5 wt% of TTFA with 0.2 wt% of surfactant and measuring the average size of the nanoparticle aggregates formed versus time. Based on this dispersion stability analysis, the non-ionic surfactants (N1, N2, N3) and anionic surfactants (A1, P1) were selected for further analysis, as given below.

10.3.3 Foam Stability with Fly Ash

10.3.3.1 Bulk Foamability Tests

Based on the above dispersion stability analysis, the screened surfactants were mixed in glass test tubes with TTFA nanoparticles (*e.g.*, 0.2 wt% and 0.5 wt%, respectively) and air foam was generated in ambient conditions by shaking the test tubes vigorously. The decay of foam height with time was monitored and half-lives were determined. For the TTFA-only dispersion, no foaming occurred, as TTFA nanoparticles were too hydrophilic. The anionic surfactant A1 showed the highest foamability with a half-life greater than 3000 minutes, while non-ionic surfactants (N1, N2, and N3) showed not as good, but decent foamability.

10.3.3.2 Foam Texture Analysis

Foam texture analysis is performed as a screening method to evaluate potential surfactants that could be used in combination with TTFA nanoparticles in stabilizing foam in the porous media. The experimental setup for foam generation is similar to that described in Section 10.2.2.1 and shown in Figure 10.5. These experiments were performed at room temperature with a backpressure of 1300 psi. The CO_2 was in a liquid state under these conditions. Mixtures of 0.5 wt% of TTFA nanoparticle and 0.2 wt% surfactant solution were co-injected with carbon dioxide at a quality of 90% at a total flow rate of 2 cc/min. Figure 10.19 shows the foam texture as seen from the view cell installed downstream of the sandpack for different surfactant-TTFA systems. Stable foam was observed in the case of surfactants N2, A1 and N3. No sign of strong foam was seen in the case of N1. The P1-TTFA system was able to stabilize large carbon dioxide-in-water bubbles, but not a fine-textured foam. Based on this screening experiment, the nonionic surfactant N2 and anionic surfactant A1 were chosen for further study on foam in porous media.

| N1 | N2 | A1 | P1 | N3 |

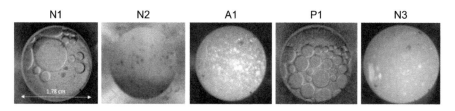

FIGURE 10.19
Foam texture as seen from view cell for different TTFA-surfactant system. From Singh et al. (2015). Copyright 2015, Society of Petroleum Engineers. Reproduced with permission of SPE. Further reproduction prohibited without permission.

10.3.4 Foam Flow Experiments

Based on the above foam stability study for formulation optimization, Singh et al. (2015) conducted foam flow experiments to investigate the synergy between the surfactant (non-ionic or anionic) and TTFA nanoparticles in stabilizing foam. The coreflood setup is similar to that shown in Figure 10.5 for silica nanoparticle-stabilized CO_2 foam coreflood experiments. A cylindrical core of Berea sandstone (1.5-inch diameter and 1-ft long) was mounted vertically in a Hassler-type core holder with a confining pressure of 1500 psi. Carbon dioxide gas and brine/surfactant solution are co-injected first through an inline filter (2 micron) and then a sandpack (0.6-inch diameter, 6-inch long; 40–70 Mesh Ottawa sand) to ensure proper mixing and foam generation. The pre-generated foam was then injected into the core. The testing of different combinations of TTFA nanoparticles and surfactant selected above is illustrated here with the case of the anionic surfactant (A1). The base case was first performed in which brine (1 wt% NaCl) and CO_2 were co-injected at 90% quality into the core. After cleaning the core, 0.4 wt% of A1 surfactant in brine was co-injected with CO_2 at similar conditions as the base case. The steady state pressure drop after about 15 PV was about 5.9 psi (Figure 10.20(a)). The core was again cleaned, and a mixture of 0.4 wt% of A1 and 0.4 wt% of TTFA in brine was co-injected with CO_2 similarly as above. The steady state pressure drop in this case was about 21.9 psi (Figure 10.20(b)) which is 3.5 times greater than the previous case. The core was again cleaned and brine permeability was measured. No permeability reduction was observed suggesting no nanoparticle plugging. The mobility reduction factor (see Equation 10.2) can be calculated using the above data which increased from 7 (surfactant case) to 25.8 (surfactant-nanoparticle mixture case). This result shows that there is

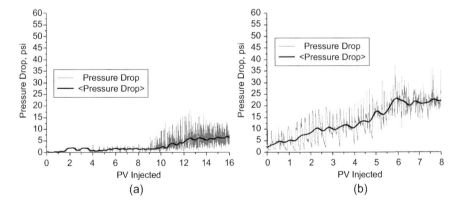

FIGURE 10.20
Pressure drop profile during foam flow experiments with A1 (a) and A1-TTFA mixture (b) as foaming agents. From Singh et al. (2015). Copyright 2015, Society of Petroleum Engineers. Reproduced with permission of SPE. Further reproduction prohibited without permission.

a significant synergy between the TTFA nanoparticles and the anionic surfactant (A1) in generating nanoparticle-stabilized CO_2 foam in porous media. Thus, the fly ash nanoparticles have potential to serve the dual-purpose of acting as gas mobility control agents in enhanced oil recovery processes as well as sequestration of both CO_2 and fly ash in subsurface formations.

The sequence of experiments described above in Sections 10.3.2 to 10.3.4 showed that fly ash nanoparticles even in low concentration (0.4 wt%) can significantly improve the stability and apparent viscosity (MRF) of CO_2 foams, when employed together with a low-concentration of surfactant, resulting in a significant synergistic benefit.

10.4 Emulsions for Improved Reservoir Sweep

In Section 10.2, the generation and characterization of CO_2 foams both in bulk state and in porous media, and their rheological and transport behaviors in reservoir rocks, are described in detail. In this section, the use of nanoparticle-stabilized emulsions for EOR and conformance control will be described. While not widely researched and developed for EOR, the use of surfactant-stabilized emulsion has been investigated for many years (see, *e.g.*, Sjoblom 2006) to provide an improved mobility control during heavy oil recovery processes. An obvious question to ask is: Why an emulsion instead of CO_2 foam and why not surfactant-stabilized emulsion instead of nanoparticle-stabilized emulsion? The answers to the questions, and thereby the potential advantages of using nanoparticle-stabilized emulsions, will be given here.

There are a number of reasons why use of light hydrocarbons, instead of CO_2, as the internal, dispersed phase is sometimes more advantageous. First, as described in Chapter 4, for a nanoparticle to serve as a good stabilizer for foams and emulsions, its surface coating chemical should provide a good hydrophilic-lipophilic-balance (HLB), or hydrophilic-CO_2 philic-balance (HCB) in the case of CO_2. However, CO_2 has a very weak van der Waals force and balancing the weak solvation of the coating chemical by CO_2 with the strong solvation by water is quite difficult. For this reason, as is well-known by the dispersion formulation chemists (Sjoblom 2006), stabilizing a hydrocarbon-in-water emulsion with nanoparticle is much easier with a wide range of nanoparticles. Other reasons are: (1) A steady source of CO_2 may not be as readily available as light hydrocarbons in many reservoir locations; (2) use of CO_2 requires protection from corrosion for oilfield facilities that contact CO_2; and (3) with the current, active development of shale gas and oil fields in the US, natural gas liquids (NGL) are now abundantly available at low cost. As described more in detail below, such use of NGL as the emulsion's internal phase brings forth an interesting advantage for heavy oil recovery: In addition to enhancing the apparent viscosity of the injection water, when the NGL droplets of emulsion contact the resident heavy oil and the NGL plops into the oil phase, it helps to lower oil viscosity.

Use of nanoparticles instead of surfactants as emulsion stabilizers has a number of benefits. First, as with the CO_2 EOR case, where the use of surfactant as a stabilizer is difficult due to its degradation at high-temperature condition, use of nanoparticle is a good alternative. Second, as discussed by Binks (2017), a variety of complex emulsions (*e.g.*, double emulsions, high-internal volume emulsion) that are stabilized by stimuli-responsive coating chemicals could be developed according to the specific oilfield needs, such as the use of stimuli-responsive emulsions for the conformance control at the injection or production wells.

10.4.1 Pickering Emulsion Re-Visited

As described in Chapter 4, nanoparticle-stabilized emulsions have triggered great interest in recent years (*e.g.*, Binks 2017). Active research efforts are on-going in many areas, especially in chemical engineering and materials science. These research efforts led to the detailed characterization of the properties of emulsions solely stabilized by nanoparticles in many aspects, *e.g.*, emulsion type, droplet size, stability, bulk viscosity, and interfacial properties. The influence of experimental conditions such as nanoparticle wettability, particle concentration, their initial location (i.e., dispersed in water or dispersed in oil), and salt concentration and pH of the aqueous phase, as well as the oil type, on the emulsion system has also been elucidated, and detailed reviews are available (*e.g.*, Binks and Horozov 2006; Ngai and Bon 2015). The most commonly used nanoparticles are spherical fumed silica particles with a diameter in the range of several to tens of nanometers. Their wettability is controlled by the coating extent of silanol groups on their surface (Binks 2002). The nanoparticles can be made hydrophilic with high percentage (over 90%) of silanol groups on the surface, and consequently they form stable oil-in-water (o/w) emulsions. On the other hand, when the silica particles are only coated about 10% on their surface by silanol groups, they are hydrophobic and yield water-in-oil (w/o) emulsions. Furthermore, when the nanoparticles are only partially coated with silanol groups (*e.g.*, 70%), they become particles with "intermediate wettability or hydrophobicity"; the stable emulsion type they generate depends on the oil polarity, i.e., formation of o/w emulsions are favored with nonpolar oils whereas w/o emulsions are preferred with polar oils.

The effect of the initial phase in which intermediate-wettability nanoparticles are dispersed on the emulsion they subsequently produce was investigated by Binks and Lumsdon (2000). They showed that with equal volumes of toluene and water, w/o emulsions were preferred when particles were originally dispersed in toluene, while o/w emulsions were produced when particles were initially dispersed in water. Experiments have been carried out to investigate the dependence of emulsion stability and rheological properties on electrolyte concentrations and pH (Horozov et al. 2007). They showed that fumed silica particle-stabilized emulsions were not very sensitive to NaCl concentration at pH of 7. Furthermore, it was also found that emulsion stability increased and the average size of emulsion droplets deceased with decreasing particle size (Hunter et al. 2008; Lopetinsky et al. 2006).

To understand and quantify the conditions for equilibrium and stability of emulsion systems stabilized with solid particles, many theoretical models have been developed (Kralchevsky et al. 2005; Reincke et al. 2006). Every nanoparticle in the emulsion may experience electrostatic repulsions, van der Waals attractions, capillary attractive forces, and so on (Binks and Horozov 2006; Bresme and Oettel 2007). The nanoparticle monolayer on a droplet surface can reach equilibrium when and only when the interactions between the nanoparticles are balanced. From the free energy considerations, the equilibrium density of the spheres at the interface was defined. Some qualitative comparisons of the model prediction with the measured data are discussed in some reviews (Reincke et al. 2006). Kralchevsky et al. (2005) also investigated the nanoparticle-stabilized emulsion stability by defining the preferred interfacial bending moment and deducing the emulsion droplet size as a function of the interfacial density of the adsorbed particles. This investigation of emulsion stability accounted for the interactions between the emulsion droplets. The well-established DLVO (Derjaguin-Landau-Verwey-Overbeek) theory, together with the viscous resistance to the thinning of the liquid film between the droplets, explained the emulsion stability reasonably well. Stability mechanisms other than the above conventional approach have also been proposed (Lopetinsky et al. 2006).

As mentioned above, nanoparticle-stabilized emulsions can avoid the deficiencies caused by the large size of Pickering emulsion droplets with colloidal particles, such as flowing through the reservoir rock pores. In addition, these nanoparticles can be made with required characteristics so that the formation and destabilization of emulsions as well as their rheological properties can be precisely controlled. These attributes of nanoparticle-stabilized emulsions offer exciting opportunities for the upstream oil industry. Examples include water-in-oil emulsions stabilized with catalytic nanoparticles for heavy oil upgrading (Thompson et al. 2008), and external control of the rheology and mobility of the drilling fluids and the flooding fluids during enhanced oil recovery processes (Melle et al. 2005).

10.4.2 Emulsion Generation and Characterization

While extensive research has been carried out on the thermodynamic and dynamic characterization of nanoparticle-stabilized emulsions (*e.g.*, Binks 2017; Ngai and Bon 2015), the oilfield applications still require additional investigation due to the unique nature of oil reservoirs, such as high salinity and high hardness. In this sub-section, such characterization work carried out by Zhang et al. (2009, 2010, 2011) will be mainly described.

10.4.2.1 *Emulsion Phase Behavior*

- *Emulsion Type Determination:* As described in Chapter 4, nanoparticles with different wettabilities may stabilize different types of emulsions (Binks 2002). Nanoparticles having a contact angle at the oil–water interface smaller than 90° are hydrophilic and stabilize oil-in-water (o/w) emulsions, while the nanoparticles having a contact angle greater than 90° are hydrophobic and yield water-in-oil (w/o) emulsions. The emulsion type produced with intermediate-wettability nanoparticles, which have a contact angle around 90°, is determined by other factors, *e.g.*, salinity, oil type, and the initial phase volume ratio. After an emulsion is generated, *e.g.*, by sonication of a mixture of oil, water, and nanoparticles, and reached equilibrium, aliquots of the generated emulsion can be added to a bulk water phase and a bulk oil phase, separately, to verify the emulsion type. If the emulsion drop disperses in oil (or water) but keeps intact as a drop in water (or oil), it is a w/o emulsion (or o/w emulsion).

- *Emulsion Phase Volume Measurements:* The emulsion stability is determined by monitoring the post-sonication emulsion state with time. Due to the density difference between oil and water, a separation of the excess phase always occurred after sonication. Oil-in-water emulsion droplets always cream to the top, while water-in-oil drops always sediment to the bottom. In an o/w emulsion system, an interface between the bulk emulsion phase and excess water phase is formed; while in a w/o emulsion system, an interface between the bulk emulsion phase and the excess oil is generated. The creaming (or sediment) movement stopped once the system reached "equilibrium." It is noted that the phase equilibrium thus attained is, strictly speaking, a pseudo-equilibrium state, because emulsion is by definition not in a thermodynamic equilibrium state. However, since the nanoparticles provide a much more robust stability to emulsions than surfactants, the phase pseudo-equilibrium serves as a useful way to characterize emulsion states.

- *Oil-in-Water Emulsions with Hydrophilic Nanoparticles:* Zhang et al. (2011) created a set of decane-in-water emulsions with different nanoparticle concentrations, salinities, and initial water/decane volume ratios (IVR), thereby mapping their phase behavior. The emulsions in Group 1 were prepared with 2 mL of water and 2 mL of decane. The nanoparticle concentration in water varied from 0.05 wt% to 5 wt%, and the salinity varied from 0 to 10 wt%, independently. When the nanoparticle concentration in water was 0.5 wt% or greater, little decane was in excess after sonication. The emulsified oil fraction at equilibrium was calculated through volume measurement. Well over half of the decane was emulsified for nearly all the samples. The fraction increased gradually with increasing nanoparticle concentration and reached a plateau close to 1 when the nanoparticle concentration was over 0.5 wt%. The fraction also changed with salinity. With a fixed nanoparticle concentration of 0.05 wt%, the fraction decreased as the salinity increased.

The dispersed phase (i.e., decane) volume fraction φ within the bulk emulsion phase was also calculated. While $\varphi = 0.64$ is the largest volume fraction possible for randomly packed spherical droplets, because the droplets could be deformed under compression, φ could even approach close to 1.0. All φ values were over 0.5, which means that the decane droplets are densely packed, some with deformation. The high values indicate droplet deformation as well as wide size-distribution of droplets. When the nanoparticle concentration increased to 0.5wt% or larger, φ dropped to around 0.7; the emulsion was still close-packed, but the droplet deformation was not significant. It appears that, with the lower nanoparticle concentration, the adsorbed nanoparticle layer at the oil/water interface is not compact and the droplet surface can be easily deformed. With the higher nanoparticle concentration, on the other hand, a compact nanoparticle layer is formed at the droplet surface whose deformation then becomes much more difficult. Also note that, for a fixed nanoparticle concentration (between 0.5 and 5 wt%), higher salinity yielded lower oil volume fraction in the emulsion phase. This is because more salt ions attached onto the nanoparticle surface with higher salinity and they increased the electrostatic repulsion as well as the separation between the droplets in the condensed emulsion phase. The most probable cause for droplet deformation in a compressed emulsion system is the creaming of oil droplets under gravity.

- *Water-in-Oil Emulsions with Hydrophobic Nanoparticles*: Zhang et al. (2011) created a set of water-in-decane emulsions with different particle concentrations, salinities, and initial decane volume fractions in the mixture, thereby mapping their phase behavior. The emulsions in Group 4 were prepared with equal volumes of decane (with hydrophobic nanoparticle loading) and brine (or de-ionized water). The nanoparticle concentration in the decane varied from 0.05 to 5 by wt%, and the salinity of water changed as 0, 0.1, 1, and 10 wt%, independently. The phase behavior suggests a threshold nanoparticle concentration exists. When the nanoparticle concentration was 0.5 wt% or higher, no excess water existed, no matter what the salinity was. Otherwise, less than 10% water was emulsified. With a nanoparticle concentration of 0.05 wt% or 0.1 wt%, the emulsified water fraction increased initially when the salinity increased from 0 to 1%, then dropped when the salinity additionally increased.

The dispersed phase (i.e., water) volume fraction within the water-in-oil (w/o) emulsion phase increased when the nanoparticle concentration increased from 0.05 to 0.5 wt%, then it decreased when the particle concentration kept increasing. When the nanoparticle loading is below 0.5 wt%, most of the water fractions are less than 0.5, and those emulsions were sparsely dispersed with water drops. This phenomenon is different from the result for the oil-in-water (o/w) emulsion with low nanoparticle concentration (see above), where the dispersed phase volume fraction was very high. The hydrophilic and hydrophobic nanoparticles thus behaved quite differently in stabilizing emulsions at small nanoparticle concentrations. Within w/o emulsions prepared with at least 0.5 wt% hydrophobic nanoparticles, the water volume fraction in the emulsion phase is large, suggesting dense packing of droplets. Moreover, when the nanoparticle concentration was fixed, the water volume fraction did not change much with varying salinity. It can be explained by the fact that the salt ions stay only inside the dispersed phase in the emulsion. Thus, they cannot significantly affect the electrostatic repulsion between the droplets. Therefore, the salinity has a different effect in the two types of emulsions.

10.4.2.2 Internal Structure of Emulsions

Once the emulsion's pseudo-equilibrium phase states are quantified, the next step is to determine the emulsion's internal structure. That is, what is the emulsion droplet size and what is the nanoparticle adsorption density at the droplet surface?

- *Emulsion Droplet Size Measurements*: The emulsion droplet size measurement equipment consists of a microscope apparatus which is connected to a PC running an imaging software, which processes emulsion microscope images for droplet size distribution. The microscopic images of decane-in-water emulsions with hydrophilic nanoparticles (shown in Figure 10.21) were taken five days after sonification. The emulsion droplets are spherical with diameters between several microns to hundreds of microns. Overall, smaller droplets were generated with higher nanoparticle concentration; especially when the nanoparticle concentration increased from 0.1 wt% to 0.5 wt%, the droplet sizes shrink dramatically. When the nanoparticle concentration was less than 0.5 wt% in the aqueous phase, the emulsion droplets exhibited a dual-mode diameter distribution, with one mode around 10 microns and the other around 100 microns in the presence of salt. In addition, the increased salinity widened the droplet size range by making the larger drops even larger. However, mono-dispersed droplets within a diameter range of 2~10 μm were observed within emulsions with a nanoparticle concentration over 0.5 wt%. The arithmetic mean of all the measured droplet radii in every microscope image was calculated and tabulated in Table 10.3. The table mostly verified the observations in the microscope images (Figure 10.21) regarding the nanoparticle concentration and the salinity effects on the emulsion droplet size.

The microscopic images of water-in-decane emulsions with hydrophobic nanoparticles were made, with different nanoparticle concentrations and different salinities (shown in Figure 10.22). Overall, those water-in-decane emulsion droplets were smaller than the oil-in-water ones. Within w/o emulsions, the water was dispersed as the inner phase, while decane was

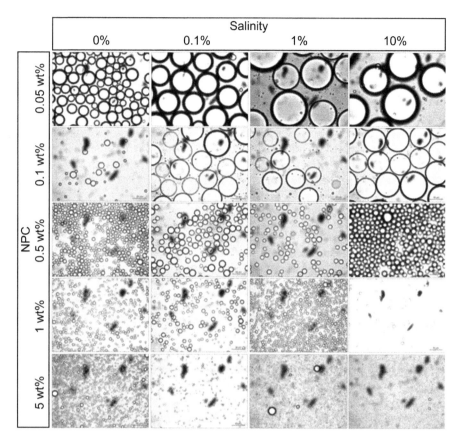

FIGURE 10.21
Droplet images for decane-in-water emulsions (stabilized with hydrophilic nanoparticles) made with varied nanoparticle concentration (each row) and salinity (each column). Initial volume ratio: water/decane = 1. From Zhang et al. (2010). Copyright 2010, Society of Petroleum Engineers. Reproduced with permission of SPE. Further reproduction prohibited without permission.

the continuous phase. When the nanoparticle concentration increased from 0.1 wt% to 5 wt%, the emulsion droplet diameters decreased gradually from tens of microns to a couple of microns. In addition, the emulsion droplet size distribution attained a narrower range with more nanoparticles when the salinity was fixed or with higher salinity when the nanoparticle concentration was fixed. Some differences in size distribution compared with o/w emulsions were observed in w/o emulsions with a low nanoparticle concentration (~0.1 wt %). First, the bimodal size distribution was not evident, as shown in Figure 10.22, compared with Figure 10.21. Second, the droplet size distribution scale changed oppositely when the salinity increased.

- *Nanoparticle Adsorption Density at Droplet Surface*: In order to determine how many nanoparticles are adsorbed at the oil/water

TABLE 10.3

Dispersed Phase Volume Fraction in Emulsion and Measured Emulsion Droplet Radii

		Salinity (wt%)							
		Dispersed Phase Vol. Fraction				Emulsion Droplet Radius (μm)			
		0.	0.1	1.	10.	0.	0.1	1.	10.
Hydrophilic NP conc.	0.05	0.80	0.75	0.80	0.80	13.1	16.	4.02	4.94
(wt%)	0.1	0.76	0.75	0.76	0.78	4.92	7.5	28.2	33.1
*Oil vol. fraction	0.5	0.72	0.72	0.67	0.68	3.75	6.54	6.77	7.02
**Oil droplet radius	1.	0.72	0.70	0.70	0.68	3.92	3.92	2.87	4.6
	5.	0.71	0.70	0.60	0.55	1.79	1.89	1.29	1.05
Hydrophobic NP	0.05	0.19	0.32	0.33	0.13				
conc. (wt%)	0.1	0.22	0.44	0.64	0.50	1.89	3.25	3.58	2.37
*Water vol. fraction	0.5	0.69	0.64	0.71	0.71	3.25	3.06	1.79	2.84
**Water droplet radius	1.	0.60	0.61	0.62	0.58	2.83	2.57	1.96	2.98
	5.	0.56	0.54	NA[1]	NA[1]	1.75	0.59	1.14	0.5

interface, we need to know how many nanoparticles are still in the fluid phases. The nanoparticle concentration in the excess aqueous phase can be determined indirectly by measuring the refractive index of the liquid using a refractometer. With a constant salinity, the nanoparticle concentration in the nanoparticle dispersion increases almost linearly with its refractive index, allowing nanoparticle concentration determination from the refractive index measurement. For the o/w emulsions, using the measured emulsion droplet radii r in the microscope images, the arithmetic means of r^2 and r^3 were calculated. Assuming that the droplet size distribution obtained from microscope analysis applies to the whole emulsion phase, and using the measured decane volume V_d in the bulk emulsion, the total surface area of all the emulsion droplets can be estimated by $S_e = 3V_d \langle r^2 \rangle / \langle r^3 \rangle$. Then the ratio of emulsion total surface area to its total volume V_e, which came from volume measurement, was calculated. When preparing emulsion samples, the initial amount of nanoparticles in the mixture was known. For the Group 1 samples, the nanoparticle amount left in the excess water after sonication was determined by measuring the refractive index of the dispersion. The difference between initial and final masses of nanoparticles in the aqueous phase is the quantity of nanoparticles attached to the droplet surfaces in the bulk emulsion. Knowing the size of each nanoparticle, the total surface area occupied by nanoparticles at droplet surfaces S_p was calculated from the nanoparticle mass. With the total emulsion droplet surface area calculated before, the

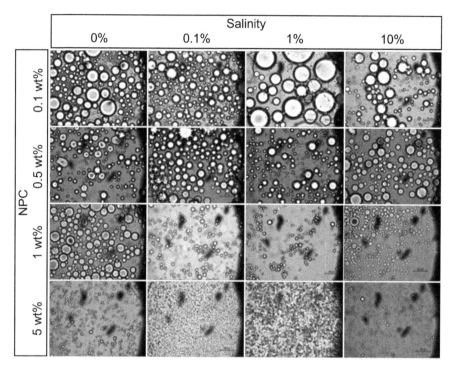

FIGURE 10.22

Water-in-decane emulsion (stabilized with hydrophobic nanoparticles) droplet images with varying nanoparticle concentration in decane (each row) and salinity (each column). Initial volume
ratio: water/decane = 1. From Zhang et al. (2010). Copyright 2010, Society of Petroleum Engineers.
Reproduced with permission of SPE. Further reproduction prohibited without permission.

nominal fraction of droplet surface covered by nanoparticles (S_p/S_e)
can be subsequently calculated. Details of the nanoparticle adsorption density calculation and analysis of the results are provided in
Zhang et al. (2010).

10.4.3 Rheology of Emulsions

The equilibrated emulsion viscosity was measured across a range of shear
rates by using the TA Instruments' Advanced Rheometric Expansion System
(ARES) LS-1 rheometer.

10.4.3.1 Oil-in-Water Emulsions with Hydrophilic Nanoparticles

The apparent viscosities of oil-in-water emulsions in Group 1 were measured for a shear rate range of $0.01\sim100$ s^{-1}. Figure 10.23(a) presents the plot
of bulk emulsion apparent viscosity versus shear rate for varied nanoparticle
concentrations without salt. The emulsions were shear-thinning across the
entire shear rate range. At a smaller nanoparticle concentration, the emulsion

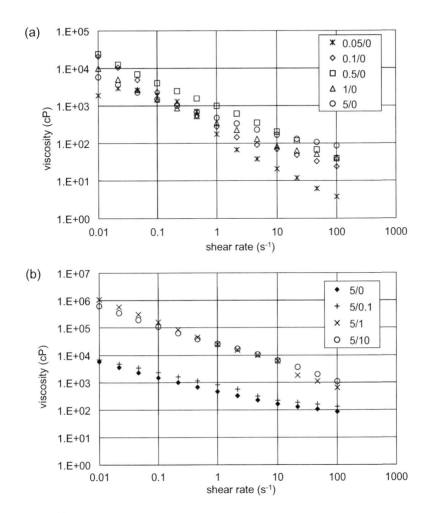

FIGURE 10.23
O/W emulsion (stabilized with hydrophilic nanoparticles) viscosity versus shear rate. (a) Varied nanoparticle concentration (0.05, 0.1, 0.5, 1., 5. wt%) in DI water. (b) Different salinities (0, 0.1, 1., 10. wt%) but with nanoparticle concentration of 5 wt%. From Zhang et al. (2010). Copyright 2010, Society of Petroleum Engineers. Reproduced with permission of SPE. Further reproduction prohibited without permission.

viscosity was larger at very low shear rate ($\sim$0.01 s^{-1}), dropped faster as the shear rate increased, and finally reached a smaller value at very high shear rate ($\sim$100 s^{-1}). The salinity effect on the o/w bulk emulsion apparent viscosity is shown in Figure 10.23(b), with a constant nanoparticle concentration of 5 wt%. The emulsion apparent viscosity might correspond to the inner structure of the emulsion, *e.g.*, the emulsion droplet sizes, the droplet surface coverage with particles, and the average distance between the drops. Some consistencies between their dependences on the nanoparticle concentration and on the salinity were found. For example, with 5 wt% nanoparticles, the emulsion

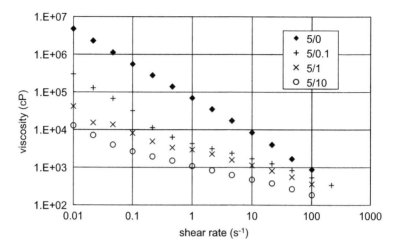

FIGURE 10.24

W/O emulsion viscosity (stabilized with hydrophobic nanoparticles) versus shear rate with different salinities (0, 0.1, 1., 10. wt%). Nanoparticle concentration in decane is 5 wt%. From Zhang et al. (2010). Copyright 2010, Society of Petroleum Engineers. Reproduced with permission of SPE. Further reproduction prohibited without permission.

prepared with 0.1% salinity and the emulsion without salt shared similar inner structures and exhibited almost the same rheological property. Two other o/w emulsion samples, which had 1% salinity and 10% salinity individually, have similar inner structures and resemble each other in rheology.

10.4.3.2 Water-in-Decane Emulsions with Hydrophobic Nanoparticles

The emulsions were shear-thinning in a shear rate range of 0.01~100 s^{-1}. Their viscosity did not change significantly when the nanoparticle concentration increased from 0.5 wt% to 5 wt%, which is different from the o/w emulsions. At a shear rate of about 0.01 s^{-1}, the emulsions exhibited very high viscosity, up to 10^7 cp, which is even higher than that of o/w emulsions. The dependence of w/o emulsion viscosity on salinity is shown in Figure 10.24 for a nanoparticle concentration in decane of 5 wt%. The emulsion viscosity decreased with a factor between 10 and 1000 as the salinity increased and the factor became smaller as the shear rate increased. The dependence of w/o emulsion apparent viscosity on salinity is opposite to that in o/w emulsions, which indicates that different mechanisms occur in o/w emulsions and in w/o emulsions.

10.4.4 Emulsions with Natural Gas Liquids for Heavy Oil Recovery

NGL is a mixture of short carbon chain alkanes often produced with light crude oil or with wet natural gas. They are relatively inexpensive and are often treated as waste products in the oil and gas industry, especially in the unconventional oil fields that are currently very actively developed in the US for light oil production. Nanoparticle-stabilized NGL emulsions show promise as a new practical method for the tertiary recovery of viscous oils (Griffith et al. 2016). As a preliminary step of concept validation, Griffith et al. (2016) carried out a study to first characterize emulsions which were generated using pentane and butane as surrogate NGL and then the oil-recovery coreflood experiments. As their work is unique and innovative while still quite preliminary, it will be described in some detail below. In the study, the effects of salinity on the stability of two nanoparticle dispersions were analyzed first. Using nanoparticle dispersions of varying salinity, pentane emulsions were generated to observe the effects of salinity on droplet size, stability, and rheology. Liquid butane emulsions of varying salinity were generated as well to assess the effects of salinity on the stability of these emulsions. Coreflooding experiments using pentane emulsions were then conducted to investigate the efficacy of using NGL emulsions for residual oil recovery.

10.4.4.1 Procedure to Make Pentane and Butane Emulsions

To make pentane emulsions, the silica nanoparticles of 46 nm and 13 nm nominal diameters were used. The former (DP9711) is from Nyacol Nano Technologies, and the latter is from 3M. Pentane and nanoparticle dispersion were co-injected at a 1:1 volume ratio into a high pressure column filled with 180 μm hydrophilic glass beads. To prevent the nanoparticles from directly contacting the elements of the pump, an accumulator was used. For emulsion stability tests, changes in emulsion volume fraction with time were observed. Immediately after an emulsion was generated, viscosity versus shear rate measurements were conducted using an AR-G2 rheometer from TA Instruments. Also, estimations of average droplet size were made using a microscope and image analysis software. These three evaluation steps have been described in the above sub-sections 10.4.2 and 10.4.3, and will not be repeated here. One operational aspect that is different from above is the generation of butane-in-brine emulsion: Because of the volatility of butane, a gas cylinder containing n-butane at 20 psi was inverted to maximize the amount of liquid butane that could be pulled from the cylinder. By inverting the cylinder, the denser liquid butane phase settled to the bottom and flow from the cylinder was driven by the gas-phase on top (due to expansion).

10.4.4.2 Procedure for Residual Oil Recovery Corefloods

A Boise sandstone core was first dried and evacuated, then saturated with brine. The core was then injected with light mineral oil until no water was being produced in the effluent, reaching the state of residual water saturation. The core was then waterflooded at incrementally increasing flowrates until no oil was produced in the effluent, which was taken to be at the residual oil saturation condition. A pentane-in-water emulsion that was stabilized with DP9711 nanoparticles (2 wt% in dispersion) and 3 wt% NaCl was first generated by co-injection into a beadpack with 180 μm hydrophilic glass beads, at a 1:1 volume ratio of pentane to nanoparticle dispersion for a total flow rate of 24 mL/min. To keep the emulsion stable, it was stored in an accumulator pressurized to 100 psi. The emulsion was then injected into the core either at 1 or 4 mL/min. The mineral oil was dyed red to distinguish how much residual mineral oil was recovered from the coreflood. Tests were also conducted where a half pore volume of the core was injected with emulsion and then was driven by a post-brine flush.

10.4.4.3 Coreflood Analysis

In oil-displacement corefloods, pentane emulsions were injected into sandstone cores at residual oil saturation with the resident oil being light mineral oil. The emulsions used in these experiments had a median droplet diameter of 69.5 μm. Griffith et al. (2016) carried out four corefloods, each of which will be briefly described. For the first experiment, pentane emulsion was injected into the core at 4 mL/min until no more light mineral oil was seen to be recovered. The injected emulsion droplets likely coalesce upon contacting the residual mineral oil in the core, becoming miscible with the residual mineral oil in the core. Guillen et al. (2012) showed that surfactant-stabilized emulsions increased mobility of resident oil and improved sweep efficiencies through a mechanism of droplet pore-blockage and displacing phase redirection. This mechanism, which may also describe nanoparticle-stabilized emulsions, results in an increase in the apparent viscosity of the displacing phase, thus increasing residual oil recovery. To assess the amount of residual mineral oil recovery, the effluent test tubes were placed under the fume hood to allow the pentane to evaporate from the samples. The estimated residual oil recovery for this coreflood experiment (69%) is encouraging given the viscosity difference between the pentane emulsion and the light mineral oil (~30 cp).

In the second coreflood experiment, the flow rate of the injected pentane emulsion was reduced to 1 mL/min. An estimated oil recovery of 81% was obtained, using a core with virtually the same porosity and permeability. Guillen et al. (2012) proposed that at a certain threshold capillary pressure, pore-blocking droplets were forced through pore throats, thus no longer aiding in the increase in mobility control and sweep efficiency. The increase in recovery seen in the slower injection rate experiment could be a result of an increase in droplets below this threshold capillary pressure needed to displace a droplet through a particular pore throat. The third coreflood conducted involved injecting a 0.50 core pore volume (PV) of pentane emulsion into the core and then injecting a post-brine flush behind the emulsion. All displacing fluids were injected into the core at 4 mL/min. The pressure drop increased throughout the duration of emulsion injection but then stabilized as the post-brine flush was injected. This stabilization of the pressure drop indicates an end to the coalescence and regeneration of emulsion. The oil recovery (82%) was equally good. The fourth coreflood was similar to the third, but the flow rate was reduced to 1 mL/min. Reducing the flow rate had a negative effect on the residual oil recovery (57%). The difference in the effect of reducing flow rates, between the second and this coreflood, is likely a result of complete emulsion coalescence and phase separation. With the injection of a finite-size bank of 0.5 PV at low flow rates, the emulsion may begin to completely coalesce and separate (reducing the chances of regeneration) when the post-brine flush commences, leading to a less continuous displacing front resulting in less residual oil recovery.

The emulsion flooding experiments by Griffith et al. (2016) show that nanoparticle-stabilized NGL emulsions can be a low-cost method for enhanced recovery of viscous residual oil phases. Finally, when compared with conventional emulsion-stabilizing materials such as surfactants, nanoparticles offer an inexpensive and robust alternative with stability over a wider range of temperature and salinity. In addition to the above examples of using silica nanoparticles as emulsion stabilizers, other kinds of nanoparticles have also been employed (see, *e.g.*, Binks 2017). A good example is the efficient use of graphene-oxide nano-platelets, which have a very large surface area per mass, to stabilize oil-in-water emulsions for high-salinity conditions (Yoon et al. 2013). Such novel use of various nanoparticles for different EOR processes offers an exciting opportunity.

References

Adkins, S. S., Gohil, D., Dickson, J. L., Webber, S. E., and Johnston, K. P. (2007) Water-in-carbon dioxide emulsions stabilized with hydrophobic silica particles. *Phys. Chem. Chem. Phys.*, 9(48), 6333–6343.

Aminzadeh, B., DiCarlo, D. A., Kianinejad, A., Chung, D. H., Bryant, S. L., and Huh, C. (2012a) Effect of Nanoparticles on Flow Alteration during CO2 Injection (SPE 160052). SPE Annual Technical Conference, Oct. 8–10, San Antonio, TX, USA.

Aminzadeh, B., DiCarlo, D. A., Roberts, M., Chung, D. H., Bryant, S. L., and Huh, C. (2012b) Effect of Spontaneous Formation of Nanoparticle Stabilized Emulsion on the Stability of a Displacement (SPE 154248). SPE Improved Oil Recovery Symposium, Apr. 14–18, Tulsa, OK, USA.

Aminzadeh, B., Kianinejad, A., Chung, D. H., Bryant, S. L., Huh, C., and DiCarlo, D. A. (2013a) CO2 leakage prevention by introducing engineered nanoparticles to the in-situ brine. *Energy Procedia*, 37, 5290–5297.

Aminzadeh, B., Chung, D. H., Zhang, X., Bryant, S. L., Huh, C., and DiCarlo, D. A. (2013b) Influence of Surface-Treated Nanoparticles on Displacement Patterns during CO2 Injection (SPE 166302). SPE Annual Technical Conference, New Orleans, LA, USA.

Aroonsri, A., Worthen, A. J., Hariz, T., Johnston, K. P., Huh, C., and Bryant, S. L. (2013) Conditions for Generating Nanoparticle-Stabilized CO2 Foams in Fracture and Matrix Flow (SPE 166319). SPE Annual Technical Conference, Sept. 30–Oct. 2, New Orleans, LA, USA.

Aroonsri, A. (2014) Nanoparticle-Stabilized Supercritical CO2 Foam for Mobility Control in CO2 Enhanced Oil Recovery. Master's Thesis, University of Texas at Austin, Austin, TX, USA.

Bennetzen, M. V., and Mogensen, K. (2014) Novel Applications of Nanoparticles for Future Enhanced Oil Recovery (IPTC 17857). International Petroleum Technology Conference, Dec. 10–12, Kuala Lumpur, Malaysia.

Binks, B. P. (2002) Particles as surfactants—similarities and differences. *Curr. Opin. Colloid Interface Sci.*, 7(1), 21–41.

Binks, B. P. (2017) Colloidal Particles at a Range of Fluid–Fluid Interfaces. *Langmuir*, 33(28), 6947–6963.

Binks, B. P., and Horozov, T. S. (2006) *Colloidal Particles at Liquid Interfaces*. Cambridge University Press, Cambridge, UK.

Binks, B. P., and Lumdson, S. O. (2000) Effect of oil type and aqueous phase composition on oil-water mixtures containing particles of intermediate hydrophobicity. *Phys. Chem. Chem. Phys.*, 2, 2959–2967.

Bouyer, F., Robben, A., Yu, W. L., and Borkovec, M. (2001) Aggregation of colloidal particles in the presence of oppositely charged polyelectrolytes: Effect of surface charge heterogeneities. *Langmuir*, 17(17), 5225–5231.

Bresme, F., and Oettel, M. (2007) Nanoparticles at fluid interfaces. *J. Phys. Condens. Matter*, 19, 413101.

Chen, K. L., and Elimelech, M. (2006) Aggregation and deposition kinetics of fullerene (C-60) nanoparticles. *Langmuir*, 22(26), 10994–11001.

Dickson, J. L., Binks, B. P., and Johnston, K. P. (2004) Stabilization of carbon dioxide-in-water emulsions with silica nanoparticles. *Langmuir*, 20(19), 7976–7983.

DiCarlo, D. A., Aminzadeh, B., Roberts, M., Chung, D. H., Bryant, S. L., and Huh, C. (2011) Mobility control through spontaneous formation of nanoparticle stabilized emulsions. *Geophys. Res. Lett.*, 38, L24404.

Eftekhari, A. A., Krastev, R., and Farajzadeh, R. (2015) Foam stabilized by fly ash nanoparticles for enhancing oil recovery. *Ind. Eng. Chem. Res.*, 54(50), 12482–12491.

Elimelech, M., and Omelia, C. R. (1990) Kinetics of deposition of colloidal particles in porous-media. *Environ. Sci. Technol.*, 24(10), 1528–1536.

Emrani, A., Ibrahim, A., and Nasr-El-Din, H. (2017) Evaluation of Mobility Control with Nanoparticle-Stabilized CO2 Foam (SPE 185551). SPE Latin America Caribbean Petroleum Engineering Conference, May 17–19, Buenos Aires, Argentina.

Emrani, A. S., and Nasr-El-Din, H. A. (2017) Stabilizing CO2 foam by use of nanoparticles. *Soc. Pet. Eng. J.*, 4, 494–504.

Espinosa, D. A., Caldelas, F. M., Johnston, K. P., Bryant, S. L., and Huh, C. (2010) Nanoparticle-Stabilized Supercritical CO2 Foams for Potential Mobility Control Applications (SPE 129925). SPE Improved Oil Recovery Symposium, Apr. 24–28, Tulsa, OK, USA.

Gauglitz, P. A., Friedmann, F., Kam, S. I., and Rossen, W. R. (2002) Foam generation in homogeneous porous media. *Chem. Eng. Sci.*, 57(19), 4037–4052.

Gonzenbach, U. T., Studart, A. R., Tervoort, E., and Gauckler, L. J. (2006a) Stabilization of foams with inorganic colloidal particles. *Langmuir*, 22(26), 10983–10988.

Gonzenbach, U. T., Studart, A. R., Tervoort, E., and Gauckler, L. J. (2006b) Ultrastable particle-stabilized foams. *Angew. Chem. Int. Ed.*, 45(21), 3526–3530.

Griffith, N., Ahmad, Y., Daigle, H., and Huh, C. (2016) Nanoparticle-Stabilized Natural Gas Liquid-in-Water Emulsions for Residual Oil recovery (SPE 179640). SPE Improved Oil Recovery Symposium, Apr. 11–13, Tulsa, OK, USA.

Guillen, V. R., Carvalho, M. S., and Alvarado, V. (2012) Pore scale and macroscopic displacement mechanisms in emulsion flooding. *Trans. Porous Media*, 94(1), 197–206.

Guo, F., and Aryana, S. (2016) An experimental investigation of nanoparticle-stabilized $CO2$ foam used in enhanced oil recovery. *Fuel*, 186, 430–442.

Guo, F., He, J., Johnson, P. A., and Aryana, S. A. (2017) Stabilization of $CO2$ foam using by-product fly ash and recyclable iron oxide nanoparticles to improve carbon utilization in EOR processes. *Sustainable Energy Fuels*, 1(4), 814–822.

Hirasaki, G. J., and Lawson, J. B. (1985) Mechanisms of foam flow in porous media: Apparent viscosity in smooth capillaries. *Soc. Pet. Eng. J.*, 4, 176–190.

Holthoff, H., Egelhaaf, S. U., Borkovec, M., Schurtenberger, P., and Sticher, H. (1996) Coagulation rate measurements of colloidal particles by simultaneous static and dynamic light scattering. *Langmuir*, 12(23), 5541–5549.

Horozov, T. S., Binks, B. P., and Gottschalk-Gaudig, T. (2007) Effect of electrolyte in silicone oil-in-water emulsions stabilized by fumed silica particles. *Phys. Chem. Chem. Phys.*, 9, 6389–6404.

Hunter, T. N., Pugh, R. J., Franks, G. V., and Jamenson, G. J. (2008) The role of particles in stabilizing foams and emulsions. *Adv. Colloid Interface Sci.*, 137, 57–81.

Jaisi, D. P., Saleh, N. B., Blake, R. E., and Elimelech, M. (2008) Transport of single-walled carbon nanotubes in porous media: Filtration mechanisms and reversibility. *Environ. Sci. Technol.*, 42(22), 8317–8323.

Jarrell, P., Fox, C., Morgan, K., Stein, M., and Webb, S. (2002) *Practical Aspects of CO_2 Flooding*. SPE Monograph Series No. 22, Richardson, TX, USA.

Jiménez, A., and Radke, C. (1961) Dynamic stability of foam lamellae flowing through a periodically constrained pore. In Borchardt, J. K., and Yen, T. F., (eds.), *Oil-Field Chemistry, Enhanced Recovery and Production Stimulation*. Amer Chemical Society, Washington, DC, USA.

Kalyanaraman, N., Arnold, C., Gupta, A., Tsau, J. S., and Ghahfarokhi, R. B. (2017) Stability improvement of CO_2 foam for enhanced oil-recovery applications using polyelectrolytes and polyelectrolyte complex nanoparticles. *J. Appl. Polym. Sci.*, 44491.

Kam, S., and Rossen, W. (2003) A model for foam generation in homogeneous porous media. *Soc. Pet. Eng. J.*, 8, 417–425.

Kaptay, G. (2003) Interfacial criteria for stabilization of liquid foams by solid particles. *Colloids Surf. A: Physicochem. Eng. Aspects*, 230(1–3), 67–80.

Kim, I., Taghavy, A., DiCarlo, D., and Huh, C. (2015) Aggregation of silica nanoparticles and its impact on particle mobility under high-salinity conditions. *J. Pet. Sci. Eng.*, 133, 376–383.

Kim, I., Worthen, A. J., Johnston, K. P., DiCarlo, D., and Huh, C. (2016a) Size-dependent properties of silica nanoparticles for pickering stabilization of emulsions and foams. *J. Nanopart. Res.*, 18(4), 82–93.

Kim, I., Worthen, A. J., Lotfollahi, M., Johnston, K. P., DiCarlo, D., and Huh, C. (2016b) Nanoparticle-stabilized Emulsions for Improved Mobility Control for Adverse-Mobility Waterflooding (SPE 179644). SPE Symposium Improved Oil Recovery, Apr. 11–13, Tulsa, OK, USA.

Kovscek, A., Patzek, T., and Radke, C. (1995) A mechanistic population balance model for transient and steady-state foam flow in Boise sandstone. *Chem. Eng. Sci.*, 50(23), 3783–3799.

Kralchevsky, P. A., Ivanov, I. B., Ananthapadmanabhan, K. P., and Lips, A. (2005) On the thermodynamics of particle-stabilized emulsions: curvature effects and catastrophic phase inversion. *Langmuir*, 21, 50–63.

Lake, L. W., Johns, R. T., Rossen, W. R., and Pope, G. A. (2014) Fundamentals of Enhanced Oil Recovery. Society of Petroleum Engineers, Richardson, TX, USA.

Legg, B. A., Zhu, M. Q., Comolli, L. R., Gilbert, B., and Banfield, J. F. (2014) Impacts of ionic strength on three-dimensional nanoparticle aggregate structure and consequences for environmental transport and deposition. *Environ. Sci. Technol.*, 48(23), 13703–13710.

Lee, D., Cho, H., Lee, J., Huh, C., and Mohanty, K. (2015) Fly ash nanoparticles as a CO_2 foam stabilizer. *Powder Technol.*, 283, 77–84.

Li, S., Li, Z., and Wang, P. (2016) Experimental study of the stabilization of CO2 foam by sodium dodecyl sulfate and hydrophobic nanoparticles. *Ind. Eng. Chem. Res.*, 55(5), 1243–1253.

Lopetinsky, R. J. G., Masliyah, J. H., and Xu, Z. (2006) Solids-stabilized emulsions: A review. In Binks, B. P., and Horozov, T. S., (eds.), *Colloidal Particles at Liquid Interfaces*. Cambridge University Press, Cambridge, UK, pp. 186–224.

Lotfollahi, M., Kim, I., Beygi, M. R., Worthen, A. J., Huh, C., Johnston, K. P., Wheeler, M. F., and DiCarlo, D. (2017) Experimental studies and modeling of foam hysteresis in porous media. *Trans. Porous Media*, 116(2), 387–406.

Luo, D., Wang, F., Zhu, J., Cao, F., Liu, Y., Li, X., Wilson, R.C., Yang, Z., Chu, C.-W., and Ren, Z. (2016) Nanofluid of graphene-based amphiphilic Janus nanosheets for tertiary or enhanced oil recovery: High performance at low concentration. *PNAS* 113(28), 7711–7716.

Ma, K., Ren, G., Mateen, K., Morel, D., and Cordelier, P. (2015) Modeling techniques for foam flow in porous media. *Soc. Pet. Eng. J.*, 6, 453–470.

Melle, S., Lask, M., and Fuller, G. G. (2005) Pickering emulsions with controllable stability. *Langmuir*, 21, 2158–2162.

Mo, D., Jia, B., Yu, J., Liu, N., and Lee, R. (2014) Study of Nanoparticle-Stabilized CO2 Foam for Oil Recovery at Different Pressure, Temperature, and Rock Samples (SPE 169110). SPE Improved Oil Recovery Symposium, Apr. 12–16, Tulsa, OK, USA.

Mo, D., Yu, J., Liu, N., and Lee, R. L. (2012) Study of the effect of different factors on nanoparticlestablized CO2 foam for mobility control (SPE 159282). SPE Annual Technical Conference, Oct. 8–10, San Antonio, TX, USA.

Murray, B. S., and Ettelaie, R. (2004) Foam stability: Proteins and nanoparticles. *Curr. Opin. Colloid Interface Sci.*, 9(5), 314–320.

Nelson, K. E., and Ginn, T. R. (2011) New collector efficiency equation for colloid filtration in both natural and engineered flow conditions. *Water Resour. Res.*, 47, W05543.

Ngai, T., and Bon, S. A. F. (2015) *Particle-Stabilized Emulsions and Colloids*. Royal Society of Chemistry, Cambridge, UK.

Nguyen, P., Fadaei, H., and Sinton, D. (2014) Pore-scale assessment of nanoparticle-stabilized CO2 foam for enhanced oil recovery. *Energy Fuels*, 28(10), 6221–6227.

Prigiobbe, V., Worthen, A. J., Johnston, K. P., Huh, C., and Bryant, S. L. (2016) Transport of nanoparticle-stabilized CO2-foam in porous media. *Trans. Porous Media,* 111(1), 265–285.

Qajar, A., Xue, Z., Worthen, A. J., Johnston, K. P., Huh, C., Bryant, S. L., and Prodanović, M. (2016) Modeling fracture propagation and cleanup for dry nanoparticle-stabilized-foam fracturing fluids. *J. Pet. Sci. Eng.,* 146, 210–221.

Ransohoff, T. C., and Radke, C. J. (1988) Mechanisms of foam generation in glass-bead packs. *Soc. Pet. Eng.,* 3(2), 573–585.

Reincke, F., Kegel, W. K., Zhang, H., Nolte, M., Wang, D., Vanmaekelbergh, D., and Mohwald, H. (2006) Understanding the self-assembly of charged nanoparticles at the water/oil interface. *Phys. Chem. Chem. Phys.,* 8, 3828–3835.

Roberts, M., Aminzadeh, B., DiCarlo, D. A., Bryant, S. L., and Huh, C. (2012) Generation of Nanoparticle-Stabilized Emulsions in Fractures (SPE 154228). SPE Improved Oil Recovery Symposium, Apr. 14–18, Tulsa, OK, USA.

Rossen, W. R. (1996) Foams in enhanced oil recovery. In Prud'homme, R. K., and Khan, S. A., (eds.), *Foams: Theory, Measurements, and Applications.* pp. 413–462. Surfactant Science Series, Marcel Dekker, New York.

Rossen, W. (2013) Numerical Challenges in Foam Simulation: A Review (SPE 166232). SPE Annual Technical Conference, Sept. 30–Oct. 2, New Orleans, LA, USA.

Rossen, W. R., and Gauglitz, P. A. (1990) Percolation theory of creation and mobilization of foams in porous media. *AIChE J.,* 36(8), 1176–1188.

Rossen, W., and Lu, Q. (1997) Effect of capillary crossflow on foam improved oil recovery (SPE 38319). SPE Western Regional Meeting, Jun. 25–27, Long Beach, CA, USA.

San, J., Wang, S., Yu, J., Liu, N., and Lee, R. (2017) Nanoparticle-stabilized carbon dioxide foam used in enhanced oil recovery: effect of different ions and temperatures. *Soc. Pet. Eng. J.,* 10, 1416–1423.

Sheng, J. J. (2011) *Modern Chemical Enhanced Oil Recovery—Theory and Practice.* Gulf Professional Pub., Burlington, MA, USA.

Singh, R., Gupta, A., Mohanty, K. K., Huh, C., Lee, D., and Cho, H. (2015) Fly Ash Nanoparticl-estabilized CO2-in-Water Foams for Gas Mobility Control Applications (SPE 175057). SPE Annual Technical Conference, Sept. 28–30, Houston, TX, USA.

Singh, R., and Mohanty, K. K. (2014) Synergistic Stabilization of Foams by a Mixture of Nanoparticles and Surfactants (SPE 169126). SPE Improved Oil Recovery Symposium, Apr. 12–16, Tulsa, OK, USA.

Singh, R., and Mohanty, K. K. (2015) Synergy between nanoparticles and surfactants in stabilizing foams for oil recovery. *Energy Fuels,* 29(2), 467–479.

Singh, R., and Mohanty, K. K. (2016) Foams stabilized by in-situ surface-activated nanoparticles in bulk and porous media. *Soc. Pet. Eng. J.,* 21(1), 121–130.

Sjoblom, J. (2006) *Emulsions and Emulsion Stability.* Surfactant Science Series. CRC Taylor & Francis.

Sun, Q., Li, Z., Li, S., Jiang, L., Wang, J., and Wang, P. (2014) Utilization of surfactant-stabilized foam for enhanced oil recovery by adding nanoparticles. *Energy Fuel,* 28, 2384–2394.

Tanzil, D., Hirasaki, G. J., and Miller, C. A. (2002) Conditions for Foam Generation in Homogeneous Porous Media (SPE 75176). SPE/DOE Improved Oil Recovery Symposium, Apr. 13–17, Tulsa, OK, USA.

Thompson, J., Vasquez, A., Hill, J. M., and Pereira-Almao, P. (2008) The synthesis and evaluation of up-scalable molybdenum based ultra dispersed catalysts: Effect of temperature on particle size. *Catal. Lett.*, 123, 16–23.

Vandesteeg, H. G. M., Stuart, M. A. C., Dekeizer, A., and Bijsterbosch, B. H. (1992) Polyelectrolyte adsorption - a subtle balance of forces. *Langmuir*, 8(10), 2538–2546.

Walstra, P., and Smulders, P. E. A. (1998) Emulsion formation. In Binks, B. P., (ed.), *Modern Aspects of Emulsion Technology*. Royal Society of Chemistry, Cambridge, UK. pp. 56–99.

Worthen, A., Taghavy, A., Aroonsri, A., Kim, I., Johnston, K., Huh, C., Bryant, S., and DiCarlo, D. (2015) Multi-Scale Evaluation of Nanoparticle-Stabilized CO2-in-Water Foams: From the Benchtop to the Field (SPE 175065). SPE Annual Technical Conference, Sept. 28–30, Houston, TX, USA.

Worthen, A. J., Bagaria, H. G., Chen, Y., Bryant, S. L., Huh, C., and Johnston, K. P. (2013a) Nanoparticle-stabilized carbon dioxide-in-water foams with fine texture. *J. Colloid Interface Sci.*, 391, 142–151.

Worthen, A. J., Bryant, S. L., Huh, C., and Johnston, K. P. (2013b) Carbon dioxide-in-water foams stabilized with nanoparticles and surfactant acting in synergy. *AIChE J.*, 59(9), 3490–3501.

Worthen, A., Bagaria, H., Chen, Y., Bryant, S. L., Huh, C., and Johnston, K. P. (2012) Nanoparticle Stabilized Carbon Dioxide in Water Foams for Enhanced Oil Recovery (SPE 154285). SPE Improved Oil Recovery Symposium, Apr. 14–18, Tulsa, OK, USA.

Xue, Z., Worthen, A., Qajar, A., Robert, I., Bryant, S. L., Huh, C., Prodanović, M., and Johnston, K. P. (2016a) Viscosity and stability of ultra-high internal phase CO2-in-water foams stabilized with surfactants and nanoparticles with or without polyelectrolytes. *J. Colloid Interface Sci.*, 461, 383–395.

Xue, Z., Worthen, A. J., Da, C., Qajar, A., Ketchum, I. R., Alzobaidi, S., Huh, C., Prodanović, M., and Johnston, K.P. (2016b) Ultradry carbon dioxide-in-water foams with viscoelastic aqueous phases. *Langmuir*, 32(1), 28–37.

Yekeen, N., Manan, M. A., Idris, A. K., Padmanabhan, E., Junin, R., Samin, A. M., Gbadamosi, A. O., and Oguamah, I. (2017a) A comprehensive review of experimental studies of nanoparticles-stabilized foam for enhanced oil recovery. *J. Pet. Sci. Eng.*, 164, 43–74.

Yekeen, N., Manan, M. A., Idris, A. K., Samin, A. M., and Risal, A. R. (2017b) Experimental investigation of minimization in surfactant adsorption and improvement in surfactant-foam stability in presence of silicon dioxide and aluminum oxide nanoparticles. *J. Pet. Sci. Eng.*, 159, 115–134.

Yoon, K. Y., An, S. J., Chen, Y., Lee, J. H., Bryant, S. L., Ruoff, R., Huh, C., and Johnston, K. P. (2013) Development of graphene-oxide-stabilized oil-in-water emulsions for high-salinity conditions. *J. Colloid Interface Sci.*, 403, 1–6.

Yu, J., Mo, D., Liu, N., and Lee, R. (2013) The Application of Nanoparticle-Stabilized CO2 Foam for Oil Recovery (SPE 164074). SPE International Symposium Oilfield Chemistry, Apr. 8–10, The Woodlands, TX, USA.

Yu, J., Wang, S., Liu, N., and Lee, R. (2014a) Study of Particle Structure and Hydrophobicity Effects on the Flow Behavior of Nanoparticle-Stabilized CO_2 Foam in Porous Media (SPE 169047). SPE Improved Oil Recovery Symposium, Apr. 12–16, Tulsa, OK, USA.

Yu, J., Khalil, M., Liu, N., and Lee, R. (2014b) Effect of particle hydrophobicity on CO_2 foam generation and foam flow behavior in porous media. *Fuel*, 126, 104–108.

Zhang, T., Roberts, M., Bryant, S. L., and Huh, C. (2009) Foams and emulsions stabilized with nanoparticles for potential conformance control applications (SPE 121744). SPE International Symposium Oilfield Chemistry, Apr. 20–22, The Woodlands, TX, USA.

Zhang, T., Davidson, D., Bryant, S. L., and Huh, C. (2010) Nanoparticle-Stabilized Emulsions for Applications in Enhanced Oil Recovery (SPE 129885). SPE Improved Oil Recovery Symposium, Apr. 24–28, Tulsa, OK, USA.

Zhang, T., Espinosa, D., Yoon, K. Y., Rahmani, A. R., Yu, H., Caldelas, F. M., Ryoo, S., Roberts, M., Prodanovic, M., Johnston, K. P. , Milner, T. E., Bryant, S. L., and Huh. C. (2011) Engineered Nanoparticles as Harsh-Condition Emulsion and Foam Stabilizers and as Novel Sensors (OTC 21212). Offshore Technology Conference, May 2–5, Houton, TX, USA.

Zhang, X., Tian, F.-G., Zhou, H., Du, H.-L., Ding, C., Wang, J., and DiCarlo, D. A. (2017) Effect of Nanoparticles on Flow Alteration and Saturation during CO2 Injection and Post Flush (SPE 187168). SPE Annual Technical Conference, Oct. 9–11, San Antonio, TX, USA.

11

Enhanced Oil Recovery: Wettability Alteration and Other Topics

11.1 Introduction

In addition to the research efforts to employ different kinds of nanoparticles as stabilizers for foams and emulsions for enhanced oil recovery (EOR) use as described in detail in Chapter 10, another important use of nanoparticles is to alter wettability of reservoir rock, which will be described in this chapter. As described in Section 7.6, a self-assembled layer of nanoparticles deposited on a solid surface can make the surface either superhydrophilic or superhydrophobic (Krause 2008a,b). With a superhydrophilic coating, any particulates (such as dirt) attached to the surface can be easily removed by gentle rinsing with water. Such a superhydrophilic coating on glass is also employed for antifogging purpose because the superhydrophilicity prevents formation of tiny water droplets, preferring to form a thin water film on the glass surface. With a superhydrophobic coating, water droplets on the solid surface, such as rain drops falling on an automobile windshield, simply roll off without wetting and leaving a stain on the surface at all. Because wettability alteration is an important topic for improved oil production as described below, the possibility of such use of nanoparticles to alter the wettability of reservoir rock is drawing considerable interest. For example, if a dilute concentration of nanoparticles is injected into a well, which then adsorb on the rock surface and alter its wettability for a meaningful duration of production or injection period, and if the specially surface-coated nanoparticles do a more effective and durable job than the current suite of specialty surfactants and polymers, such application will be a significant technical innovation for production and/or injection stimulation.

One important application of such wettability alteration is to induce oil out of the oil-wet pores of tight matrix zones of naturally fractured reservoirs (Gupta and Mohanty 2011). The conventional method of injecting water to displace oil does not work, because the water simply zips through the fracture networks, bypassing the matrix pores where the bulk of oil resides. By injecting a dilute concentration of hydrophilic surfactant which then diffuses into the matrix pores, their wettability can be altered to water-wet condition, thus allowing the injected water to imbibe into the pores and expel the resident oil to the fracture and subsequently to the production well. If the nanoparticles can do a better job than the surfactant, *e.g.*, for high-temperature reservoirs where the surfactant would easily degrade, it will bring a significant business benefit. Another reason for the significant interest in the use of aqueous dispersions of nanoparticles for EOR is the "dynamic spreading" concept proposed by Wasan and Nikolov (2003), as further described below. According to their theory, the nanoparticles at the "wedge-like" solid/liquid/fluid-phase contact line zone form structured layers, which creates a disjoining pressure and makes the "wedge" advance forward, i.e., enhances the spreading of the nanoparticle-containing fluid phase.

Because of the above reasons, there have been quite active research efforts to use aqueous nanoparticle dispersions to enhance oil recovery by way of wettability alteration. For recent reviews on the subject, see Idogun et al. (2016), Negin et al. (2016), and Sun et al. (2017). These investigations will be described in detail in the next section. In Section 11.3, the recent research efforts on different uses of nanoparticles for the surfactant/polymer EOR processes are described. In Section 11.4, the use of silica nanoparticles to enhance the generation of CO_2 hydrates and subsequent development of highly porous hydrate structure is described. While the topic is not directly related to EOR, it is closely related to the use of silica nanoparticles for CO_2 foam stabilization and could be potentially employed for low-cost and secure CO_2 sequestration at deep subsea conditions.

11.2 Wettability Alteration and Imbibition Enhancement

Prompted by the recent extensive research efforts to modify the wettability of a solid surface by depositing a self-assembling layer of nanoparticles, the possibility of adding some nanoparticles in the injection water so that the wettability of reservoir rock is altered to improve oil recovery has recently been actively investigated. A number of researchers, especially the Norwegian University of Science and Technology group, studied how injection of aqueous dispersion of different nanoparticles (SiO_2, other metal oxides, and their mixtures) into rock cores improves oil recovery, as described more in detail below. To better interpret the coreflood results, they investigated the effects of nanoparticles on interfacial tension and contact angle, and observed oil displacement behavior in micromodels. They also demonstrated that improved oil recovery generally resulted from the alteration of the rock wettability to a more hydrophilic condition.

11.2.1 Wettability Alteration Experiments

Wettability of rock is a key factor that governs oil saturation distribution in rock pores, its displacement, and ultimate recovery, as it controls the capillary pressure and relative permeabilities of immiscible fluid phases in reservoir rock. To determine the effect of nanoparticles on reservoir rock's wettability, a number of researchers either measured the contact angle change or carried out the Amott test. For the measurement of the contact angle, a small piece of rock with a surface that is smooth, plane, and clean is immersed in the test brine; a drop of oil is placed on top of the rock surface and the contact angle is measured. Because the rock is in reality a porous medium, the contact angle is not a true measure of wettability as the specimen's surface consists of not only solid area but also void area representing rock pores. Nevertheless, the changes in contact angle are often employed to quantify the wettability alteration. The Amott test, also known as a spontaneous imbibition test, is probably the most commonly employed technique to quantify the wettability alteration because the consequence of the alteration can be directly obtained in terms of the amount of oil that percolates out from the rock sample, as a result of enhanced imbibition of water. Some of the more notable studies on the wettability alteration experiments, reported in the literature, are briefly discussed here.

Karimi et al. (2012) studied the wettability alteration by aqueous dispersions of zirconium oxide (ZrO_2) in carbonate cores. They reported that the contact angle of n-heptane/water on calcite plates aged in crude oil was 180°. A surfactant solution without nanoparticles altered the rock's wettability from oil-wet ($\theta \sim 180°$) to an intermediate state ($\theta \sim 90°$). Introducing the nanoparticles into the solution resulted in a strongly water-wet condition. They also carried out oil-displacement corefloods as briefly described below. Zhang et al. (2014a) conducted spontaneous imbibition tests in water-wet Berea sandstone core and in glass capillary tubes using a reservoir crude oil with a viscosity of 24.6 cp at 50°C and a high salinity reservoir brine. A 10 wt% dispersion of SiO_2 (nominal diameter of 20 nm) in deionized (DI) water at pH = 9.7 showed a 91% (16 to 1.4 dynes/cm) reduction in interfacial tension and a 98% decrease in contact angle (74° to 1.2°) in oil/water systems. Nazari et al. (2015) investigated the effectiveness of eight different nanoparticles for wettability alteration from oil-wet to water-wet conditions in carbonate rocks and also carried out oil-displacement corefloods as described in Section 11.2.2.2.

Li and Torsaeter (2015a,b) studied two different silica nanoparticles, namely, nanostructured particles (NSP) and colloidal nanoparticles (CNP). TEM images revealed that NSP has an average primary particle size of 7 nm but they tend to aggregate to form bigger particles of ~100 nm; and CNP has average particle size of 18 nm and does not aggregate easily. Spontaneous imbibition tests in sandstone rock samples, which had been aged in a mixture of pentane and siliconizing fluid to make it oil-wet, were carried out using three different concentrations of NSP and CNP dispersed in 3 wt% NaCl brine. Figure 11.1 shows the imbibition volume versus time for different nanoparticle concentrations of NSP and CNP. Their results reveal that hydrophilic nanoparticles can alter the oil-wet reservoir rock to water-wet state, and the alteration is more pronounced with higher nanoparticle concentration.

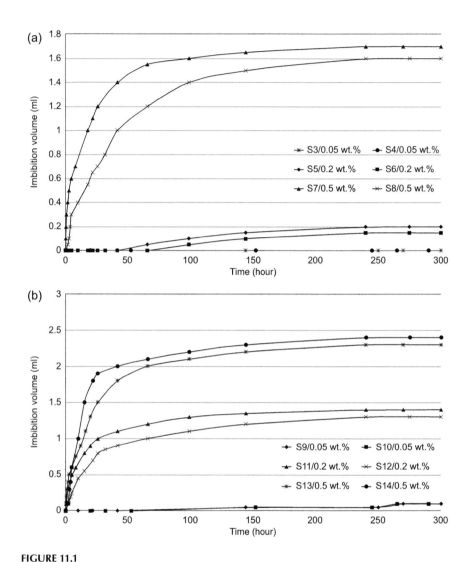

FIGURE 11.1
Spontaneous imbibition of silica nanoparticle-containing water (3 wt% NaCl) into a sandstone rock sample which had been aged to make it oil-wet. Two dynamic imbibition experiments each for different nanoparticle concentrations: (a) NSP (~100 nm aggregates) and (b) CNP nanoparticle (18 nm). From Li and Torsaeter (2015b).

Roustaei and Bagherzadeh (2015) carried out a series of Amott tests to study the impact of SiO_2 nanoparticles on the wettability of a carbonate rock, which showed that the nanoparticle dispersion indeed helped to make the oil-wet rock to more water-wet condition. Al-Anssari et al. (2016) found that the SiO_2 nanoparticles helped alter the oil-wet condition of fractured limestones to water-wet condition. Nwidee et al. (2017) investigated the effectiveness of altering wettability by injecting zirconium oxide (ZrO_2) and nickel oxide

(NiO) nanoparticles into oil-wet fractured limestone. ZrO_2 demonstrated a better efficiency by altering strongly oil-wet ($\theta = 152°$) calcite substrates into a strongly water-wet ($\theta = 44°$) state, while NiO changed wettability to an intermediate wet condition ($\theta = 86°$) at 0.05 wt% nanoparticle concentration. The contact angles decrease with the increase in nanoparticle concentration (0.004–0.05 wt%), time, and salinity, showing a similar decreasing trend.

11.2.2 Nanoparticle-Assisted Oil Displacement Waterfloods

Motivated by the nanoparticle's potential for EOR as described above, many researchers carried out oil-displacement coreflood experiments in which aqueous dispersions of various nanoparticles (often called "nanofluid") are injected into reservoir rock core samples. Some of the notable results and observations are discussed here.

11.2.2.1 Use of Silica Nanoparticles

The most commonly employed nanoparticles are SiO_2 particles. As noted earlier, the research group at the Norwegian University of Science and Technology carried out a systematic study on the effectiveness of adding various metal oxide nanoparticles in injection water for EOR (Cheraghian and Hendraningrat 2016a,b). Hendraningrat et al. (2013a) and Li et al. (2013) found that the oil recovery from waterflooding with addition of SiO_2 nanoparticles during the secondary flooding stage was better than the addition during the tertiary flooding stage, with ~8% higher ultimate oil recovery. Hendraningrat et al. (2013b) further investigated the effects of nanoparticle size, rock permeability, initial rock wettability, and temperature on oil recovery. They found that smaller nanoparticle size and higher temperature resulted in higher incremental oil recovery. Increasing permeability did not show a similar trend for incremental oil recovery. Hendraningrat et al. (2013c) studied the effectiveness of using SiO_2 dispersion for EOR in sandstones (9–400 md) and concluded that they are indeed effective in water-wet sandstone, even though the magnitude of incremental oil recovery is not substantial. They also found that nanoparticle concentration is an important factor for flooding in low-permeability water-wet sandstones because higher concentrations tend to block pore throats. Li et al. (2015) conducted coreflood experiments to study the effect of hydrophilic silica nanostructure particles and colloidal particles (NSP and CNP, discussed above) on oil recovery. Figure 11.2 shows their coreflood results for two different nanoparticle concentrations (a) for NSP and (b) for CNP. They observed the same results that injecting both types of nanoparticles could increase oil recovery even with a very low concentration.

In addition to the spontaneous imbibition tests, Zhang et al. (2014a) conducted oil-displacement corefloods with a reservoir crude oil and a high-salinity brine in water-wet Berea sandstone cores and observed an increase

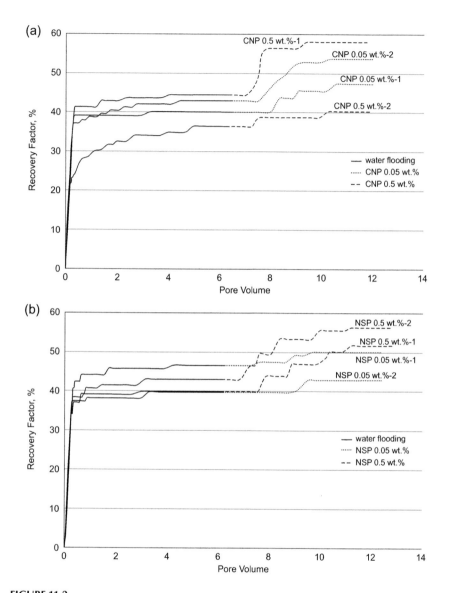

FIGURE 11.2
Oil-displacement coreflood with injection of silica nanoparticle-containing water (3 wt% NaCl) into a sandstone rock core which had been aged to make it oil-wet. Two corefloods each for different nanoparticle concentrations: (a) NSP (~100 nm aggregates) and (b) CNP nanoparticle (18 nm). From Li et al. (2015).

in oil recovery by about 13% OOIP. El-Diasty (2015) evaluated the effect of nanoparticle size (5 to 60 nm) on SiO_2 nanofluid flooding in Egyptian sandstone cores, reaching the same conclusions as Hendraningrat et al. (2013a,b,c). Roustaei and Bagherzadeh (2015) reported that aging of nanofluid in cores contributed to increased oil recovery. The experimental results showed that oil recovery factors increased by 16% and 17% when the nanofluid was aged for 24 hrs in two core samples. Oil recovery experiments by Ragab and Hannora (2015a,b) are also of note.

11.2.2.2 Use of Metal Oxides Nanoparticles Other Than Silica

Hendraningrat and Torsaeter (2014a,b) employed hydrophilic metal-oxide nanoparticles, Al_2O_3 and TiO_2, with a primary size in the range of 40–60 nm to enhance oil recovery. Nanoparticle dispersions of 0.05 wt% were prepared in synthetic seawater. Berea sandstones core samples with average porosity and permeability of 15% and 60 md, respectively, were used, and the viscosity of the crude oil ranged from 5 to 50 cp. The results showed that metal oxides altered the wettability to more water-wet (the contact angle decreased from 54° to 21°) and enhanced the oil recovery, ~ 24% higher than the nanoparticle-free case. Bayat et al. (2014) evaluated the performance of three metal oxide nanoparticles, Al_2O_3 (40 nm), TiO_2 (10–30 nm), and SiO_2 (20 nm), on EOR in limestone cores at different temperatures. They found that nanoparticles mobilized the residual oil droplets, probably through wettability alteration. Al_2O_3 performed better than TiO_2 and SiO_2 for EOR at all temperatures studied.

Nazari et al. (2015) also carried out oil-displacement experiments employing eight different nanoparticles in carbonate rock cores. The results showed an increase, about 8–9%, in oil recovery with injection of $CaCO_3$ and SiO_2 nanoparticles (~35 to 40 nm diameter). Ogolo et al. (2012) and Onyekonwu and Ogolo (2010) investigated the performance of eight different nanoparticles (oxides of aluminum, zinc, magnesium, iron, zirconium, nickel, tin, and silicon) and four base fluids (distilled water, brine, ethanol, and diesel oil) for enhanced oil recovery. The results showed that Al_2O_3 nanoparticles dispersed in brine and distilled water showed improved oil recovery. Magnesium oxide and zinc oxide in distilled water and in brine resulted in permeability problems due to their large diameter. Ehtesabi et al. (2013) found that TiO_2 nanoparticles in low concentration could enhance heavy oil recovery in laboratory flooding experiments. Karimi et al. (2012), whose work on the contact angle change upon addition of ZrO_2 was described above, also carried out oil-displacement corefloods which showed improved oil recovery with the nanoparticle addition.

Hu et al. (2016) conducted oil-displacement experiments in water-wet Berea sandstone cores to study the effectiveness of rutile TiO_2 nanoparticles. Adding TiO_2 nanoparticles could enhance oil recovery (measured at water breakthrough) significantly. At lower nanoparticle concentrations, the oil recovery increases with the nanoparticle concentration, reaching the maximum value, 39.8%, at about 20 ppm. This represents about 9.5% incremental increase in the oil recovery at water breakthrough, compared with the nanoparticle-free waterflooding (30.3%). The general dependence of the ultimate oil recovery on nanoparticle concentration was similar to the water-breakthrough oil recovery, but the peak value occurred at a nanoparticle concentration of 10 ppm instead of 20 ppm, with a total oil recovery of 41.8% of OOIP, representing an 11.5% increase in oil recovery compared with the nanoparticle-free case. They also observed that nanoparticle transport is strongly dependent on its concentration.

11.2.2.3 Use of Mixture of Metal Oxides Nanoparticles

Seeking synergy among different nanoparticles, Alomair et al. (2015) performed heavy oil recovery coreflood experiments using Berea sandstone cores, injecting aqueous dispersions of mixtures of different nanoparticles. Their results showed that a mixture of SiO_2 and Al_2O_3 at 0.05 wt% had the highest incremental oil recovery among all mixtures tested. They believed this is due to its capability to hinder asphaltene precipitation and reduce the interfacial tension values. Tarek and El-Banbi (2015) showed that the use of nanoparticle mixtures (40 wt% Fe_2O_3 + 35% Al_2O_3 + 25% SiO_2) resulted in additional oil recovery beyond what a single nanoparticle could. Then, Tarek (2015) conducted several experiments with different concentrations of nanoparticle mixtures in a high-permeability core starting directly with tertiary recovery. They found that the optimum nanoparticle mixture concentration depended on both the fluid and rock properties.

11.2.2.4 Use of Other Nanoparticles

In addition to the metal-oxide nanoparticles described above, use of other nanomaterials for EOR were reported in literature, such as cellulose nanocrystals (CNC) and micro-gel nanospheres. Molnes et al. (2016) investigated the injectivity of CNC in a high-permeable sandstone core and found that CNC has potential as additives in injection water for EOR. Heggset et al. (2017) found that CNC has temperature stability that was stable after heating to 140°C for 3 days. From their micromodel experiments, Wei et al. (2016) also believed that the nanocrystals are promising EOR additives, resulting in sweep improvement with increase in effective viscosity of injection water due to emulsification. Wang et al. (2010) prepared a polyacrylamide microgel nanospheres and used them to enhance the recovery of Zhuangxi heavy oil (its viscosity is 238 mPa·s at 55°C) in a sand-pack model. They confirmed experimentally the effectiveness of nanospheres for EOR.

One interesting application of metal-oxide nanoparticles is a new EOR method proposed by Haroun et al. (2012) called electrokinetics (EK)-assisted nanofluid flooding. In this process, a direct current with about 2 V/cm voltage gradient was applied, together with nanoparticle dispersion, to increase oil recovery in carbonate cores from Abu Dhabi. They compared the performance of several nanoparticles (FeO, CuO and NiO) with EK applied after the waterflooding stage and found that CuO produced better results than FeO and NiO, presumably due to higher electrical conductivity.

11.2.3 Possible EOR Mechanism(s) with Nanoparticle Addition to Waterflood

In the above sub-section, the recent literature on the effects of adding nanoparticles to the waterflood injection water was described. The key assumption there is that, just as the surfactant adsorption on pore wall alters reservoir rock wettability, the nanoparticle deposition on the pore wall also alters wettability, and thereby increases oil recovery. The above review suggests that, while the injection of carefully chosen nanoparticles indeed enhances the oil recovery, the incremental recovery is not substantial. As discussed in the introduction of this chapter, one other mechanism of EOR by nanoparticle addition is Wasan-Nikolov (2003)'s "super-spreading" concept (Nikolov and Wasan 2014). In a series of related publications on theoretical modeling (Chengara et al. 2004; Liu et al. 2012; Wasan et al. 2011), spreading experiments (Kondiparty et al. 2012; Lim et al. 2015; Zhang et al. 2014a,b) and oil-displacement coreflood experiments (Zhang et al. 2016), they showed that, due to structure formation by nanoparticles at the three-phase contact line zone which creates a disjoining pressure for spreading, the addition of nanoparticles can significantly enhance the spreading of one fluid (*e.g.*, water), displacing another immiscible fluid, on solid. McElfresh et al. (2012) also showed that the nanofluids significantly increased oil recovery even though they did not reveal the nature of their injection formulation.

While Wasan-Nikolov's "super-spreading" concept prompted active research efforts on the use of nanofluids for EOR, a couple of difficulties need to be resolved. First, the recent studies by Lu et al. (2013, 2015) and Wang et al. (2007) showed that the addition of nanoparticles in fact inhibited the spreading of the particle-containing liquid. Apparently, how the nanoparticles form structure and generate the disjoining pressure is not a straightforward matter and the details of the structure-formation and spreading mechanism further need to be defined (Lu 2016). The second difficulty is that, according to the theoretical calculations of Liu et al. (2012), to generate any meaningful disjoining pressure for spreading, the nanoparticle concentration has to be larger than 15 wt%. For EOR application, such a high concentration is probably too costly and may have difficulty for long-distance transport in the oil reservoir.

11.3 Use of Nanoparticles for Surfactant/Polymer EOR

In addition to the use of nanoparticles as stabilizers for foams and emulsions (as described in Chapter 10) and as an additive to injection water for wettability alteration (as described above), there have been efforts to use nanoparticles for other EOR applications. One notable approach is to add nanoparticles to the surfactant/polymer EOR formulations to seek synergy between nanoparticles and mainly polymer in increasing the polymer's viscosifying ability. Another potentially important application is the use of magnetic nanoparticles (1) to remove the hardness from the water to be used for chemical EOR injection formulation; and (2) to remove the remnant chemicals (polymer or surfactant) from the water that was produced from the chemical EOR process implementation. Just as the "green" use of magnetic nanoparticles to remove the micron-size oil droplets from the oilfield produced water (see Section 8.2.1), such applications of magnetic nanoparticles show a good promise as a way of implementing environmentally responsible oilfield operations.

11.3.1 Addition of SiO₂ Nanoparticles for Surfactant/Polymer EOR

Sharma et al. (2016) investigated the effect of adding silica nanoparticles to polymer EOR and surfactant/polymer EOR processes, respectively, in terms of wettability alteration and incremental oil recovery, at varying nanoparticle concentrations and temperatures. They found that, when the silica nanoparticles were added (1.0 wt%), the viscosities of both the polymer solution and the injectant surfactant/polymer formulation increased by about 3-fold from nanoparticle-free cases. The increase in viscosity of nanofluids was greater at higher temperatures. The pressure drops during nanoparticle-added flooding cases were higher than those during nanoparticle-free flooding cases, in accordance with the viscosity increases due to particle addition. With the chase brine injection, the pressure drop relaxed back to its initial brine injection values (0.05–0.07 MPa) for nanoparticle-free polymer and SP flooding cases, but not for nanoparticle-added polymer and SP flooding cases. This suggests that the permeability of the sand pack decreased during the transport of nanoparticle-added chemical formulations due to nanoparticle adsorption/retention. For both polymer and surfactant/polymer EOR processes, the addition of SiO₂ nanoparticles enhanced the oil recovery. They observed that the incremental oil recovery from nanoparticle-added surfactant/polymer flooding was greater than that from nanoparticle-added polymer flooding. The higher the temperature was, the greater the recovery enhancement.

Maghzi et al. (2013) experimentally investigated the effect of adding silica nanoparticles to polyacrylamide polymer solutions on the rheological behavior and enhanced heavy oil recovery. When 0.1 wt% of nanoparticles were added to 1000 ppm polymer solution, significant increases in viscosity, especially at low shear rate, were observed, and in oil-displacement experiments in glass micromodels, 10% more oil was recovered with addition of silica nanoparticles from the polymer flooding.

11.3.2 Conversion of Hard Brine to Soft Brine for Polymer EOR

For the polymers (such as the partially hydrolyzed polyacrylamide) that are used for chemical EOR processes, their viscosifying ability decreases significantly if the salinity of the brine in which the polymer is dissolved becomes higher. Also, if the hardness of the brine increases, the polymer viscosity decreases, mainly because of its precipitation which results from the association of the multi-valent cations with anions on polymer chain and the contraction of the chains. It is therefore advantageous to use a polymer solution that has a low salinity, preferably without any divalent ions. In some reservoir applications, it is difficult to obtain fresh, soft water, especially at the offshore platforms where the use of seawater is the only cost-effective option. If the salinity of the available water could be reduced at a low cost, or at least the divalent ions could be removed, the effectiveness of the polymer flooding would improve significantly. Ayirala et al. (2008) recently introduced the "designer water process," which is a two-step process including both nanofiltration (NF) and reverse osmosis (RO). NF membranes will allow only monovalent ions to pass through, while divalent ions such as sulfate, calcium, and magnesium are removed. Depending on the requirements for the particular polymer flooding application, only the NF may be employed, or the RO process is subsequently employed to lower the salinity of the water to be used. The latter scheme, of course, adds more cost.

Applications of nano-adsorbents for multi-valent ion removal to soften water have been significantly increasing due to their large surface area, a great number of active sites, and low diffusion resistance. In a manner similar to the removal of micron-scale oil droplets from oilfield produced water utilizing the surface-functionalized superparamagnetic nanoparticles (SPM-NPs) (see Section 8.2.1), Wang et al. (2014) investigated the feasibility of using surface-functionalized SPM-NPs to remove Ca^{2+} from hard brine to convert it to "soft" brine. The schematics of the process is shown in Figure 11.3, in which the Ca^{2+} cation is selectively removed using polyacrylic acid (PAA), where its carboxylic group readily attach divalent cations. In Section 2.6 on the description of different steps needed to properly characterize newly synthesized nanoparticles, Wang et al. (2017)'s PAA-attached superparamagnetic nanoparticles (PAA-SPM-NPs) were described in detail. Once a sufficient amount of Ca^{2+} ions is attached to PAA chains grafted to the SPM-NPs, the particles can be collected utilizing the high-gradient magnetic separation technique, as was done in Section 8.2.1 for oil droplet removal. If the NaCl salinity of the Ca^{2+} ion-containing water is low, the Ca^{2+} adsorption capacity of PAA-SPM-NPs as high as 57.3 mg/g at pH 7 from the 400 mg/L Ca^{2+} solution was obtained (Wang et al. 2014). The adsorption capacity at high salinity conditions is not as good because the high salinity screens the negative charges on the surface of PAA-MNPs, resulting in the formation of nanoparticle aggregates. Use of a copolymer of PAA and poly 2-acrylamido-2-methyl-1-propanesulfonic acid (PAMPS) as the MNP surface coating is promising (Wang et al. 2017), as it brings the synergy between the ability of carboxyl anions along PAA's flexible C-C chain to selectively capture Ca^{2+} and the PAMPS ligand's ability to prevent the chain contraction in a high-NaCl environment.

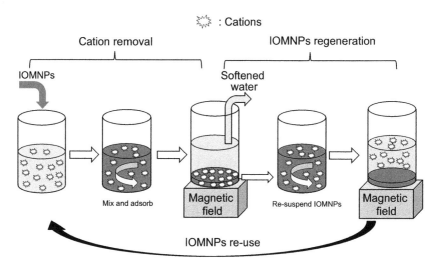

FIGURE 11.3

Process schematics for the selective adsorption of Ca^{2+} cations to the anionic adsorption sites along the PAA molecular chain attached to the iron-oxide superparamagnetic nanoparticles; and their magnetic separation, and regeneration and re-use.

11.3.3 Removal of Remnant EOR Polymer from Produced Water

When the remnant of the partially hydrolyzed polyacrylamide (HPAM) is produced from surfactant/polymer or polymer EOR operations, its removal from the produced water is required before the water is either re-used for injection or disposed. The use of SPM-NPs to remove this polymer was described in Section 8.2.3. Virtually 100% removal of HPAM from water was feasible, depending on the reaction conditions. The regeneration of spent SPM-NPs, using pH adjustment to recover the reactive sites, maintained above 90% removal efficiency for three-time repetitive usages. The electrostatic attraction between negatively charged HPAM polymer and positively charged SPM-NPs controls the attachment of SPM-NPs to the HPAM molecular chain; and the subsequent aggregation of the now neutralized SPM-NP-attached HPAM plays a critical role for accelerated and efficient magnetic separation (Ko et al. 2017).

For the field-scale application development for the selective attachment of "contaminants" (oil droplets in Section 8.2.1; and Ca^{2+} and left-over EOR polymer in the above and this sub-section, respectively) to SPM-NPs and the collection of the "contaminant"-attached magnetic nanoparticles by the well-known technique of high-gradient magnetic separation (HGMS) technique, an associated modeling work is required. Preliminary modeling attempts on the competitive adsorption of Ca^{2+} and other cations on anionic adsorption sites along the PAA molecular chain (Wang et al. 2017), and on the dynamics of magnetic separation of "contaminant"-attached SPM-NPs (Prigiobbe et al. 2015a,b), are available.

11.4 Nanoparticle-Stabilized CO_2 Hydrate Structure for Improved CO_2 Sequestration

Even though the use of nanoparticles to enhance the formation of CO_2 hydrates is unrelated to the EOR theme of this chapter, in view of its potential importance for improved CO_2 sequestration, a recent work on the generation of a highly porous CO_2 hydrate structure will be briefly described in this section. First, brief background information: at the deep sea's near-freezing temperature and high hydrostatic pressure conditions, CO_2 hydrate can be naturally formed from a mixture of CO_2 and seawater (House et al. 2006; Ohgaki et al. 1993). Exploiting this phenomenon, in-situ generation of CO_2 hydrate at deep sea-beds has been studied extensively as an economic way of sequestering anthropogenic CO_2 (Linga et al. 2007; Tajima et al. 2004; Tohidi et al. 2010). Being solid with its density slightly higher than that of seawater, the CO_2 hydrate can potentially stay for a long time at the deep sea-bed. This highly attractive concept, however, has two shortcomings. First, the hydrate formation is known to occur at the interface between CO_2 and water and, once a hydrate layer is formed at the CO_2/water interface (Lehmkuhler et al. 2009), the layer serves as a barrier to further contact between CO_2 and water. Consequently, subsequent hydrate generation is considerably slowed down. Therefore, a vigorous mixing of CO_2 and water, thus a significant input of mechanical energy, is needed to generate as much CO_2/water interfacial area as possible. Second, CO_2 hydrate generally consists of 5.8~7.8 molecules of water per CO_2 molecule (Kyung et al. 2015), which means that in order to store a meaningful volume of CO_2 in hydrate form, an inordinate amount of water also has to be processed.

Kim et al. (2017) found that a highly porous CO_2 hydrate structure can be generated by employing high internal-volume CO_2-in-water foam that is stabilized with hydrophilic silica nanoparticles (NPs). The concept was derived from the earlier work (Xue et al. 2016), as described in Section 7.3.2, to develop nanoparticle-stabilized CO_2-in-water foams that can be employed as an almost water-less fracturing fluid for oil and gas production. The generation of such high internal-phase volume fraction, stable foam, which has its continuous external phase in the form of very thin lamellae, require nanoparticles or colloidal particles to bring robust stability to the lamellae. According to Kim et al. (2017), the CO_2 hydrate particulates generated next to nanoparticles (that straddle the CO_2/water interface; see Figure 11.4) appear to provide certain structural rigidity for the lamella's integrity. Earlier studies (Carter et al. 2010; Farhang et al. 2014a,b) found that, when a water-in-CO_2 dispersion is generated by way of "dry water" (Binks and Murakami 2006) and is subsequently brought to the hydrate-generating condition, hydrate generation kinetics are significantly enhanced. Such enhancement is in line with earlier observations that the presence of silica surface enhances CO_2

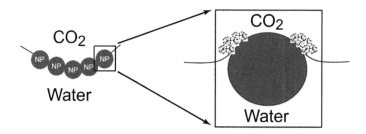

FIGURE 11.4
The presence of hydrophilic nanoparticles at the CO_2-water interface increases the length of the three-phase contact line, improving CO_2 hydrate nucleation. From Kim et al. (2017).

hydrate nucleation and growth. Importantly, the enhancement may result not only from the large solid surface area available from silica NPs, but also from the generation of tremendous lengths of three-phase (CO_2/water/silica) contact line on the NPs adsorbed at the CO_2/water interfaces of the dispersion. Recent molecular dynamics simulations (Bai et al. 2012) showed that the CO_2/water/silica contact line zone serves as a highly favorable nucleation site for CO_2 hydrates.

Figure 11.5(a) from Kim et al. (2017) shows a SEM image of the hydrate generated from gentle mixing of CO_2 and seawater that contains hydrophilic silica nanoparticles in a tumbler. At ambient conditions, such a gentle mixing cannot produce stable CO_2-in-water foam, because the adsorption of nanoparticles to the CO_2/water interface requires to overcome a certain energy barrier (Binks and Horozov 2006). Brought to the hydrate-generating condition, however, it appears that the CO_2-in-seawater foam generation was possible even with only moderate energy input from synergy between the newly generated hydrate at the CO_2/water interface and the nanoparticles that were attached to the hydrates and thus more easily brought to the CO_2/water interface. The adsorbed nanoparticles, in turn, would enhance further hydrate generation at three-phase contact lines. The nanoparticle-stabilized CO_2 hydrate structure, even generated with gentle mixing only, appears to be able to hold a substantial volume of CO_2 throughout the numerous lamellae. The nanoparticles added to water thus appear to contribute to formation of a highly porous CO_2 hydrate structure by way of generation of CO_2 foams with highly stable, nanoparticle-stabilized lamellae. Figure 11.5(b) shows the CO_2 hydrate generated from gentle mixing of CO_2 and seawater, but without nanoparticles. The planar surface indicates that there was little possibility to form stable CO_2 bubbles without the nanoparticles. As a result, the CO_2/seawater interface as a large plane served as the CO_2 hydrate generation sites, leading to a planar growth. This type of surface structure was also found with the CO_2 hydrate generated by gentle mixing of CO_2 and water. This also suggests that the stability of the CO_2 hydrate-induced CO_2/seawater

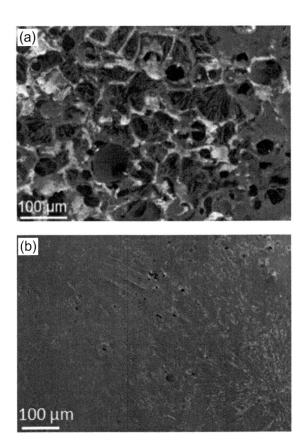

FIGURE 11.5
SEM images of CO_2 hydrate via gentle mixing of CO_2 and seawater (a) with nanoparticles and (b) without nanoparticles. From Kim et al. (2017).

dispersion in the absence of nanoparticles is rather tenuous, and the ability of such hydrate structure to hold CO_2 is quite uncertain.

Figure 11.6 shows a nanoscale SEM image of CO_2 hydrate generated by gentle mixing of CO_2 and seawater with NPs (see Figure 11.5(a)). The skeletal structure, which is the vestige of the CO_2 foam lamellae, was clearly captured, displaying its thickness around 200–500 nm. The swollen three-way junction appears to represent a plateau border with the formerly continuous aqueous phase. This SEM image suggests simultaneous CO_2 foam generation and hydrate generation. Such a highly porous CO_2 hydrate structure, generated in-situ either in geological formations or at subsea and containing CO_2 in its void space, could potentially be employed for secure sequestration of CO_2 in quantity, in the shallow subsurface of deep-sea sediments which are abundantly available worldwide.

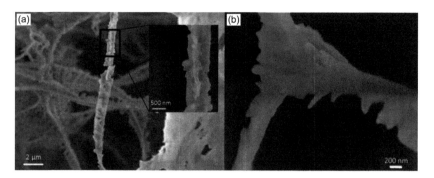

FIGURE 11.6
Nanoscale SEM image of CO_2 hydrate formed by gentle mixing of CO_2 and seawater with nanoparticles: Focusing on (a) a rod and (b) a three-way junction. From Kim et al. (2017).

11.5 Concluding Remarks

As discussed in Section 11.2.3, while the addition of nanoparticles to waterflood injection water definitely enhances the waterflood oil recovery, the incremental recovery is so far shown not to be substantial. A couple of possible reasons are suggested here with the hope of finding ways to further increase the oil recovery with the use of nanoparticles. One reasonably accepted mechanism for EOR with addition of nanoparticles is the deposition of nanoparticles on the pore wall to alter the wettability of reservoir rock, just like the addition of a dilute concentration of surfactant for wettability alteration, especially in naturally fractured reservoirs. As well-recognized with the use of surfactant, wettability is generally known to be altered with the diffusion of surfactant through rock pores and fractures, which is a rather slow process. Since a nanoparticle is still a finite-size solid even if it is very small, the Brownian diffusion of nanoparticles would be much slower than that of surfactant; and accordingly, the nanoparticle-based wettability alteration appears to be a very slow and tedious process, and some ways to accelerate the deposition of nanoparticles on rock pore surface need to be found. Wasan-Nikolov's "super-spreading" concept is a promising way to accomplish such acceleration. As discussed in Section 11.2.3, however, the nanoparticle addition sometimes actually inhibits spreading and the exact way to achieve the nanoparticle-induced, accelerated spreading is still unknown.

An interesting analogy is the mechanism for the well-known "coffee-ring effect" (Deegan et al. 1997; Shen et al. 2010), which is the accumulation of colloidal particles at the three-phase contact line region ("ring" formation) when a particle-containing liquid drop placed on a solid surface is evaporated. Even though the particles were initially uniformly dispersed in the liquid, due to the temperature-driven surface-tension gradient, the particles are drawn to the contact line zone, causing their accumulated deposition there. Somewhat related to this phenomenon, one possible scheme to bring the colloidal or nanoparticles to the three-phase contact line zone is to utilize the recent finding by Binks et al. (2006) that, when the particles adsorbed at a liquid surface exert a sufficient lateral repulsion and consequently a strong surface pressure, the particles result in the spontaneous film climbing up the container's wall. This novel observation warrants a further study for possible exploitation for the nanoparticle-based EOR.

Another nanoparticle-based EOR approach is the controlled association and dissociation of EOR polymer molecules with use of nanoparticles as a "linker." As reviewed in Section 11.3.1, the addition of nanoparticles to the EOR polymer can significantly modify the polymer's rheological behavior and consequently the oil-displacement behavior during the surfactant/ polymer and polymer flooding processes. Much improved understanding of the mechanisms for the nanoparticle-polymer association and dissociation, and the subsequent development of optimally functionalized nanoparticles, have the potential to significantly improve the performance of surfactant/ polymer and polymer EOR processes.

An intriguing and exciting development in wettability alternation is through manipulation of surface topology (Forbes 2008). This is a drastic departure from the traditional approach of altering the wettability by the chemical modification of the solid surface. Researchers at Oak Ridge National Laboratory (Krause 2008a,b) and others formed well-organized, nanoscale surface structures ("protrusions") that utilize the strong capillary force to prevent the wetting of the solid surface not only by hydrophilic liquids (*e.g.*, water) but also by lipophilic liquids (*e.g.*, hydrocarbons). Due to the large surface tension between liquid and air, the structured nanoscale protrusions can hold air in the interstices between the protrusions and make the solid surface nonwetting to *any* liquids. Other interesting new developments in nano-based surface treatment technology are also described by Forbes (2008). Treatments of the surfaces of oilfield facilities with such coatings could potentially reduce corrosion and scaling. In view of the frequently very harsh oilfield conditions, however, whether such rather delicate, nanoscale surface structures can be effectively maintained for long-term use is still to be demonstrated.

References

Al-Anssari, S., Barifcani, A., Wang, S., Maxim, L., and Iglauer, S. (2016) Wettability alteration of oil-wet carbonate by silica nanofluid. *J. Colloid Interface Sci.*, 461, 435–442.

Alomair, O. A., Matar, K. M., and Alsaeed, Y. H. (2015) Experimental study of enhanced heavy-oil recovery in Berea sandstone cores by use of nanofluids applications. *SPE Reservoir Eval. Eng.*, 8, 387–399.

Ayirala, S., Doe, P., Curole, M., and Chin, R. (2008) Polymer Flooding in Saline Heavy Oil Environments. (SPE 113396), SPE Improved Oil Recovery Symposium, Apr. 19–23, Tulsa, OK, USA.

Bai, D., Chen, G., Zhang, X., and Wang, W. (2012) Nucleation of the CO2 hydrate from three-phase contact lines. *Langmuir*, 28, 7730–7736.

Bayat, E. A., Junin, R., Samsuri, A., Piroozian, A., and Hokmabadi, M. (2014) Impact of metal oxide nanoparticles on enhanced oil recovery from limestone media at several temperatures. *Energy & Fuels*, 28(10), 6255–6266.

Binks, B. P., and Horozov, T. S. (2006) *Colloidal Particles at Liquid Interfaces*. Cambridge University Press, Cambridge, UK.

Binks, B. P., and Murakami, R. (2006) Phase inversion of particle-stabilized materials from foams to dry water. *Nat. Mater.*, 5, 865–869.

Binks, B. P., Clint, J. H., Fletcher, P. D. I., Lees, T. J. G., and Taylor, P. (2006) Growth of gold nanoparticle films driven by the coalescence of particle-stabilized emulsion drops. *Langmuir*, 22, 4100–4103.

Carter, B. O., Wang, W., Adams, D. J., and Cooper, A. I. (2010) Gas storage in "dry water" and "dry gel" clathrates. *Langmuir*, 26(5), 3186–3193.

Chengara, A., Nikolov, A. D., Wasan, D. T., Trokhymchuk, A., and Henderson, D. (2004) Spreading of nanofluids driven by the structural disjoining pressure gradient. *J. Colloid Interface Sci.*, 280(1), 192–201.

Cheraghian, G., and Hendraningrat, L. (2016a) A review on applications of nanotechnology in the enhanced oil recovery part A: Effects of nanoparticles on interfacial tension. *Int. Nano Lett.*, 6(2), 129–138.

Cheraghian, G., and Hendraningrat, L. (2016b) A review on applications of nanotechnology in the enhanced oil recovery part B: Effects of nanoparticles on flooding. *Int. Nano Lett.*, 6(1), 1–10.

Deegan, R. D., Bakajino, D. D. F., Huber, G., Nagel, S. R., and Witten, T. A. (1997) Capillary flow as the cause of ring stains from dried liquid drops. *Nature*, 389, 827–829.

Ehtesabi, H., Ahadian, M. M., Taghikhani, V., and Ghazanfari, M. H. (2013) Enhanced heavy oil recovery in sandstone cores using TiO2 nanofluids. *Energy & Fuels*, 28(1), 423–430.

El-Diasty, A. I., and Aly, A. M. (2015) Understanding the Mechanism of Nanoparticles Applications in Enhanced Oil Recovery. (SPE 175806), SPE North Africa Technical Conference, Sept. 14–16, Cairo, Egypt.

Farhang, F., Nguyen, A. V., and Hampton, M. A. (2014a) Influence of sodium halides on the kinetics of CO2 hydrate formation. *Energy & Fuels*, 28, 1220–1229.

Farhang, F., Nguyen, A. V., and Sewell, K. B. (2014b) Fundamental investigation of the effects of hydrophobic fumed silica on the formation of carbon dioxide gas hydrates. *Energy & Fuels*, 28, 7025–7037.

Forbes, P. (2008) Self-cleaning materials. *Sci. Am.*, 299, 88–95.

Gupta, R., and Mohanty, K. K. (2011). Wettability alteration mechanism for oil recovery from fractured carbonate rocks. *Transp. Porous Media*, 87(2), 635–652.

Haroun, M. R., Alhassan, S., Ansari, A. A., Al Kindy, N. A. M., Abou Sayed, N., Abdul Kareem, B. A., and Sarma, H. K. (2012). Smart Nano-EOR Process for Abu Dhabi Carbonate Reservoirs. Abu Dhabi International Petroleum Conference and Exhibition, November 11–14, Abu Dhabi, UAE.

Heggset, E. B., Chinga-Carrasco, G., and Syverud, K. (2017). Temperature stability of nanocellulose dispersions. *Carbohydr. Polym.*, 157, 114–121.

Hendraningrat, L., Li, S., and Torsater, O. (2013a) Enhancing Oil Recovery of Low-Permeability Berea Sandstone Through Optimised Nanofluids Concentration. (SPE 165283), SPE Enhanced Oil Recovery Conference, Jul. 2–4, Kuala Lumpur, Malaysia.

Hendraningrat, L., Li, S., and Torsater, O. (2013b) Effect of Some Parameters Influencing Enhanced Oil Recovery Process using Silica Nanoparticles: An Experimental Investigation. (SPE 165955), SPE Reservoir Characterisation and Simulation Conference, Sept. 16–18, Abu Dhabi, UAE.

Hendraningrat, L., Li, S., and Torsæter, O. (2013c) A coreflood investigation of nanofluid enhanced oil recovery. *J. Pet. Sci. Eng.*, 111, 128–138.

Hendraningrat, L., and Torsæter, O. (2014a) Understanding Fluid-Fluid and Fluid-Rock Interactions in the Presence of Hydrophilic Nanoparticles at Various Conditions. (SPE 171407), SPE Asia Pacific Oil and Gas Conference and Exhibition, Oct. 14–16, Adelaide, Australia.

Hendraningrat, L., and Torsaeter, O. (2014b) Unlocking the Potential of Metal Oxides Nanoparticles to Enhance the Oil Recovery. (OTC 24696), SPE Offshore Technology Conference, Mar. 25–28, Kuala Lumpur, Malaysia.

House, K. Z., Schrag, D. P., Harvey, C. F., and Lackner, K. S. (2006) Permanent carbon dioxide storage in deep-sea sediments. *Proc. Natl. Acad. Sci. USA*, 103, 12291–12295.

Hu, Z., Azmi, S. M., Raza, G., Glover, P. W. J., and Wen, D. (2016) Nanoparticle-assisted water-flooding in Berea sandstones. *Energy Fuels*, 30(4), 2791–2804.

Idogun, A. K., Iyagba, E. T., Ukwotije-Ikwut, R. P. and Aseminaso, A. (2016) A Review Study of Oil Displacement Mechanisms and Challenges of Nanoparticle Enhanced Oil Recovery. (SPE 184352), SPE Nigeria Annual International Conference and Exhibition, Aug. 2–4, Lagos, Nigeria.

Karimi, A., Fakhroueian, Z., Bahramian, A., Pour Khiabani, N., Darabad, J. B., Azin, R., and Arya, S. (2012) Wettability alteration in carbonates using zirconium oxide nanofluids: EOR implications. *Energy & Fuels*, 26(2), 1028–1036.

Kim, I., Nole, M., Jang, S., Ko, S., Daigle, H., Pope, G. A., and Huh, C. (2017) Highly porous CO_2 hydrate generation aided by silica nanoparticles for potential secure storage of CO_2 and desalination. *RSC Adv.*, 7, 9545–9550.

Ko, S., Lee, H., and Huh, C. (2017) Efficient removal of EOR polymer from produced water using magnetic nanoparticles and regeneration/re-use of spent particles. *SPE Prod. Oper.*, 32, 374–381.

Kondiparty, K., Nikolov, A. D., Wasan, D., and Liu, K. L. (2012) Dynamic spreading of nanofluids on solids. Part I: Experimental. *Langmuir*, 28(41), 14618–14623.

Krause, C. (2008a) Extremely Waterproof. ORNL Review 41, Oakridge National Laboratory.

Krause, C. (2008b) Miraculous Coatings. ORNL Review 41, Oakridge National Laboratory.

Kyung, D., Lim, H.-K., Kim, H., and Lee, W. (2015) CO_2 hydrate nucleation kinetics enhanced by an organo-mineral complex formed at the montmorillonite-water interface. *Environ. Sci. Technol.*, 49, 1197–1205.

Lehmkuhler, F., Paulus, M., Sternemann, C., Lietz, D., Venturini, F., Gutt, C., and Tolan, M. (2009) The carbon dioxide-water interface at conditions of gas hydrate formation. *J. Am. Chem. Soc.*, 131, 585–589.

Li, S., Hendraningrat, L., and Torsaeter, O. (2013) Improved Oil Recovery by Hydrophilic Silica Nanoparticles Suspension: 2-Phase Flow Experimental Studies. (IPTC 16707), International Petroleum Technology Conference, Mar. 26–28, Beijing, China.

Li, S., Genys, M., Wang, K., and Torsæter, O. (2015) Experimental Study of Wettability Alteration during Nanofluid Enhanced Oil Recovery Process and Its Effect on Oil Recovery. (SPE 175610), SPE Reservoir Characterization and Simulation Conference, Sept. 14–16, Abu Dhabi, UAE.

Li, S., and Torsaeter, O. (2015a) The Impact of Nanoparticles Adsorption and Transport on Wettability Alteration of Intermediate Wet Berea Sandstone. (SPE 172943), SPE Middle East Unconventional Resources Conference, Jan. 26–28, Muscat, Oman.

Li, S., and Torsaeter, O. (2015b) Experimental Investigation of the Influence of Nanoparticles Adsorption and Transport on Wettability Alteration for Oil Wet Berea Sandstone. (SPE 172539), SPE Middle East Oil Gas Conference, Mar. 8–11, Manama, Bahrain.

Lim, S., Horiuchi, H., Nikolov, A. D., and Wasan, D. (2015) Nanofluids alter the surface wettability of solids. *Langmuir*, 31(21), 5827–5835.

Linga, P., Kumar, R., and Englezos, P. (2007) Gas hydrate formation in a variable volume bed of silica sand particles. *J. Hazard. Mater.*, 149, 625–629.

Liu, K. I., Kondiparty, K., Nikolov, A. D., and Wasan, D. T. (2012) Dynamic spreading of nanofluids on solids. Part II: Modeling. *Langmuir*, 28, 16274–16284.

Lu, G., Duan, Y. Y., and Wang, X. D. (2015) Experimental study on the dynamic wetting of dilute nanofluids. *Colloids Surf. A*, 486, 6–13.

Lu, G., Hu, H., Duan, Y. Y., and Sun, Y. (2013) Wetting kinetics of water nano-droplet containing non-surfactant nanoparticles. *Appl. Phys. Lett.*, 103, 253104.

Lu, G. (2016) *Dynamic Wetting by Nanofluids.* Springer, Berlin.

McElfresh, P. M., Holcomb, D. L., and Ector, D. (2012) Application of Nanofluid Technology to Improve Recovery in Oil and Gas Wells. (SPE 154827), SPE International Oilfield Nanotechnology Conference and Exhibition, Noordwijk, Jun. 12–14, the Netherlands.

Maghzi, A., Mohebbi, A., Kharrat, R., and Ghazanfari, M. H. (2013) An experimental investigation of silica nanoparticles. Effect on the rheological behavior of polyacrylamide solution to enhanced heavy oil recovery. *Pet. Sci. Technol.*, 31, 500–508.

Molnes, S. N., Torrijos, I. P., Strand, S., Paso, K. G., and Syverud, K. (2016) Sandstone injectivity and salt stability of cellulose nanocrystals (CNC) dispersions— Premises for use of CNC in enhanced oil recovery. *Ind. Crops Prod.*, 93, 152–160.

Nazari, M. R., Bahramian, A., Fakhroueian, Z., Karimi, A., and Arya, S. (2015) Comparative study of using nanoparticles for enhanced oil recovery: Wettability alteration of carbonate rocks. *Energy & Fuels*, 29(4), 2111–2119.

Negin, C., Ali, S., and Xie, Q. (2016) Application of nanotechnology for enhancing oil recovery–A review. *Petroleum*, 2(4), 324–333.

Nikolov, A., and Wasan, D. (2014) Wetting-dewetting films: The role of structural forces. *Adv. Colloid Interface Sci.*, 206, 207–221.

Nwidee, L. N., Al-Anssari, S., Barifcani, A., Sarmadivaleh, M., Lebedev, M., and Iglauer, S. (2017) Nanoparticles influence on wetting behaviour of fractured limestone formation. *J. Pet. Sci. Eng.*, 149, 782–788.

Ogolo, N. A., Olafuyi, O. A., and Onyekonwu, M. (2012) Enhanced Oil Recovery Using Nanoparticles. (SPE 160847), SPE Annual Technical Conference, Apr. 8–11, Al-Khobar, Saudi Arabia.

Ohgaki, K., Makihara, Y., and Takano, K. (1993) Formation of CO_2 hydrate in pure and sea waters. *J. Chem. Eng. Japan.*, 26, 558–564.

Onyekonwu, M. O., and Ogolo, N. A. (2010) Investigating the Use of Nanoparticles in Enhancing Oil Recovery. (SPE 140744), Nigeria Annual International Conference, Jul. 31–Aug. 7, Tinapa-Calabar, Nigeria.

Prigiobbe, V., Ko, S., Wang, Q., Huh, C., Bryant, S. L., and Bennetzen, M. V. (2015a) Magnetic Nanoparticles for Efficient Removal of Oilfield Contaminants: Modeling of Magnetic Separation and Validation. (SPE 173786), SPE International Symposium Oilfield Chemistry, Apr. 13–15, The Woodlands, TX.

Prigiobbe, V., Ko, S., Huh, C., and Bryant, S. L. (2015b) Measuring and modeling the magnetic settling of superparamagnetic nanoparticle dispersions. *J. Colloid Interface Sci.*, 447, 58–67.

Ragab, A. M. S., and Hannora, A. E. (2015a) A Comparative Investigation of Nano Particle Effects for Improved Oil Recovery—Experimental Work. (SPE 175395), SPE Kuwait Oil and Gas Show and Conference, Oct. 11–14, Mishref, Kuwait.

Ragab, A. M. S., and Hannora, A. E. (2015b) An Experimental Investigation of Silica Nano Particles for Enhanced Oil Recovery Applications. (SPE 175829), SPE North Africa Technical Conference, Sept. 14–16, Cairo, Egypt.

Roustaei, A., and Bagherzadeh, H. (2015) Experimental investigation of SiO2 nanoparticles on enhanced oil recovery of carbonate reservoirs. *J. Pet. Explor. Prod. Technol.*, 5, 27–33.

Sharma, T., Iglauer, S., and Sangwai, J. S. (2016) Silica nanofluids in an oilfield polymer polyacrylamide: Interfacial properties, wettability alteration and applications for chemical enhanced oil recovery. *Ind. Eng. Chem. Res.*, 55, 12387–12397.

Shen, X. Y., Ho, C. M., and Wong, T. S. (2010) Minimal size of coffee ring structure. *J. Phys. Chem. B*, 114, 5269–5274.

Sun, X., Zhang, Y., Chen, G., and Gai, Z. (2017) Application of nanoparticles in enhanced oil recovery: A critical review of recent progress. *Energies*, 10, 345.

Tajima, H., Yamasaki, A., Kiyono, F., and Teng, H. (2004) New method for ocean disposal of CO2 by a submerged kenics-type static mixer. *AIChE J.*, 50, 871–878.

Tarek, M., and El-Banbi, A. H. (2015) Comprehensive Investigation of Effects of Nano-Fluid Mixtures to Enhance Oil Recovery. (SPE-175835), SPE North Africa Technical Conference, Sept. 14–16, Cairo, Egypt.

Tarek, M. (2015) Investigating Nano-Fluid Mixture Effects to Enhance Oil Recovery. (SPE-178739-STU), SPE Annual Technical Conference, Sept. 14–16, Houston, TX, USA.

Tohidi, B., Yang, J. H., Salehabadi, M., Anderson, R., and Chapoy, A. (2010) CO_2 Hydrates Could Provide Secondary Safety Factor in Subsurface Sequestration of CO2. *Environ. Sci. Technol.*, 44, 1509–1514.

Wang, Q., Prigiobbe, V., Huh, C., Bryant, S. L., Mogensen, K., and Bennetzen, M. V. (2014) Removal of Divalent Cations from Brine Using Selective Adsorption onto Magnetic Nanoparticles. (IPTC 17901), International Petroleum Technology Conference, Dec. 10–12, Kuala Lumpur, Malaysia.

Wang, Q., Prigiobbe, V., Huh, C., and Bryant, S. L. (2017) Alkaline earth element adsorption onto PAA-coated magnetic nanoparticles. *Energies*, 10(2), 223.

Wang, X. D., Lee, D. J., Peng, X. F., and Lai, J. Y. (2007) Spreading dynamics and dynamic contact angle of non-Newtonian fluids. *Langmuir*, 23, 8042–8047.

Wang, L., Zhang, G. C., Ge, J. J., Li, G. H., Zhang, J. Q., and Ding, B. D. (2010) Preparation of Microgel Nanospheres and Their Application in EOR. (SPE-130357), CPS/SPE International Oil Gas Conference, Jun. 8–10, Beijing, China.

Wasan, D. T., and Nikolov, A. D. (2003) Spreading of nanofluids on solids. *Nature*, 423, 156–159.

Wasan, D., Nikolov, A., and Kondiparty, K. (2011) The wetting and spreading of nanofluids on solids: Role of the structural disjoining pressure. *Curr. Opin. Colloid Interface Sci.*, 16(4), 344–349.

Wei, B., Li, Q., Jin, F., Li, H., and Wang, C. (2016) The potential of a novel nanofluid in enhancing oil recovery. *Energy Fuels*, 30(4), 2882–2891.

Xue, Z., Worthen, A., Qajar, A., Robert, I., Bryant, S. L., Huh, C., Prodanovic, M., and Johnston, K. P. (2016) Viscosity and stability of ultra-high internal phase CO2-in-water foams stabilized with surfactants and nanoparticles with or without polyelectrolytes. *J. Colloid Interface Sci.*, 461, 383–395.

Zhang, H., Nikolov, A., and Wasan, D. (2014a) Enhanced oil recovery (EOR) using nanoparticle dispersions: Underlying mechanism and imbibition experiments. *Energy & Fuels*, 28(5), 3002–3009.

Zhang, H., Nikolov, A., and Wasan, D (2014b) Dewetting film dynamics inside a capillary using a micellar nanofluid. *Langmuir*, 30(31), 9430–9435.

Zhang, H., Ramakrishnan, T. S., Nikolov, A., and Wasan, D. (2016) Enhanced oil recovery driven by nanofilm structural disjoining pressure: Flooding experiments and microvisualization. *Energy & Fuels*, 30(4), 2771–2779.

12

Heavy Oil Recovery

12.1 Introduction

All around the world and especially in Alberta, Canada, and the Orinoco Belt of Venezuela, there are still trillions of barrels of heavy oil to be recovered if the economics of their production is right. The utilization of this tremendous resource with the development of a new production technology is actually an achievable possibility if we consider the facts that the presence of the huge reserves of shale gas and oil in the US had been known for decades, but their economic production had been considered to be practically impossible; and that the persistent development and field application of the combined technology of horizontal well and multiple hydraulic fracturing brought the recent explosive activity in producing shale oil and gas.

Because of the extreme high viscosity of bitumen, the normal methods of fluid injection to displace resident oil, such as waterflooding, are not feasible; and the only way to produce heavy oil is somehow to reduce its viscosity drastically. Accordingly, different ways to reduce the heavy oil viscosity have been extensively researched during the last 100 years or so. Because the heavy oil viscosity rapidly decreases with temperature as shown Figure 12.1, various ways to efficiently introduce heat energy into the reservoir have been extensively investigated, among which steam injection is the most effective and most widely employed technique. As can be seen from Figure 12.1, the viscosity of Alberta's Athabasca bitumen at room temperature is ~1,000,000 cp so that it is practically a solid. When the temperature is raised to 200°C, its viscosity is lowered to ~10 cp, allowing it to flow in the reservoir rock pores for production, and in the pipeline for transport to the refinery. The viscosity of the oil is, therefore, a key parameter in selecting the most appropriate technique for production from heavy oil reservoirs. Figure 12.2 shows various heavy oil recovery techniques that are demonstrated to be effective in producing heavy oils of different viscosities. The techniques include not only thermal methods but also chemical methods such as the use of polymers to raise the viscosity of water that is injected to displace heavy oils with moderate viscosities (< ~5,000 cp).

There exists a huge literature on steam-assisted gravity drainage (SAGD), cyclic steam stimulation (CCS), and other steam-based heavy oil recovery

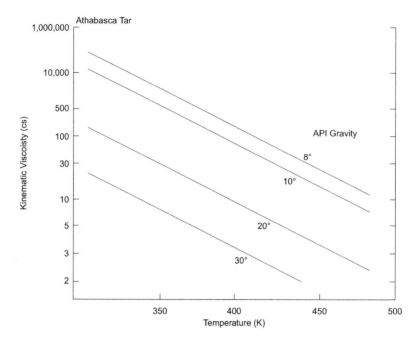

FIGURE 12.1
Dependence of viscosity of different heavy oils on temperature.

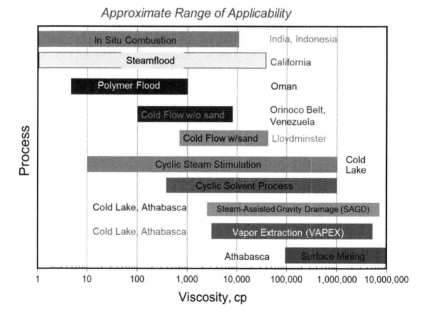

FIGURE 12.2
Applicability of various heavy oil recovery processes in terms of oil viscosity and the typical oil field for their application.

techniques, for which there are excellent reviews and monographs (Butler 1997; Lake et al. 2014). Other important techniques that have been researched extensively to raise temperature in the reservoir are in-situ combustion and electromagnetic (EM) heating. Again, excellent reviews on in-situ combustion (Greaves and Xia 2000; Shah et al. 2010) and EM heating (Bera and Babadagli 2015) are available. Because raising temperature by steam injection or EM heating is expensive, in order to help reduce the heavy oil viscosity more efficiently, addition of various low-cost solvents to steam have been studied which proved to provide excellent synergy with steam and which can also be separated from the produced oil and recycled. Before we describe the various use of nanoparticles to enhance recovery of heavy oil, a reader is also referred to recent reviews by Iskandar et al. (2016), Maity et al. (2010), and Muraza and Galadima (2015) on development and utilization of nano-catalysts for heavy oil recovery.

12.1.1 Brief Introduction to In-Situ Catalytic Upgrading of Heavy Oil

While use of steam is proven to be very effective from extensive commercial-scale applications, it is nevertheless expensive and also generates significant amounts of greenhouse gas. Researchers at the Alberta Research Council (Clark and Hyne 1984; Clark et al. 1990b; Hyne et al. 1982) earlier pointed out that, when the reservoir oil is heated to reduce its viscosity, the chemical nature of oil is also changed in the presence of steam and some mineral components of the reservoir rock which serve as reaction catalysts. A typical reaction described by Hyne et al. (1982) is the breakage of the C–S bond in sulfur-bearing hydrocarbon species which are often significant constituents in bitumen:

$$RCH_2CH_2SCH_3 + 2H_2O \leftrightarrow RCH_3 + CO_2 + H_2 + H_2S + CH_4 \qquad (12.1)$$

where R is an alkyl chain of a typical hydrocarbon molecule. Such reactions which involve water (steam) produce a variety of gases as shown in the above equation. Hyne et al. (1982) defined the breakage of heavy oil molecules by chemical reactions that involve heat, water, and some catalysts, broadly as "aquathermolysis." Notably, they proposed that breaking the asphaltene and resin molecules and their aggregates, with the help of catalysts, is another important way of lowering heavy oil viscosity. This is because the molecular networks formed by asphaltenes and resins are known to be mainly responsible for the extremely high viscosity of heavy oil. Built upon the above series of early work, an extensive research has since been carried out, as further discussed in Sections 12.3 and 12.4.

In the approaches to lower the heavy oil viscosity by chemically breaking oil molecules, as suggested above, in-situ processing at downhole where a high-temperature condition is already available has several advantages over conventional surface upgrading technology. First, in-situ upgrading can be applied on a well-by-well basis, so that large volumes of production needed for surface processing are not required. This is because there is no need to build large and costly high-temperature, high-pressure reaction vessels, as the reservoir formation serves as the reactor. In-situ upgrading can be applied both on land and offshore, in remote locations, and in places where a surface upgrader would be inappropriate. On the other hand, because in-situ processing needs to treat the whole crude oil, and not just specific boiling range fractions as is commonly done in refineries, the upgrading efficiency would inevitably be low. Reactants such as hydrogen or synthesis gas can be injected into the production well, so as to pass over a catalyst bed, or generated by in-situ combustion or thermal decomposition of the heavy oil in place. Production of CO and H_2 by the latter processes has been demonstrated.

12.1.2 Near-Wellbore Application of In-Situ Catalytic Upgrading

Weissman et al. (1996) and Weissman (1997) describe a number of downhole near-wellbore upgrading options, in which oil is passed over a fixed bed of catalyst that is placed at the gravel pack around wellbore or at the proppant pack in a hydraulic fracture. As most heavy oil reservoirs have a matrix consisting of unconsolidated sand, gravel packing to prevent sand production is usually a necessity, which could be exploited to place the catalyst bed. The presence of brine and the need to provide heat and reactant gases in a downhole environment present challenges not present in the conventional surface processing. Since the upgrading reactions need a high temperature even with the help of catalysts, Weissman et al. (1996) and Weissman (1997) suggested three different near-wellbore upgrading schemes, depending on how the catalytic bed formed at the near-wellbore zone is heated, as schematically shown in Figures 12.3 through 12.5. For all three schemes, the catalytic bed is established at the production well.

In the scheme shown in Figure 12.3, a catalyst-bearing zone is first established either in the gravel or proppant pack as mentioned above, or at the near-wellbore reservoir zone (as shown in the figure) by injecting the catalyst material and making it adsorb/adhere to the reservoir pore walls. Steam can be injected not only to heat up the catalytic bed but also to supply the water for aquathermolysis. Alternatively, an electric heater could be installed downhole to heat up the catalytic bed, over which the heavy oil being produced is flown so that the upgrading reactions can occur. Hydrogen gas or hydrogen-donor chemicals can also be injected to the catalytic bed to facilitate the upgrading reactions. In the schemes shown in Figures 12.4 and 12.5, instead of steam or electrical heating, the catalytic bed is heated by the hot gas that was generated by the wet combustion process (Hart et al. 2015a). For Figure 12.4, water and air are injected into a well and are ignited to start a partial combustion of reservoir oil. Once a sufficient heat is generated in the reservoir, the well is back-produced so that the heated oil and the hot gas generated from the burning can flow over the catalytic bed for upgrading

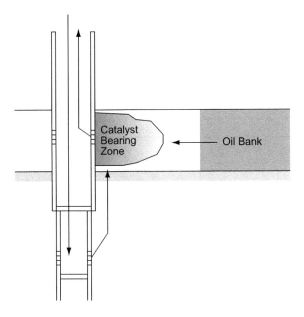

FIGURE 12.3
Downhole near-wellbore upgrading of heavy oil. As the oil with viscosity is lowered, *e.g.*, by earlier steam injection, it can be upgraded by passing through the catalyst bed installed at the annular gravel pack. Electromagnetic heating can be applied to develop the upgrading reaction condition and the hydrogen donor chemical can also be supplied to enhance the upgrading efficiency, as schematically shown. This also can be carried out as a huff-n-puff process. Adapted with permission from Weissman et al. (1996). Copyright 1996 American Chemical Society.

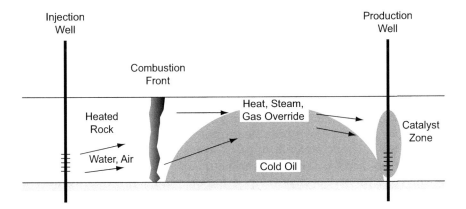

FIGURE 12.4
Downhole near-wellbore upgrading of heavy oil. Similar to the case of Figure 12.3 except that the oil viscosity reduction is achieved by the wet combustion process. Adapted with permission from Weissman et al. (1996). Copyright 1996 American Chemical Society.

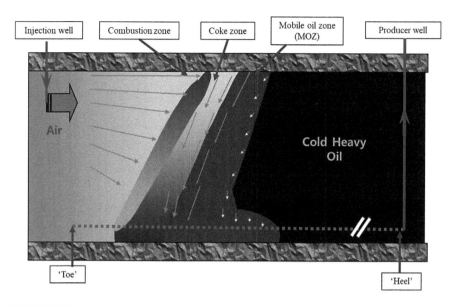

FIGURE 12.5
Schematic of the toe-to-heel air injection (THAI) in-situ combustion process. In its extension known as THAI-CAPRI process which also includes the downhole near-wellbore upgrading, the hot oil from the combustion process can be upgraded by passing through the catalyst bed installed at the annular gravel pack along the horizontal wellbore.

reactions. The scheme shown in Figure 12.5 only differs from the huff-n-puff process of Figure 12.4 in that the wet combustion is initiated at a nearby well, and the heated oil and hot gas generated are pushed to a producer, around which the catalytic bed has been established. A refined version of this scheme, as developed by Malcom Greaves' group at Bath University will be described in Section 12.3.5.

12.1.3 Effective Catalyst Delivery and Upgrading Deep in the Reservoir

Generation of catalytic sites on a solid surface, onto which oil is flown at the conditions for upgrading reactions, is a well-established and widely employed technique in the oil refining industry. While the catalytic solid is generally a fixed bed for refining purposes, such is obviously not feasible because the catalytic material must be injected, from the surface, into the porous reservoir rock. Therefore, various nanoparticles whose surface serves as catalytic sites have been developed and their upgrading effectiveness has been investigated. As described in detail below, while the laboratory batch experiments for heavy oil upgrading show significant reductions in oil viscosity, the upgrading experiments by injecting the catalytic nanoparticles (NPs) into a heavy oil-containing porous column such as a sand-pack do not appear to produce as good results. As further described below, the limited number of nanocatalyst injection field tests showed that the upgrading efficiency is far below the batch experiment results. This suggests that the delivery of the catalytic NPs to the resident oil and the subsequent creation of the effective environment for upgrading reactions there have not been properly established; and that the NP delivery and the creation of the catalytic reaction conditions in the reservoir are as important tasks as the development of good catalysts.

Because the only way, probably, of delivering the NPs to the oil deep in the reservoir is to inject them as a dispersion in an injection fluid such as steam (or more likely in the liquid water portion of a high-quality steam), the NP surface coating requirements that were discussed in Section 2.1 are relevant here: (1) long-term dispersion stability; (2) long-distance transport with minimal retention; and (3) targeted delivery to the resident oil phase. Because of the high-temperature requirement for catalytic reactions, however, fulfilling the above requirements is quite difficult. For example, stably dispersing the NPs in the low-density, gas-phase steam and making them flow in reservoir formation would be difficult. For an effective NP delivery as well as to create the required high temperature, the injection steam needs to be supercritical which then requires a high pressure or an addition of solvent (Butler 1997).

A number of plausible options for NP surface coating that satisfy the above requirements will be briefly discussed here. If the upgrading reactions could be effectively carried out in the absence of water (but may need H_2 donor such as decalin) and require only the NPs with catalytic surface sites, the surface coating requirements are: (1) NP dispersion stability in supercritical steam (or hot water) so that they can be injected and carried to where the oil is; and (2) when the NPs meet the oil, they go into the oil and the surface coating peels off, exposing the NP surface with catalytic sites. As employed for biomedical applications to transfer the NPs that had been synthesized in an oleic solvent to the water phase (Prakash et al. 2009; Shen et al. 1999), a bilayer adsorption of surfactant on the NP surface could potentially satisfy the requirement. With the bilayer adsorption with hydrophilic moiety outside, the NPs will be stable in water. When they contact the oil/water interface of the oil in the reservoir, the outer surfactant layer will peel off and the NPs now with the inner coating layer with lipophilic moiety will plop into the oil phase. Due to the high temperature in the reservoir, the inner surfactant layer would degrade exposing the bare, catalytic solid surface. A shortcoming of such an approach is that, because the surfactant is simply adsorbed on the NP surface, they could be prematurely desorbed while the NPs are transported in the reservoir. To prevent such desorption, a sufficient concentration of the surface coating surfactant needs to be included in the injection water.

One innovative surface coating method that has been employed to overcome the above shortcoming (Yoon et al. 2011) is the use of a polymer that can provide the NP's dispersion stability and its long-distance transport, as well as its affinity to the oil/water interface. More importantly, the polymer can form a network that wraps around the NP but that does not need to be attached to the NP surface. Because the polymer network completely surrounds the NP, it still serves as an effective surface coating. Once the NP is delivered to the oil phase, the polymer coating could again be made to degrade in the high-temperature environment. A detailed example on how such a polymeric coating material that satisfies all the requirements has been developed is described in Section 2.3.

If the upgrading reactions need water, probably the most effective way to create the catalytic reaction environment is to generate, in-situ in the reservoir, the NP-stabilized oil-in-steam (or water) emulsion, or more preferably, the steam (or water)-in-oil emulsion. By generating finely dispersed emulsions with NPs at their interface, a tremendous length of contact line can be produced for intimate catalytic reactions. Because the 3-phase contact line brings the oil, steam (or water), and the catalytic sites on the NP surface very closely together, the upgrading reactions could occur more effectively. We note that for the above case of catalytic reactions without water, the NP surface coating should be able to deliver the NPs to entirely inside the oil phase. With the present case of reactions that need water, however, the NP surface coating should deliver the NPs to the oil/steam interface to form stable emulsion droplets.

12.1.4 Oil Recovery Enhancement and Oil Quality Enhancement

At this point, it is important to make a distinction between the (mainly physical) techniques to enhance the oil recovery efficiency and the (mainly chemical) techniques to enhance the quality of oil. Steam injection to lower oil viscosity, thereby drastically reducing the displacement mobility ratio, is a representative "recovery enhancement" technique. Breaking the asphaltene and resin molecules and their molecular networks, with combined use of thermal energy and catalyst, as mentioned above and will further be described below, is a representative "quality enhancement" technique. It is noted here that another important "quality enhancement" technique is the physical removal of asphaltenes and resins from oil, especially when their mass fraction is not large. Such removal is important because it is believed that even a small amount of such molecular networks causes a significant increase in oil viscosity. This approach is described in more detail in Section 12.3.4.

12.2 Nanoparticle-Based Oil Recovery Enhancement

While the main focus of the NP usage for heavy oil recovery is as oil-upgrading catalysts, the NPs can also be employed as foam and emulsion stabilizers and wettability alteration agents, as used for lighter-oil EOR and described in detail in Chapters 10 and 11.

12.2.1 Nanoparticle-Stabilized Steam Foam

Just as CO_2 foam is employed to improve the sweep efficiency for the EOR with CO_2 injection, steam foam has been employed to improve the sweep efficiency for the steam-based heavy oil recovery processes. An excellent review on steam foam is available (Hirasaki 1989). Again, just as using nanoparticles to stabilize CO_2 foam brings a number of advantages over the use of surfactant, the steam foam could be stabilized with nanoparticles to bring some unique advantages. A key difficulty in using surfactants as a steam foam stabilizer is its thermal degradation at high temperature. Because the nanoparticle is a solid and accordingly is much less susceptible to degradation, its use as a steam foam stabilizer has great potential. Furthermore, the nanoparticle can be utilized as an oil-upgrading catalyst as described in some detail below. Recognizing these advantages, Khajehpour et al. (2016) recently carried out an investigation on the stability of high-temperature gas foam, employing N_2 as a surrogate for high-pressure steam. They also carried out coreflood experiments with Ottawa sandpacks to measure the in-situ foam mobility while flowing in the porous medium. While silica nanoparticle is quite effective in stabilizing CO_2 and N_2 foams at low-temperature conditions, it is not as effective at the high-temperature conditions; however, a combined use of nanoparticles with surfactants produced a synergy in generating foam and developing an apparent high viscosity for a good mobility control during the steam-based oil recovery process.

12.2.2 Nanoparticle-Stabilized Emulsion

In Section 10.4, the use of nanoparticle-stabilized emulsions for improved reservoir sweep is described and, in particular, the use of an emulsion whose internal phase is the natural gas liquid (NGL) which is abundantly available in the US due to the current, active development of shale oil reservoirs. As described there, such nanoparticle-stabilized emulsions, which can generate a high apparent viscosity, while remaining stable during their flow in the generally high-permeability heavy oil reservoirs, offer an alternative option to steam for effective recovery of heavy oil where viscosity is not extremely high (<10,000 cp). With such an objective, Arab et al. (2018) studied the nanoparticle-fortified emulsification of heavy oil, employing different surfactants and surface-modified silica nanoparticles. They found a significant synergy between surfactant and nanoparticles in developing a long-term stability of in-situ generated oil-in-water emulsions.

12.2.3 Wettability Alteration

In Section 11.2, various research efforts to use nanoparticles to alter wettability of reservoir rocks, thereby enhancing recovery of mainly light oil, have been described. Since most of the heavy oil reservoir rocks are strongly water-wet (Butler 1997) so that displacing the oil (as long as the viscosity of oil is sufficiently reduced) with the help of condensed, hot water can be efficiently achieved, altering the reservoir wettability is deemed not to be a critical requirement, unlike some other kinds of reservoirs such as the naturally fractured, oil-wet reservoirs. While not as extensive, efforts have been made to alter the wettability of heavy oil reservoir rocks. Cao et al. (2017) studied the effects of adding silica (SiO_2), alumina (Al_2O_3), and zirconia (ZrO_2) nanoparticles to water on the contact angle made by heavy oil and water on various rock surfaces.

12.3 Development of Catalysts for In-Situ Heavy Oil Upgrading

Before the use of various nanoparticles as upgrading catalysts is described, different kinds of upgrading catalysts for potential use deep in the reservoir are briefly described, namely: (1) water-soluble catalysts; (2) oil-soluble catalysts; and (3) dispersed solid catalysts. Some metals and metallic ions naturally available in the reservoir were found to serve as upgrading catalysts (Clark et al. 1990a; Fan et al. 2004; Hyne et al. 1982), but their effectiveness at the temperature and pressure conditions of the currently practiced, steam-based oil recovery processes is believed to be marginal. Since the theme of this book is the utilization of solid nanoparticles, *e.g.*, as heavy oil upgrading catalysts, the review of literature on the water-soluble and oil-soluble catalysts, given in Section 12.3.1, will be quite limited while the main focus is on literature on the development and use of nanoparticle catalysts.

For upgrading of crude oil in the refinery, catalysts are placed on fixed-bed support. For the catalyst to be effective deep in the reservoir, either it is prepared at the surface and transported to the target location (i.e., where oil is) as a stable dispersion in the injection fluid as described above, or the constituent chemical components that will generate the needed catalyst in-situ in the reservoir need to be injected. As the latter option is difficult to achieve, the former option, which is the use of nanoparticle catalysts, has been extensively investigated in recent years. There are four different kinds of nanoparticle catalysts: (1) metal or metal oxide nanoparticles whose surface directly serves as catalytic sites; (2) catalyst chemical which is in nano-size solid form and can slowly dissolve either in water or oil; (3) nanoparticles on whose surface catalyst chemical(s) is attached; and (4) nanoparticles on whose surface another smaller catalyst nanoparticles (such as the above three) are attached. These will be described in some detail below.

12.3.1 Water- and Oil-Soluble Catalysts

12.3.1.1 Water-Soluble Catalysts

As described in the introduction to this chapter, by carrying out a comprehensive series of aquathermolysis experiments, Hyne and Clark's group at the Alberta Research Council in Canada found that many transition metal cations serve as catalysts in breaking not only model compounds such as thiophene and tetrahydrothiophene, but also some bitumen. Clark and Kirk (1994) reported that Fe^{++} and Ru^{+++} served as effective catalysts in lowering bitumen viscosity from 2140 cp to 520 cp by aquathermolysis at high temperatures of 375–415°C. Zhong et al. (2003) reported that Fe^{++}, Co^{++}, Mo^{++}, Ni^{++}, and Al^{++} cations reduced viscosity and lowered the molecular weight of Liaohe extra-heavy oil by aquathermolysis reactions, when a hydrogen donor (tetralin) was added. The reactions were carried out at temperatures of 160–280°C, pressure of 10–25 MPa, and 24–240 hours of reaction time. The bitumen viscosity was reduced by up to 90%. Fe^{++}, Co^{++}, and Mo^{++} gave almost similar viscosity reduction (by ~60%).

One notable recent aquathermolysis study employing water-soluble catalysts is by Chen et al. (2009), who employed a kegging heteropoly acid salt, $K_3PMo_{12}O_{40}$, as a catalyst. Due to their extremely strong acidity and oxidizing power, kegging heteropoly acids are well-known catalysts (Kozhevnikov 1998) and Chen et al. (2009) showed that the heteropoly acid catalyst helped to reduce the viscosity of heavy oil by 92% at 280°C of reaction temperature. Based on NMR and other characterization studies, they believe that the aquathermolysis reactions did not alter the structure of asphaltenes in heavy oil, but the structure of resins was significantly changed with the opening of rings of aromatic molecules in resins. The laboratory testing of water-soluble metallic salt (Fe^{3+}) as an additive to improve in-situ combustion performance is also of note (He et al. 2005).

12.3.1.2 Oil-Soluble Catalysts

While the water-soluble catalysts have the key advantage that they can be added to the injection water and be carried to the reservoir where the oil is, they still need to contact the oil phase in rock pores readily to serve as effective upgrading catalysts, which could be sometimes difficult. Realizing this shortcoming, use of oil-soluble catalysts has also been developed and tested by researchers, even though ways to deliver them from the surface to the oil deep into the reservoir still have to be developed. As previously described, one possible way is to incorporate the oil-soluble catalyst chemical(s) into the oil droplets of the oil-in-water emulsion that is then injected into the reservoir. As an example of oil-soluble catalyst development work, Wen et al. (2007) prepared the molybdenum oleate as catalyst (with 24.9 wt% Mo content) and used it for viscosity reduction of Liaohe heavy oils. The upgrading reaction was carried out in an autoclave by mixing 75 g of oil, 0.4 g of catalyst, and 25 g of water at 240°C for 24 hrs. During the upgrading experiment, a large amount of gas (CO_2, H_2S, light hydrocarbons of C_2–C_7) was generated and the oil viscosity was reduced by up to 90%.

Mohammad and Mamora (2008) used an organometallic iron catalyst, $Fe(CH_3COCHCOCH_3)$, which is highly soluble in tetralin, to upgrade a Jobo (Venezuelan) heavy oil. Tetralin (1,2,3,4-tetrahydronaphthalene, $C_{10}H_{12}$) is a double-ring hydrocarbon which is widely used as a hydrogen donor for heavy oil upgrading reactions. At the temperature of 273°C, the combined use of the oil-soluble catalyst and tetralin was highly effective in reducing the viscosity of oil. Yufeng et al. (2009) compared the effectiveness of the water-soluble and oil-soluble catalysts in reducing the viscosity of Liaohe heavy oil. For their experiments, they separated out the asphaltene and some other portions from the oil and used them for the reaction and measured the degree of conversion of asphaltene and resin. They employed $NiSO_4$ and $FeSO_4$ as water-soluble catalysts; and Ni naphthenate and Fe naphthenate as oil-soluble catalysts. The conversion of asphaltene was 3.8–14.9% and that of resin was 8.1–22.9%; and the order of conversion effectiveness is Fe naphthenate >Ni naphthenate >$FeSO_4$ >$NiSO_4$.

12.3.2 Catalysts Supported by Solid Substrate

Solid catalyst materials found effective for upgrading heavy crude oil, vacuum residual oil, and similar heavy petroleum fractions should be applicable for in-situ upgrading by making them as nanoparticles. Catalysts consisting of molybdenum oxide with cobalt or nickel oxides on high-surface-area alumina or silica-alumina supports are well known and used in large quantities in refining processes, including hydrotreating and hydrocracking (Speight 1991). Exposing conventional hydroprocessing catalysts to a high-pressure water/steam environment, downhole in the reservoir, apparently does not affect the dispersion state of the catalyst metals, but overall catalytic activity is known to decrease (Laurent and Delmon 1994). Among many hydroprocessing catalysts, Ni-Mo and Co-Mo catalysts had similar sulfur and nitrogen removal activities from heavy gas oils. Commercial Ni-Mo and Co-Mo catalysts were found to be more or less equivalent for hydrodesulfurization of Athabasca oil sands coker gas oil.

Wang et al. (2012) prepared zirconia-supported tungsten oxide and used it as a superacid catalyst to decrease the viscosity of heavy oil without water through mild cracking of heavy hydrocarbons. The catalyst loading of 20 wt% tungsten prepared by the hydrothermal method can reduce the viscosity of Liaohe heavy oil from 5740 cp to 1020 cp under reaction conditions of 220°C, 6 hr., and 2 MPa. The upgrading reaction involved is likely to follow the ionic mechanism, which is different from the free radical mechanism in conventional aquathermolysis, with which the viscosity of reacted oil is sometimes regressed to a higher value, even though not to the initial, very high value.

12.3.3 Nanoparticle Catalysts

As discussed in Section 12.1.3, the most practical way of placing the heavy oil upgrading catalysts in the reservoir where the oil exists is probably by injecting the catalyst nanoparticles that are stably dispersed in the injection fluid. For this reason, extensive laboratory studies have been recently carried out in testing the effectiveness of various nanoparticles as catalysts. Very promising and active developments are reported in the literature on the use of various nanoparticles as upgrading catalysts, where the large surface area per mass offers a large number of catalytic sites. Furthermore, its nano size allows their delivery to the oil deep in the reservoir, with proper surface coating to allow their long-term dispersion stability and minimal retention in the reservoir rock, as pointed out earlier. Many different kinds of nanoparticles were employed for testing: Ni (Li et al. 2007; Wu et al. 2013), α-Fe_2O_3 together with aromatic sulfonic acid (Chen et al. 2008; Wang et al. 2010), NiO, Fe_3O_4 and Co_3O_4 (Nassar et al. 2011a, 2011b), NiO (Noorlaily et al. 2013), zeolite-supported α-Fe_2O_3 and Fe_3O_4 (Nurhayati et al. 2013), Fe_3O_4 (Iskandar et al. 2014; Nugraha et al. 2013), and TiO_2 (Ehtesabi et al. 2013). Use of dispersed catalysts such as molybdenum, with addition of methane as hydrogen source, by Ovalles et al. (1998, 2003) is also notable. Some of these activities are briefly discussed below.

Using microemulsion method, Li et al. (2007) prepared nickel nanoparticle catalyst which is in spheroidal form with mean particle size of 6.3 nm. They used it for their investigation to reduce the viscosity of Liaohe extra-heavy oil by aquathermolysis. During the upgrading experiments carried out at 280°C, some of C–S bonds in the extra-heavy oil are destroyed, sulfur content is reduced, and the component with high molecular weight is partly converted into smaller molecules. The sulfur content changed from 0.45% to 0.23%, the content of resin and asphaltene was reduced by 15.83% and 15.33%, respectively. With respect to the original crude oil, the viscosity of the upgraded sample is changed from 139800 mPa·s to 2400 mPa·s at 50°C, an approximately 98.90% reduction. Li et al. (2013) further investigated the upgrading performance of $Fe(C_7H_7O_3S)_3$ and $Cu(C_7H_7O_3S)_2$ catalysts prepared from Fe_2O_3 and CuO powders, respectively, which were employed for catalytic aquathermolysis of six different heavy oils with quite positive results. Figure 12.6 from Li et al. (2013) shows a typical aquathermolysis reaction catalyzed by the Fe^{3+} ion.

Wu et al. (2013) prepared nano-nickel catalyst in a microemulsion system and used it to reduce the viscosity of extra-heavy oil by aquathermolysis. The viscosity of extra-heavy oil (San56-13-19) was reduced by 90.36% at 200°C. The analysis of structure and group composition of the heavy oil before and after the upgrading reactions showed that the pyrolysis of asphaltene played a key role in viscosity reduction, the average molecular weight of asphaltene decreasing by 28.06%. The central aromatic framework was broken, which included the opening of the heterocyclic ring and the breaking of the

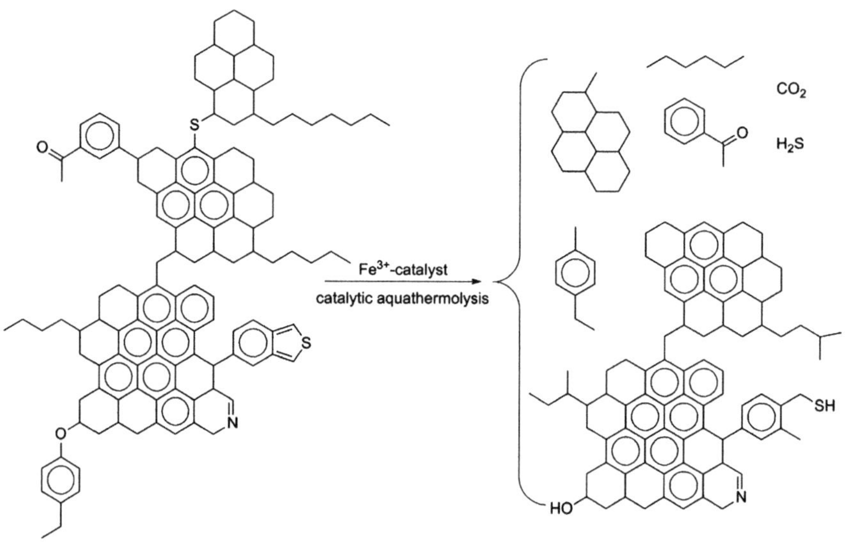

FIGURE 12.6
An example of aquathermolysis reaction catalyzed by Fe^{3+} ion. Reprinted with permission from Li et al. (2013). Copyright 2013 American Chemical Society.

cycloalkane bridge bonds and alkyl side chains. The aggregation of heavy components became more disperse in the heavy oil.

Hamedi-Shokrlu and Babadagli (2010, 2011, 2013) and Hamedi-Shokrlu et al. (2013) carried out experiments to study the effect of micro- and nano-sized metal particles on the enhancement of upgrading reactions within the oil phase. The experiments showed that even at low temperatures (<100°C), the particles reduce the heavy oil viscosity after being mixed with the oil phase. The amount of the viscosity reduction is a function of the concentration of the particles and there exists an optimum concentration of particles yielding a maximum amount of viscosity reduction. Also, the trend of viscosity versus concentration of the particles is a function of the type and size of the nanoparticles. The second series of experiments revealed that the same trend of viscosity versus concentration of particles is observed at steam injection conditions (300°C), but at a much higher degree of reduction in viscosity, compared with the low temperature, water-free experiments. The amount of viscosity reduction and the optimum concentration of the metal particles strongly depend on the oil sample composition, especially asphaltene content, and the metal type and size of nanoparticles. The coordination reactions of transition metal particles with the heavy molecules of oil may dominate the exothermic reactions in longer times. They also suggest that attachment of asphaltene molecules to the metal particles, mainly due to the Ostwald ripening process, can be considered as the non-chemical viscosity reduction mechanism.

Yi et al. (2018) studied the effects of adding nickel nanoparticles (of 40–70 nm size) on oil recovery during the late cycles of cyclic-steam-stimulation (CSS) process. Two Mexican heavy oils (of viscosities of 159,000 and 45,590 cp at 25°C) with high resin and asphaltene contents, 24.36 and 24.57 wt%, respectively, were employed for their CSS experiments with clean silica sandpacks (with porosity of 0.344). The upgrading experiments were carried out at two different temperatures (150 and 220°C) and with varying nickel nanoparticle (Ni NP) concentration of 0.05–0.5 wt%. They found that Ni NPs acted as an effective catalyst, mainly in breaking C–S bonds of constituent molecules in resins and asphaltenes; that the best oil recovery was attained at NP concentration of 0.2 wt% at 220°C; and that the contribution of Ni NPs for improving oil recovery originated mainly from their distribution near the injection port. This last finding suggests either that the injected NP dispersion did not propagate properly all through the sandpack, or that the effectiveness of Ni NP as an upgrading catalyst quickly diminished once the NPs contact some components of the oil or mineral.

While the use of Ni nanocatalysts was discussed in some detail, as an example, use of other metallic nanocatalysts, *e.g.*, Fe_3O_4 and Co_3O_4, was also investigated as cited at the beginning of this sub-section. Below, the use of multi-metallic nanocatalysts will be described.

Ultra-Dispersed Catalysts:

An active research effort is ongoing at the University of Calgary to upgrade Alberta bitumen by employing ultradispersed catalysts. Unlike the use of nanoparticles to scavenge H_2S in the oil or to adsorb off some of the asphaltenes from the oil, as described below, the hydrocracking requires a high temperature (>300°C). Unless it is employed together with the in-situ combustion process as described below in Section 12.3.5, its implementation as a part of steam-based recovery processes will be quite difficult.

The catalysts tested for hydrocracking were usually nanoparticles made of bimetallic alloys such as NiMo, CoMo, or NiW. Galarraga and Pereira-Almao (2010) used a trimetallic alloy catalyst (Ni+W+Mo) for superior hydrocracking efficiency. The nanocatalysts are usually employed together with a hydrogen donor, which not only increases the upgrading efficiency but also helps to prevent excessive generation of coke. Hashemi et al. (2013a, 2013b) further investigated the upgrading of Athabasca bitumen using the (Ni+W+Mo) catalyst nanoparticles in an oil-sandpack column. The experiments were conducted at a pressure of 3.5 MPa, temperature from 320 to 340°C, and with hydrogen flow. Figure 12.7 shows the increase in API gravity and the decrease of oil viscosity as a function of reaction time.

Nassar et al. (2011b) studied the effectiveness of Co_3O_4 nanoparticles (of 22 nm average size) in reducing the asphaltene content of a heavy oil and found that 32 wt% reduction could be achieved at 300°C. Their analysis of asphaltene cracking activation energy suggested a significant reduction in

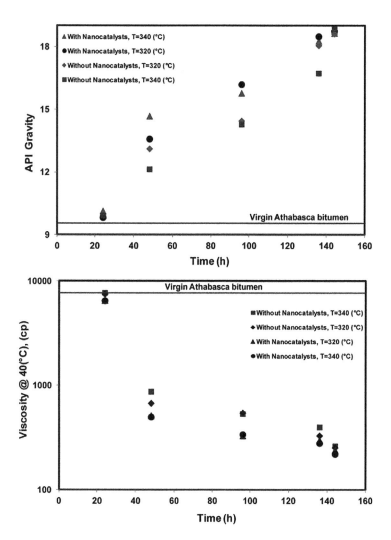

FIGURE 12.7
Athabasca bitumen upgrading versus reaction time. Trimetallic nanoparticle catalyst (Mo = 0.627, Ni = 0.181, W = 0.192) dispersed in VGO was injected into a bitumen-containing sandpack. Increase in API gravity; and decrease in oil viscosity. Reprinted with permission from Hashemi et al. (2013b). Copyright 2013 American Chemical Society.

activation energy in the presence of the nanoparticle catalyst. They also suggested that such asphaltene cracking reactions could be feasible at lower temperatures that are practically achievable in steam-based oil recovery processes such as SAGD.

Nassar and Husein (2010) and Nassar et al. (2011c) explored a novel method of using nanoparticles which carry out the dual functions of (1) preferentially adsorbing asphaltene on their surface and (2) serving as catalyst for steam cracking of adsorbed asphaltenes. They believed that catalytic steam cracking/gasification would allow for oil upgrading and production of H_2 which further helps the upgrading reactions. Catalytic steam gasification of adsorbed asphaltenes over nanoparticles was carried out and studied using a simultaneous thermogravimetric analysis/differential scanning calorimetry (TGA/DSC), which showed the extent of asphaltene gasification in terms of its mass loss. For virgin asphaltene steam gasification/cracking, the mass loss profiles can be divided into three regions: (1) a low-temperature range up to 360°C; (2) a mid-temperature range from 360 to 500°C; and (3) a high-temperature range beyond 500°C. Because asphaltenes are heavy fractions of oil, little mass change is observed during pyrolysis/gasification in the low-temperature range. Steam gasification/cracking of virgin asphaltenes without nanoparticles in the presence of steam occurs above 500°C, as evidenced by the presence of an exotherm that follows the mass loss profile shown in Figure 12.8. The mass loss profile of asphaltenes, after adsorption over different nanoparticles (Figure 12.8), shows that steam gasification and/or cracking of asphaltenes occur at a much lower temperature. From the peak temperatures of the mass loss during steam gasification/cracking reactions, and from Figure 12.9 showing the % conversion of asphaltenes, it appears that the presence of NiO, Co_3O_4, and Fe_3O_4 nanoparticles lowered the reaction temperature from ~500 to 317, 330, and 380°C, respectively.

Hashemi et al. (2014) provide a comprehensive review of the University of Calgary's UDC research and related works by other institutions.

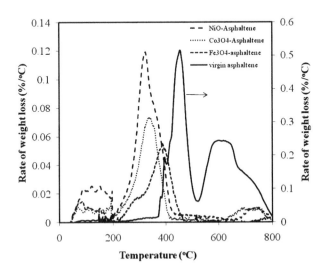

FIGURE 12.8
Thermogravimetric/differential thermal analysis (TG/DTA) curves for asphaltene gasification with and without metal oxide nanoparticles. Heating rate = 10°C/min; air flow = 100 cm³/min. Reprinted with permission from Nassar et al. (2011c). Copyright 2011 American Chemical Society.

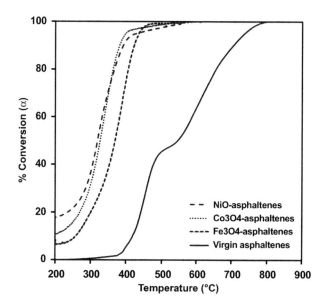

FIGURE 12.9
Percent conversion of asphaltenes in the presence and absence of different metal oxide nanoparticles. Reprinted with permission from Nassar et al. (2011c). Copyright 2011 American Chemical Society.

12.3.4 Nanoparticle-Based Removal of Asphaltenes and H_2S from Oil

One approach to make the oil in the reservoir to attain a lower viscosity is to remove some of the asphaltenes from the oil (Abu Tarboush and Husein 2012; Shayan and Mirzayi 2015), as the asphaltene content is the key contributor for the high viscosity of heavy oil (Luo and Gu 2006). This could be achieved by generating nano catalyst dispersions in the reservoir, onto which the asphaltenes are preferentially adsorbed. Nassar et al. (2012, 2011a, 2011b) investigated the kinetics of asphaltene adsorption on the commercially available alumina and other metal oxide nanoparticles. They found that a significant amount of asphaltenes was adsorbed onto nanoparticles in a very short time. Figure 12.10 from Nassar et al. (2011b) shows the asphaltene adsorption isotherm for different nanoparticles. As the asphaltenes are removed from the oil, its viscosity will be reduced. By creating appropriate reaction conditions, the asphaltene adsorbed at the nanoparticles could potentially be upgraded, the nanoparticles serving as catalyst.

Another important use of nanoparticles is to remove the hydrogen sulfide dissolved in crude oil (not necessarily the heavy oil only). Because it has been known that the oxides of zinc, iron, and other metals are good scavengers of hydrogen sulfide, the highly dispersed metal oxide nanoparticles will be effective adsorbents because of their large surface area. Nassar et al. (2011a, 2011b)

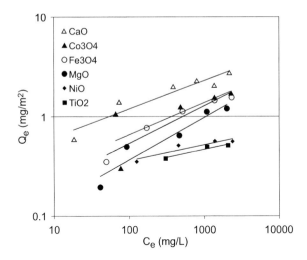

FIGURE 12.10
Comparison of asphaltene adsorption and oxidation for different metal oxide nanoparticles. Reprinted with permission from Nassar et al. (2011b). Copyright 2011 American Chemical Society.

studied the thermodynamics and kinetics of H_2S adsorption on iron oxide and iron hydroxide nanoparticles.

Nanoparticles with a design surface coating could be employed to selectively attach the asphaltenes and resins to their surface, which could then be separated from the oil, either in-situ in the reservoir and left there or after their production to the surface. As described earlier, magnetic nanoparticles (MNP) could be employed for the required attachment in-situ in the reservoir; magnetic separation of the (MNP + asphaltene, resin) aggregates after their production; and separation of MNPs from the asphaltenes and resins so that the particles can be regenerated to restore their surface catalytic sites and reuse.

12.3.5 Use of Nanoparticles for In-Situ Combustion Processes

One key difficulty in carrying out in-situ upgrading during steam-based heavy oil recovery processes is that the creation of the high temperature required for upgrading reactions is difficult. As a very high temperature is known to develop during the burning process, the in-situ combustion process offers the possibility of creating an effective environment for successful upgrading reactions. Toe-to-heel air injection (THAI) and its catalytic add-on process (CAPRI), as developed by Malcolm Greaves and his team (Al-Marshed et al. 2015a, 2015b; Greaves and Xia 2000; Hart et al. 2015a, 2015b, 2016, 2017; Shah et al. 2010), combine in-situ combustion with catalytic upgrading using an annular catalyst-pack placed around the horizontal producer well. This technique potentially offers higher recovery levels than the steam-based methods due to its ability to achieve the required upgrading reaction temperatures.

THAI is an in-situ combustion process where a fraction of the oil in the reservoir is intentionally burned with injection of air, and thereby the otherwise immobile oil is progressively mobilized along the horizontal well, with an increase of the reservoir temperature (Greaves and Xia 2000), as schematically shown in Figure 12.5. Thus, combining the in-situ combustion process with the advanced horizontal well technique, the oil whose viscosity has been reduced not only due to the temperature increase but also by thermal cracking can be more effectively produced. CAPRI is the catalytic extension to THAI whose objective is to achieve further in-situ upgrading of heavy oil by placing an annular layer of catalyst around the perforated horizontal producer well to create a downhole catalytic reactor. Oil upgrading is believed to occur by a combination of carbon-rejection reactions (thermal cracking) and hydrogen addition at the surface of a hydroconversion or hydrotreating catalyst. In this sub-section, the results from a comprehensive series of experiments carried out to further advance the THAI-CAPRI concept (Al-Marshed et al. 2015a, 2015b; Hart et al. 2015a, 2015b, 2016, 2017; Shah et al. 2011) are briefly described. As the desired upgrading reactions with the help of catalysts are expected to occur with the oil already mobilized due to combustion, the above experiments focused on the effects of different catalysts on upgrading, thereby decoupling the catalytic upgrading process from the combustion process. The mobile oil zone in THAI, the starting state for the catalytic upgrading reaction experiments, contains oil, water (steam), and combustion gases (including some CO and unconsumed O_2) at temperatures of ~500 to 600°C and at reservoir pressures of ~30 to 50 bar.

Shah et al. (2011) reported the upgrading experiments with a microreactor containing 10 g of catalyst, with oil flow of 1 mL/min and gas flow of 0.5 L/min, under different temperatures, pressures, and gas environments. Catalysts tested included alumina-supported CoMo, NiMo, and ZnO/CuO. At 500°C and 20 bar condition, the oil with 12.8° API was upgraded by an average of 6.1° API but catalyst lifetime was limited to 1.5 hours. When the temperature was lowered to 420°C, while the upgrading was by an average of 1.6° API (sometimes up to 3° API), the catalyst lifetime was extended to 77.5 hours. A significant shortcoming in applying the catalyst as an annulus of stationary pelleted pack around the wellbore is that, the coke generated as a result of the upgrading reactions is deposited on the catalyst pellets, thus blocking the access of the oil to the catalyst surface. Hart et al. (2015a) carried out a comparison study of the fixed-bed catalyst and the dispersed particulate catalyst during the upgrading reactions. The fixed-bed CoMo/Al$_2$O$_3$ catalyst (1.2 mm diameter with 2–5 mm length) was tested at a temperature of 425°C, pressure of 20 bar, and residence time of 10 min. The same catalyst in dispersed particulate form ($d = 2.6$ μm) was tested in a batch reactor, but the residence time, catalyst-to-oil ratio, and the Reynolds number were kept the same to ensure dynamic similitude. An oil with 13.8 API was upgraded by 5.6 API for the fixed-bed case while the API gravity increased by 8.7 for the dispersed particle case. The dispersed particles provided high surface area to volume ratio, allowing more accessible reaction sites per unit mass. Furthermore, the reduction of sulfur of 38.6% and (Ni + V) content of 85.2% in the produced oil show greater heteroatom removal compared with 29% (sulfur) and 45.6% (Ni + V) observed in the product from the fixed-bed.

As a part of a continuing effort to further develop the THAI-CAPRI process, the use of upgrading catalysts in the form of dispersed nanoparticles was investigated by Al-Marshed et al. (2015a) with Mo, Ni, and Fe transition metal catalysts, and by Al-Marshed et al. (2015b) with a Fe_2O_3 iron-oxide catalyst. Hydroconversion of heavy oil was carried out in a stirred batch reactor at 425°C, 50 bar (initial H_2 pressure), 900 rpm, and 60 min reaction time using transition metal (Mo, Ni, and Fe) nanoparticle catalysts. The levels of API gravity and viscosity of the upgraded oils with use of the nanoparticles were approximately 21°API and 108 cp, compared with thermal cracking alone (24°API and 53.5 cp). This moderate upgrade with nanoparticles is due to the lack of cracking functionality offered by the fixed-bed supports such as zeolite, alumina, or silica. The presence of dispersed nanoparticles, however, significantly suppressed coke formation: 4.4 wt% (MoS_2), 5.7 wt% (NiO), and 6.8 wt% (Fe_2O_3), compared with 12 wt% obtained with thermal cracking alone. The results also showed that with the dispersed nanoparticles in sulfide form, the middle distillate (177–343°C) of the upgraded oil was improved, particularly with MoS_2, which gave 50 wt% relative to 43 wt% (thermal cracking) and 28 wt% (feed oil).

Al-Marshed et al. (2015b) further investigated the use of the Fe_2O_3 dispersed nanoparticle catalyst for a range of upgrading reaction conditions. The optimum combinations of reaction parameters were found to be: temperature = 425°C, initial hydrogen pressure = 50 bar, and reaction time = 60 min, at 400 rpm agitation and iron–metal loading of 0.1 wt%. The properties of upgraded oil at the optimum condition are API gravity 21.1°, viscosity 105.75 cp, sulfur reduced by 37.54%, metals (Ni + V) reduced by 68.9%, and naphtha plus middle distillate fractions (IBP: 343°C) increased to 68 wt% relative to the feed oil (12.8°API, 1482 cp, sulfur content 3.09 wt%, metals [Ni + V] content 0.0132 wt%, and naphtha plus middle distillate fractions 28.86 wt%).

12.4 Development of In-Situ Upgrading Processes

While the concept of breaking the asphaltene-resin molecules and their molecular networks in-situ deep in the reservoir is very attractive, there are a number of difficult challenges that must be overcome. The first and foremost is how to create the appropriate temperature and pressure conditions in the reservoir so that the cracking/upgrading reactions with a meaningful conversion efficiency could be carried out. Hyne et al. (1982) showed that, without the help of catalysts, any meaningful conversion requires a temperature higher than 300°C, which is difficult to achieve in the usual steam-based heavy oil recovery processes such as SAGD. It is therefore critical to develop catalysts, and other additives such as a hydrogen donor, which will bring the meaningful upgrading reaction conditions down to those that are achievable by the current steam-based processes. As will be described below in some detail, significant progresses have been made during the last ten years, especially in the lab; however, their application in the field has not been satisfactory, mainly because there are other, associated challenges that need to be overcome. These will also be discussed below.

As described earlier, the conversion of heavy oil (or bitumen) to a much lower-viscosity oil by breaking asphaltene and resin molecules and their networks, and by deoxygenation, desulfurization, and denitrogenation at a high temperature in the presence of steam or hot water, is loosely and broadly called "aquathermolysis"; and the detailed reaction mechanisms and reaction pathways are still not known. Invariably, catalysts and/or hydrogen donors (*e.g.*, decalin, tetralin) are also employed to improve the upgrading efficiency of conversion reactions. For background information on aquathermolysis and recent progress, reviews of Maity et al. (2010) and Muraza and Galadima (2015) are available. Based on their extensive research on catalytic aquathermolysis at the China University of Geosciences (Wuhan), Wang et al. (2010) provide an organized description of different classes of upgrading reactions.

12.4.1 Heavy Oil Upgrading Reaction Kinetics and Modeling

In order to properly evaluate any heavy oil upgrading process, the modeling of the upgrading reactions and the modeling of transport of nanocatalysts in the reservoir and production of the upgraded products from the reservoir will be an important aspect. The heavy oil reaction modeling work carried out by Loria et al. (2011) and Galarraga and Pereira-Almao (2010), as a part of the University of Calgary's ultra-dispersed catalyst (UDC) development effort, is briefly reviewed in this sub-section. Hydro-processing reactions generally present three phases: solid (coke), liquid (unconverted residue and liquid products), and gas (hydrogen and product gases). In the case of all hydro-processing experiments performed with ultra-dispersed catalysts, only traces of coke were observed. In their reaction model, therefore, coke was not taken into consideration. Excess of hydrogen allows a rapid attainment of equilibrium between the gas and liquid phases and keeps a constant hydrogen concentration in the liquid phase; this was the case of the UDC experiments. Reactions for hydro-processing of bitumen can be assumed to be kinetically controlled.

12.4.1.1 Reaction Kinetics Modeling

The upgrading reaction kinetic model employed by Galarraga et al. (2011), Hassanzadeh et al. (2009), and Loria et al. (2011) is adapted from the high severity hydrocracking of heavy oils proposed by Sanchez et al. (2005). While the upgrading reactions generally produce not only the upgraded hydrocarbon liquids and associated gases but also some coke (solid) and residue (liquid), in the model, coke was not taken into account as described above. Also, with use of excess hydrogen for their upgrading experiments, a constant hydrogen concentration could be maintained in the liquid phase, which is another simplifying assumption made in the model. Figure 12.11 shows the proposed kinetic model. A further simplifying assumption in the model is that all fluids are represented by five lumped groups of components: unconverted residue (R), vacuum gas oil (VGO), distillates (D), naphta (N), and gases (G); and ten reactions. For each reaction, a kinetic expression (r_i) was formulated as a function of the lump weight percent (wt%), composition (y_i), and kinetic rate constants (k_n). All reactions were assumed to be first-order one, since hydroprocessing of hydrocarbons is of first order with respect to the hydrocarbon molecule. The reaction rates are given by the following expressions:

$$\text{Residue}: r_R = -(k_1 + k_2 + k_3 + k_4)C_R \tag{12.2}$$

$$\text{VGO}: r_{VGO} = k_1 C_R - (k_5 + k_6 + k_7)C_{VGO} \tag{12.3}$$

$$\text{Distillates}: r_D = k_2 C_R + k_5 C_{VGO} - (k_8 + k_9)C_D \tag{12.4}$$

$$\text{Naphta}: r_N = k_3 C_R + k_6 C_{VGO} + k_8 C_D - k_{10} C_N \tag{12.5}$$

$$\text{Gases}: r_G = k_4 C_R + k_7 C_{VGO} + k_9 C_D + k_{10} C_N \tag{12.6}$$

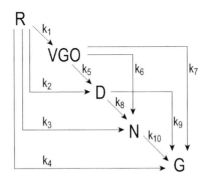

FIGURE 12.11

Modeling scheme for in-situ catalytic upgrading reactions for Athabasca bitumen. The heavy oil and its upgrading products are lumped into five pseudo components: Residue (R), vacuum gas oil (VGO), distillate (D), naphtha (N), and gases (G). The reaction routes and the corresponding reaction kinetic constants are shown. Adapted with permission from Loria et al. (2011). Copyright 2011 American Chemical Society.

where

  r_j is the rate of reaction
  C_j is the concentration of component j normalized by its initial concentration.

The dimensions for the variables present in Equations (12.2) through (12.6) were defined as r_i = wt %/time , k_n = 1/time , and y_i = wt % , where the subscript i represents the different lumps (residue, VGO, distillates, naphta, and gases); and the subscript n is the reaction number according to Figure 12.11.

The kinetic model was incorporated into a plug-flow reactor model. This model assumes an isothermal operation where reactions are assumed irreversible of the first-order. The reaction rate for each component i can be expressed by its generic mass balance:

$$\frac{dF_i}{dV} = Q\frac{dy_i}{dV} = \frac{dy_i}{d(V/Q)} = \frac{dy_i}{d\tau} \tag{12.7}$$

where

  F_i is the wt% flow rate
  V is the reactor volume
  Q is the volumetric flow rate
  τ is the residence time.

The above reaction rate equation is identical to that for a batch reactor, except that residence time replaces the batch reaction time.

Recent modeling attempt by Kapadia et al. (2013) which focuses more on the generation of various gas components from aquathermolysis reactions is also of note.

12.4.1.2 Nanoparticle Transport Experiments and Modeling

One key requirement in developing an effective in-situ upgrading process is the delivery and placement of nanocatalyst at target locations deep in the heavy oil reservoir. While there exists an extensive literature on the transport of nanoparticles in porous media as described in detail in Chapter 3, most of the earlier studies are focused on the nanoparticle adsorption and transport in the aqueous phase, and do not account for the presence of oil and gas and the importance of heat energy generation and transport. Furthermore, the transport of nanoparticle is affected by its role as a catalyst in the upgrading reactions, as described in the above section.

In this sub-section, the nanoparticle transport modeling and associated experimental work carried out by researchers at the University of Calgary (Hashemi et al. 2012; Zamani et al. 2010, 2012) will be mainly described. In their envisioned process, nanoparticle dispersion and hydrogen (or hydrogen-donor chemical) could be injected into the reservoir, *e.g.*, during the steam-assisted gravity drainage (SAGD) process. A key requirement for such an in-situ upgrading process is the placement of the nanoparticles, right at the steam chamber deep underground, so that the nanoparticles can properly serve as the upgrading reaction catalyst. Hashemi et al. (2012) and Zamani et al. (2010, 2012) showed that it is feasible to propagate the catalyst nanoparticles in the sandpack prepared from the reservoir sands without major permeability reduction. The sand media retained about 14–18% of the injected nanoparticles, mainly at the bed entrance. Carrying out the experiments at high pressure and temperature of SAGD conditions, they investigated (1) the transport, deposition, and retention of multimetallic nanoparticles through one-dimensional (1D) columns with two different permeabilities; and (2) the effects of the temperature on the transport behavior of nanoparticles.

For the nanoparticle transport experiments in sandpacks, four different metal-oxides nanoparticles dispersed in water were tested: oxides of nickel (Ni), molybdenum (Mo), tungsten (W), and an alloy of Mo, Ni, and W with metal molar fractions of 0.627, 0.181, and 0.192, respectively. The average nanoparticle diameter was ~34 nm. Hashemi et al. (2012) found that, after 1.0 PV of injection, nanoparticle concentrations sharply increased reaching a plateau, but they never reached their feed concentration, suggesting that the trapping of the nanoparticles occurred in the sandpack. Also, the level of the plateau in the low-permeability (~9 Darcy) sands is lower than the plateau level for the high-permeability (~250 Darcy) sands, again suggesting that the trapping of the nanoparticles occurred in the sandpack. The number of nanoparticles deposited on sand grains was approximately two times higher than that for high-permeability oil sands.

Another test was carried out to investigate the effect of temperature on the transport behavior of nanoparticles in an 8.9-Darcy sandpack. The experimental conditions for this test were similar to the above test, except that the temperature was 320°C instead of 300°C. The normalized effluent curves for different nanoparticles again revealed that the plateau levels did not reach the original concentration; furthermore, the plateau level for the 320°C case is lower than the 300°C case for all nanoparticles. Mass balance calculation revealed that 36.11 wt% of the injected nanoparticles was retained in the sands for the 320°C case, which is slightly higher than that for the 300°C case. The higher retention could be attributed to particle collision and aggregation because of the decrease in heavy oil viscosity in response to the temperature increase.

12.4.2 Inhibition of Formation Damage

In Section 12.3.4, it was shown that removal of asphaltenes with use of dispersed nanoparticles is a practical and potentially economic way of lowering the heavy oil viscosity, as the asphaltene content of heavy oil is generally high and the asphaltenes contribute significantly to the extremely high viscosity of heavy oil. One important application of nanoparticles is the inhibition of asphaltene deposition to avoid formation damage within the porous media (Franco et al. 2013; Mohammadi et al. 2011; Zabala et al. 2014). Precipitation of asphaltenes at reservoir rock pores during heavy oil recovery processes can significantly reduce the process performance (Akbarzadeh et al. 2007). Nanoparticles prevent asphaltene deposition by its stabilization and thus avoiding its flocculation and precipitation (Franco et al. 2013). For the purpose of recovery enhancement, in-situ adsorption of asphaltenes within the producing wellbore by nanoparticles has been demonstrated as a way not only to upgrade oil but also to reduce clogging of rock pores and well damage (Franco et al. 2013). Nanoparticles can adsorb the asphaltenes from crude oil, thus delaying the asphaltenes precipitation behavior with changes in pressure, temperature, and composition. The typical pore diameters in heavy oil reservoirs are in the order of micrometers and the nanoparticles, even with asphaltene aggregates attached to them, can flow through the reservoir without any pore plugging problem.

As pointed out in Section 12.3.5 on the use of nanocatalysts for the THAI-CAPRI in-situ combustion process, the generation of coke during the thermal cracking of heavy oil molecules and its deposition on the catalyst surface is a serious factor in lowering the effectiveness of the catalytic upgrading. As described, the use of dispersed nanoparticle catalysts suffer less of the coke deposition problem since, as the nanoparticles are freely flowing in the rock pores, they have less probability of coke deposition on their surface. When the coke deposition problem becomes severe, it can even block pore throats, hampering the production of the upgraded oil.

12.5 Field Testing of Nanocatalysts

A limited number of small-scale field testing of the nanocatalysts that showed good upgrading capability have been implemented, mainly in China. A good example of the field testing at the Henan oilfield in China is described by Chen et al. (2008). After the laboratory upgrading experiments showed ~91% viscosity reduction of EX35 oil (reaction carried out at 200°C and 6–7 MPa), the optimized injection formulation with aromatic sulfonic iron was injected into 2 wells (G61012 and G6606) after steam injection and soaking periods of 3 and 14 days, respectively. The viscosity of the oil subsequently produced from the wells was reduced by ~80% and 82%, respectively. Another example of the field testing at the Xinjiang oilfield in China is reported by Chao et al. (2012). After the laboratory upgrading tests showed ~91% viscosity reduction (from 181,000 cp) of Shengli oil when reacted at 240°C and 3 MPa, using 0.3 wt% of alkyl ester sulfonate copper, the catalyst was injected into a well (F10223) together with 1200 m^3 of superheated steam, before closing the well for 6 days. The viscosity of the oil was reduced by ~85% and the compositions of the saturated and aromatic hydrocarbons increased from 31.21% and 20.85% to 49.29% and 24.50%, respectively. Their characterization data revealed that the increase in saturates and aromatics was attributable to C–C bond cleavages of the side chains and the hydrogenation of the derived olefins.

Even though the field results were not as good as the laboratory batch reactor or sandpack results, they are still quite good in view of the difficulties of delivering the nanocatalysts to where the oil exists, creating appropriate upgrading reaction conditions (T, p, etc.), maintaining the catalyst's effectiveness for a sufficiently long time (*e.g.*, coke deposition mitigation), and ensuring that such favorable upgrading conditions are developed and maintained in as a large pore volume of reservoir as possible. Even though the details of the above and other Chinese field tests are not known, the facts that their laboratory results show good upgrading efficiency and that the field results, while small-scale, also show good performance raises the hope that well-designed nanocatalysts and their implementation may bring large-scale in-situ upgrading and oil recovery from the tremendous bitumen resources to a reality.

12.6 Concluding Remarks

As reviewed above in this chapter, active research efforts have been made on the possible use of nanoscale catalysts for the upgrading of heavy and extra-heavy oil, in-situ in the reservoir. In view of the huge amounts of the heavy oil reserve available in Alberta, Canada, and elsewhere in the world, and because the oil that is usually produced by thermal means still requires upgrading either on site or at the refinery, in-situ upgrading, if successful, will have a tremendous societal and business impact. Despite many advantages of in-situ upgrading in the reservoir, the challenges to overcome are daunting. While the conventional hydroprocessing catalysts may be applicable to the in-situ upgrading processes, the improved versions of catalysts need to have significant catalytic activity in the presence of aqueous brine and at temperatures, pressures, and hydrogen partial pressures that are generally outside of their current ranges of effectiveness. One key challenge for the in-situ upgrading reactions is the presence of steam, which is generally employed to recover the extremely high-viscosity oil. To achieve the temperature and the hydrogen partial pressure that allow the upgrading reactions in the presence of steam, the pressure in the reservoir needs to be very high, which may be difficult to maintain. Another significant drawback is that once in place, the catalyst will be difficult to retrieve for re-activation. It will likely be significantly deactivated due to the deposition of coke, asphaltenes, and metals from the crude.

As mentioned at the beginning of this chapter, the persistent development and field trials of the combined technology of horizontal well and hydraulic fracturing brought the amazing shale oil and gas production in the US and Canada, which for many decades was considered to be impossible to achieve. While a large-scale economic development of extra-heavy oil currently looks difficult, the above shale example suggests that, with a new technology such as the development of drastically improved nanocatalysts and their effective implementation in the field, heavy oil resource development may see a bright light in the near future.

References

Abu Tarboush, B. J., and Husein, M. M. (2012) Adsorption of asphaltene from heavy oil onto in situ prepared NiO nanoparticles. *J. Colloid Interface Sci.*, 378(1), 64–69.

Akbarzadeh, K., Hammami, A., Kharrat, A. et al. (2007) Asphaltenes – Problematic but rich in potential. *Oilfield Rev.*, Summer, 22–43.

Al-Marshed, A., Hart, A., Leeke, G., Greaves, M., and Wood, J. (2015a) Effectiveness of different transition metal dispersed catalysts for in situ heavy oil upgrading. *Ind. Eng. Chem. Res.*, 54, 10645–10655.

Al-Marshed, A., Hart, A., Leeke, G., Greaves, M., and Wood, J. (2015b) Optimization of heavy oil upgrading using dispersed nanoparticulate iron oxide as a catalyst. *Energy & Fuels*, 29, 6306–6316.

Arab, D., Kantzas, A., and Bryant, S. L. (2018) Nanoparticle-Fortified Emulsification of Heavy Oil. SPE-190377, SPE EOR Conference Oil and Gas West Africa, Mar. 26–28, Muscat, Oman.

Bera, A., and Babadagli, T. (2015) Status of electromagnetic heating for enhanced heavy oil/bitumen recovery and future prospects: A review. *Appl. Energy*, 151, 206–226.

Butler, R. M. 1997. *Thermal Recovery of Oil and Bitumen*, GravDrain, Alberta, Canada.

Cao, N., Mohammed, M. A., and Babadagli, T. (2017) Wettability Alteration of Heavy-Oil-Bitumen-Containing Carbonates by Use of Solvents, High-pH Solutions, and Nano/Ionic Liquids. *SPE Reservoir Evaluation Eng.*, May, 363–371.

Chao, K., Chen, Y., Liu, H., Zhang, X., and Li, J. (2012) Laboratory experiments and field test of a difunctional catalyst for catalytic aquathermolysis of heavy oil. *Energy & Fuels*, 26, 1152–1159.

Chen, Y., Wang, Y., Wu, C., and Xia, F. (2008) Laboratory experiments and field tests of an amphiphilic metallic chelate for catalytic aquathermolysis of heavy oil. *Energy & Fuels*, 22, 1502–1508.

Chen, Y., Wang, Y., Lu, J., and Wu, C. (2009) The viscosity reduction of nano-keggin K3PMo12O40 in catalytic aquathermolysis of heavy oil. *Fuel*, 88(8), 1426–1434.

Clark, P. D., Clarke, R. A., Hyne, J. B., and Lesage, K. L. (1990a) Studies on the effect of metal species on oil sands undergoing steam treatments. *AOSTRA J. Res.*, 6(1), 53–64.

Clark, P. D., Clarke, R. A., Hyne, J. B., and Lesage, K. L. (1990b) Studies on the chemical reactions of heavy oils under steam stimulation conditions. *AOSTRA J. Res.*, 6(1), 29–39.

Clark, P. D., and Hyne, J. B. (1984) Steam-oil chemical reactions: Mechanisms for the aquathermolysis of heavy oils. *AOSTRA J. Res.*, 1(1), 15–20.

Clark, P. D., and Kirk, M. J. (1994) Studies on the upgrading of bituminous oils with water and transition metal catalysts. *Energy & Fuels*, 8, 380–387.

Ehtesabi, H., Ahadian, M. M., Taghikhani, V., and Ghazanfari, M. H. (2013) Enhanced heavy oil recovery in sandstone cores using TiO2 nanofluids. *Energy & Fuels*, 28(1), 423–430.

Fan, H., Zhang, Y., and Lin, Y. (2004) The catalytic effects of minerals on aquathermolysis of heavy oils. *Fuel*, 83(14–15), 2035–2039.

Franco, C. A., Nassar, N. N., Ruiz, M. A., Pereira-Almao, P. R., and Cortes, F. B. (2013) Nanoparticles for inhibition of asphaltenes damage: Adsorption study and displacement test on porous media. *Energy & Fuels*, 27, 2899–2907.

Galarraga, C. E., and Pereira-Almao, P. (2010) Hydrocracking of Athabasca bitumen using submicronic multimetallic catalysts at near in-reservoir conditions. *Energy & Fuels*, 24(4), 2383–2389.

Galarraga, C. E., Scott, C., Loria, H., and Pereira-Almao, P. (2011) Kinetic models for upgrading Athabasca bitumen using unsupported NiWMo catalysts at low severity conditions. *Ind. Eng. Chem. Res.*, 51, 140–146.

Greaves, M., and Xia, T. X. (2000) Upgrading Athabasca Tar Sand Using THAI—Toe-To-Heel Air Injection. SPE 65524, SPE/CIM International Conference Horizontal Well Technology, November 6–8, Calgary, Canada.

Hamedi-Shokrlu, Y., and Babadagli, T. (2010) Effects of Nano Sized Metals on Viscosity Reduction of Heavy Oil/Bitumen during Thermal Applications. SPE 137540, SPE Canadian Unconventional Resources and International Petroleum Conference, October 19–21, Calgary, Canada.

Hamedi-Shokrlu, Y., and Babadagli, T. (2011) Transportation and Interaction of Nano and Micro Size Metal Particles to Improve Thermal Recovery of Heavy Oil. SPE 146661, SPE Annual Technical Conference and Exhibition, October 30–November 2, Denver, Colorado, USA.

Hamedi-Shokrlu, Y., and Babadagli, T. (2013) In-situ upgrading of heavy oil/bitumen during steam injection by use of metal nanoparticles: A study on in-situ catalysis and catalyst transportation. *SPE Reservoir Eval. Eng.*, 16(3): 333–344.

Hamedi-Shokrlu, Y., Maham, Y., Tan, X., Babadagli, T., and Gray, M. (2013) Enhancement of the efficiency of in-situ combustion technique for heavy-oil recovery. *Fuel*, 105, 397–407.

Hart, A., Greaves, M., and Wood, J. (2015a) A comparative study of fixed-bed and dispersed catalytic upgrading of heavy crude oil using CAPRI. *Chem. Eng. J.*, 282, 213–223.

Hart, A., Lewis, C., White, T., Greaves, M., and Wood, J. (2015b) Effect of cyclohexane as hydrogen-donor in ultradispersed catalytic upgrading of heavy oil. *Fuel Process. Technol.*, 138, 724–733.

Hart, A., Omajali, J. B., Murray, A. J., Macaskie, L. E., Greaves, M., and Wood, J. (2016) Comparison of the effects of dispersed noble metal (Pd) biomass supported catalysts with typical hydrogenation (Pd/C, Pd/Al2O3) and hydrotreatment catalysts (CoMo/Al2O3) for in-situ heavy oil upgrading with Toe-to-Heel Air Injection (THAI). *Fuel*, 180, 367–376.

Hart, A., Wood, J., and Greaves, M. (2017) In situ catalytic upgrading of heavy oil using a pelletized Ni-Mo/ Al2O3 catalyst in the THAI process. *J. Pet. Sci. Eng.*, 156, 958–965.

Hashemi, R., Nassar, N., and Almao, P. (2013a) Enhanced heavy oil recovery by in situ prepared ultradispersed multimetallic nanoparticles: A study of hot fluid flooding for athabasca bitumen recovery. *Energy & Fuels*, 27(4), 2194–2201.

Hashemi, R., Nassar, N., and Almao, P. (2013b) In situ upgrading of athabasca bitumen using multimetallic ultradispersed nanocatalysts in an oil sands packed-bed column: Part 1. Produced liquid quality enhancement. *Energy & Fuels*, 28, 1338–1350.

Hashemi, R., Nassar, N. N., and Pereira Almao, P. (2012) Transport behavior of multimetallic ultradispersed nanoparticles in an oil-sands-packed bed column at a high temperature and pressure. *Energy & Fuels*, 26, 1645–1655.

Hashemi, R., Nassar, N. N., and Pereira Almao, P. (2014) Nanoparticle technology for heavy oil in-situ upgrading and recovery enhancement: Opportunities and challenges. *Appl. Energy*, 133, 374–387.

Hassanzadeh, H., Galarraga, C. E., Abedi, J., Scott, C. E., Chen, Z., and Almao, P. P. (2009) Modelling of Bitumen Ultradispersed Catalytic Upgrading Experiments in a Batch Reactor. Canadian International Petroleum Conference, June 16–18, Calgary, Canada.

He, B., Chen, Q., Castanier, L. M, Kovscek, A. R. (2005) Improved In-Situ Combustion Performance with Metallic Salt Additives. SPE 93901, SPE Western Regional Meeting, Mar. 30–Apr. 1, Irvine, CA, USA.

Hirasaki, G. J. (1989) The steam-foam process. *J. Pet. Technol.*, May, 449–456.

Hyne, J. B., Greidanus, J. W., Tyrer, J. D., Verona, D., Rizek, C., Clark, P. D., Clarke, R. A., and Koo, J. (1982) Aquathermolysis of Heavy Oils. Presented at the Second International Conference on Heavy Crudes and Tar Sands, UNITAR, February 7–17, Caracas, Venezuela.

Iskandar, F., Fitriani, P., Merissa, S., Mukti, R. R., Khairurrijal, K., and Abdula, M. (2014) Fe3O4/zeolite nanocomposites synthesized by microwave assisted coprecipitation and its performance in reducing viscosity of heavy oil. *AIP Conf. Proc.*, 1586, 132–135.

Iskandar, F., Dwinanto, E., Abdullah, M., Khairurrijal, K., and Muraza, O. (2016) Viscosity reduction of heavy oil using nanocatalyst in aquathermolysis reaction. *KONA Powder Part. J.*, 33, 3–16.

Kapadia, P. R., Kallos, M. S., and Gates, I. D. (2013) A new reaction model for aquathermolysis of athabasca bitumen. *Can. J. Chem. Eng.*, 91(3): 475–482.

Khajehpour, M., Etminan, S. R., Goldman, J., and Wassmuth, F. (2016) Nanoparticles as Foam Stabilizer for Steam-Foam Process. SPE 179826, SPE EOR Conference West Asia, Mar. 21–23, Muscat, Oman.

Kozhevnikov, I. V. (1998) Catalysis by heteropoly acids and multicomponent polyoxometalates in liquid-phase reactions. *Chem. Rev.*, 98, 171–198.

Lake, L. W., Johns, R. T., Rossen, W. R., and Pope, G. A. (2014) *Fundamentals of Enhanced Oil Recovery*. Society of Petroleum Engineering., Richardson, TX.

Laurent, E., and Delmon, B. (1994) Influence of water in the deactivation of a sulfide NiMoy-Al$_2$O$_3$ catalyst during hydrodeoxigenation. *J. catalysis*, 146, 281–285, 288–291.

Li, W., Zhu, J., and Qi, J. (2007) Application of nano-nickel catalyst in the viscosity reduction of liaohe-heavy oil by aquathermolysis. *J. Fuel Chem. Technol.*, 35(2), 176180.

Li, J., Chen, Y., Liu, H., Wang, P., and Liu, F. (2013) Influences on the aquathermolysis of heavy oil catalyzed by two different catalytic ions: Cu2+ and Fe3+. *Energy & Fuels*, 27, 2555–2562.

Loria, H., Ferrer, G., Stull, C., and Almao, P. (2011) Kinetic modeling of bitumen hydroprocessing at in-reservoir conditions employing ultradispersed catalysts. *Energy & Fuels*, 25(4), 1364–1372.

Luo, P., and Gu, Y. (2006) Effects of asphaltene content on the heavy oil viscosity at different temperatures. *Fuel*, 86(7–8): 1069–1078.

Maity, S. K., Ancheyta, J., and Marroquin, G. (2010) Catalytic aquathermolysis used for viscosity reduction of heavy crude oils: A review. *Energy & Fuels*, 24, 2809–2816.

Mohammad, A. A. A., and Mamora, D. D. (2008) In Situ Upgrading of Heavy Oil under Steam Injection with Tetralin and Catalyst. SPE-117604, International Thermal Operations Heavy Oil Symposium, Oct. 20–23, Calgary, Canada.

Mohammadi, M., Akbari, M., Fakhroueian, Z., Bahramian, A., Azin, R., and Arya, S. (2011) Inhibition of asphaltene precipitation by TiO2, SiO2, and ZrO2 nanofluids. *Energy Fuels*, 25, 3150–3156.

Muraza, O., and Galadima, A. (2015) Aquathermolysis of heavy oil: A review and perspective on catalyst development. *Fuel*, 157, 219–231.

Nassar, N. N., Hassan, A., Carbognani, L., Lopez-Linares, F., and Pereira-Almao, P. (2012) Iron oxide nanoparticles for rapid adsorption and enhanced catalytic oxidation of thermally cracked asphaltenes. *Fuel*, 95, 257–262.

Nassar, N. N., Hassan, A., and Pereira-Almao, P. (2011a) Effect of surface acidity and basicity of aluminas on asphaltene adsorption and oxidation. *J. Colloid Interface Sci.*, 360(1), 233–238.

Nassar, N. N., Hassan, A., and Pereira-Almao, P. (2011b) Metal oxide nanoparticles for asphaltene adsorption and oxidation. *Energy & Fuels*, 25(3), 1017–1023.

Nassar, N. N., and Husein, M. M. (2010) Ultradispersed particles in heavy oil: Part I, preparation and stabilization of iron oxide/hydroxide. *Fuel Process. Technol.*, 91(2), 164–168.

Nassar, N. N., Hassan, A., and Pereira-Almao, P. (2011) Application of nanotechnology for heavy oil upgrading: Catalytic steam gasification/cracking of asphaltenes. *Energy & Fuels*, 25, 1566–1570.

Noorlaily, P., Nugraha, M. I., Khairurrijal, Abdullah, M., and Iskandara, F. (2013) Ethylene glycol route synthesis of nickel oxide nanoparticles as a catalyst in aquathermolysis. *Mater. Sci. Forum*, 737, 93–97.

Nugraha, M. I., Noorlaily, P., Abdullah, M., Khairurrijal, and Iskandara, F. (2013) Synthesis of NixFe3-xO4 nanoparticles by microwave-assisted coprecipitation and their application in viscosity reduction of heavy oil. *Mater. Sci. Forum*, 737, 204–208.

Nurhayati, T., Iskandara, F., Abdullah, M., and Khairurrijal. (2013) Synthesis of hematite (α-Fe2O3) nanoparticles using microwave-assisted calcination method. *Mater. Sci. Forum*, 737, 197–203.

Ovalles, C., Filgueiras, E., Morales, A., Rojas, I., de Jesus, J. C., and Berrois, I. (1998) Use of a dispersed molybdenum catalysts and mechanistic studies for upgrading extra-heavy crude oil using methane as source of hydrogen. *Energy & Fuel*, 12, 379–385.

Ovalles, C., Filgueiras, E., Morales, A., Scott, C. E., Gonzalez-Gimenez, F., and Embaid, B. P. (2003) Use dispersed catalysts for upgrading extra-heavy crude oil using methane as source of hydrogen. *Fuel*, 82, 887–892.

Prakash, A., Zhu, H., Jones, C. J., Benoit, D. N., Ellsworth, A. Z., Bryant, E. L., and Colvin, V. L. (2009) Bilayers as phase transfer agents for nanocrystals prepared in nonpolar solvents. *ACS Nano*, 3, 2139–2146.

Sanchez, S., Rodriguez, M. A., and Ancheyta, J. (2005) Kinetic model for moderate hydrocracking of heavy oils. *Ind. Eng. Chem. Res.*, 44, 9409.

Shah, A., Fishwick, R., Wood, J., Leeke, G., Rigby, S., and Greaves, M. (2010) A review of novel techniques for heavy oil and bitumen extraction and upgrading. *Energy Environ. Sci.*, 3(6), 700–714.

Shah, A., Fishwick, R. P., Leeke, G. A., Wood, J., Rigby, S. P., and Greaves, M. (2011) Experimental optimization of catalytic process in situ for heavy-oil and bitumen upgrading. *J. Can. Pet. Tech.*, November/December, 33–46.

Shayan, N. N., and Mirzayi, B. (2015) Adsorption and removal of asphaltene using synthesized maghemite and hematite nanoparticles. *Energy & Fuels*, 29(3), 1397–1406.

Shen, L., Laibinis, P. E., and Hatton, T. A. (1999) Bilayer surfactant stabilized magnetic fluids: Synthesis and interactions at interfaces. *Langmuir*, 15, 447–453.

Speight, J. G. (1991) *The Chemistry and Technology of Petroleum*, 2nd ed., Marcel Dekker, New York.

Wang, Y., Chen, Y., He, J., Li, P., and Yang, C. (2010) Mechanism of catalytic aquathermolysis: Influences on heavy oil by two types of efficient catalytic ions: Fe3+ and Mo6+. *Energy & Fuels*, 24, 1502–1510.

Wang, H., Wu, Y., He, L., and Liu, Z. (2012) Supporting tungsten oxide on zirconia by hydrothermal and impregnation methods and its use as a catalyst to reduce the viscosity of heavy crude oil. *Energy & Fuels*, 26, 6518–6527.

Weissman, J. G. (1997) Review of processes for downhole catalytic upgrading of heavy crude oil. *Fuel Process. Tech.*, 50, 199–213.

Weissman, J. G., Kessler, R. V., and Sawicki, R. A. (1996) Down-hole catalytic upgrading of heavy crude oil. *Energy & Fuels*, 10, 883–889.

Wen, S., Zhao, Y., Liu, Y., and Hu, S. (2007) A Study on Catalytic Aquathermolysis of Heavy Crude Oil during Steam Stimulation. SPE-106180, SPE International Symposium Oilfield Chemistry, February 28–March 2, Houston, TX.

Wu, C., Su, J., Zhang, R., Lei, G., and Cao, Y. (2013) The use of nano-nickel catalyst for upgrading extra-heavy oil by an aquathermolysis treatment under steam injection conditions. *Pet. Sci. Technol.*, 31, 2211–2218.

Yi, S., Babadagli, T., and Li, H. A. (2018) Use of nickel nanoparticles for promoting aquathermolysis reaction during cyclic steam stimulation. *Soc. Pet. Eng. J.*, 23(1), 145–156.

Yoon, K. Y., Kotsmar, C., Ingram, D. R., Huh, C., Bryant, S. L., Milner, T. E., and Johnston, K. P. (2011) Stabilization of superparamagnetic iron oxide nanoclusters in concentrated brine with cross-linked polymer shells. *Langmuir*, 27, 10962–10969.

Yufeng, Y., Shuyuan, L., Fuchen, D., and Hang, Y. (2009) Change of asphaltene and resin properties after catalytic aquathermolysis. *Pet. Sci.*, 6(2), 194–200.

Zabala, R., Mora, E., Botero, O., Cespedes, C., Guarin, L., Franco, C. A., Cortes, F. D., Patino, J. E., and Ospina, N. (2014) Nanotechnology for Asphaltenes Inhibition in Cupiagua South Well. Paper at International Petroleum Technology Conference, January 19-22, Doha, Qatar.

Zamani, A., Maini, B., and Pereira-Almao, P. (2010) Experimental study on transport of ultra-dispersed catalyst particles in porous media. *Energy & Fuels*, 24, 4980–4988.

Zamani, A., Maini, B., and Pereira-Almao, P. (2012) Flow of nanodispersed catalyst particles through porous media: Effect of permeability and temperature. *Can. J. Chem. Eng.*, 90, 304–314.

Zhong, L. G., Liu, Y. J., and Fan, H. F. (2003) Liaohe extra-heavy crude oil underground aquathermolytic treatments using catalyst and hydrogen donors under steam injection conditions. SPE International Improved Oil Recovery Conference Asia Pacific, October 20–21, Kuala Lumpur, Malaysia.

13

Conclusions and Future Directions

13.1 Conclusions

Throughout this book, we have seen how the various unique properties of nanoparticles may be exploited to provide innovative solutions to problems in the oil and gas industry. Many of these solutions involve manipulating the surface chemistry of nanoparticles through some sort of pre-treatment or surface coating; others rely on the intrinsic properties of the nanoparticle core material; while others simply rely on the shape and morphology of the nanoparticle. What these solutions all have in common is that they cannot be achieved with other chemical substances; as an example, the magnetic hyperthermia exploited for flow assurance in Chapter 8 is only possible in nano-sized fragments of iron oxide, since the Néel relaxation of the magnetic domains is necessary for the process to work. Small, iron-bearing compounds would not work. Overall, we hope to have shown that the term "nano" is more than a buzzword, but rather that it refers to a particular class of materials with unique properties that can accomplish things that other materials cannot.

The applications presented in this book have in some cases solved problems, but in most cases, the applicability in solutions has been suggested or shown in the laboratory, but more work remains to be done to bring the solution to fruition. In addition, there are several overarching challenges in the oil and gas industry that will need to be addressed continually for the foreseeable future. Here, we briefly describe these challenges and discuss the advances in nanoparticle science that will be necessary to meet them.

13.2 Future Challenges in the Oil and Gas Industry

Minimizing the environmental footprint of oil and gas exploration and production is a key component of producers' ability to retain a license to operate, which refers to societal acceptance of their activities. Hydraulic fracturing and offshore drilling tend to be some of the most contentious components of oil and gas operations, and oil and gas companies constantly strive to mitigate risks associated with these activities as doing so is vital to their maintaining the license to operate. Some specific challenges in this regard include minimizing the amount of water and chemicals used during drilling and completion, and responsibly treating and disposing of water and chemicals at the end of operations. We have covered several nanotechnology solutions to these problems, but further work is necessary to bring them to field readiness.

A second but equally important issue is that of safety. Many safety risks exist in oil and gas extraction, and many lives have been lost throughout the history of the industry as a result of these risks. As exploration and production activities extend to deeper, hotter, and higher-pressure environments, maintaining the safety of operations requires new materials for activities such as controlling pressure and maintaining formation integrity during drilling, safely isolating producing intervals with cement, and preventing blockages in pipeline flows. The recent advances in the oil and gas industry related to these challenges mirror those in the broader materials science community, with new nanomaterials being developed with superior performance at high pressure and temperature. The key for moving forward in nanotechnology-based safety improvements are identifying those components that require improved performance and determining the optimal nanomaterial formulations to give the desired behavior.

A final challenge is that of maximizing oil and gas production while minimizing the associated costs. This has been a particular concern during the period of low commodity prices in the mid-2010s, but economically efficient recovery is desirable even during times of high commodity prices. Reducing costs can include tertiary recovery efforts in fields with existing infrastructure, minimizing new construction costs; shortening the amount of time required to drill a well through better bottomhole assembly and drilling fluid design; reducing the number and duration of interventions in subsea pipelines; and reducing the volume of chemicals required in completion operations. As we have shown, there are nanotechnology-based solutions for all these efforts. Sustained research and development will help increase the market for nanoparticles and ultimately reduce their cost as they are produced at larger and larger scales.

13.3 Challenges in Nanotechnology Implementation

While nanotechnology applications in problems in the oil and gas industry are manifold, there are still some significant challenges that need to be overcome on the side of nanoscience. First, and probably most significant, is a persistent gap in understanding of the fundamental governing principles underlying these applications. Many times throughout this book, we have commented that such-and-such study has yielded compelling results, but we still do not understand the mechanisms behind them. Studies often slip into *post hoc* assessments of behavior ("we found this result, and it was probably because of this") or somewhat myopic extensions of results obtained in different applications ("Carbon nanotubes improve lubricity in motor oil and we found that they also do in drilling fluid. QED."). We have often heard nanoparticles referred to colloquially as "special sauce," "fairy dust," or (our favorite) "foo-foo powder." In future research, much more attention needs to be paid to the questions of why and how nanotechnology applications work, rather than simply focusing on the fact that they *do* work. We hope that this book serves as a foundation for such future studies.

Related to this challenge is the parallel challenge of process optimization. Ideally, oil and gas applications should use as few nanoparticles as possible to reduce cost and complexity. However, many studies fail to take this into account, either reporting behavior at excessive nanoparticle concentrations (>15 wt%) or not investigating behavior at a range of concentrations. An ideal study of a nanotechnology application would report nanoparticle size and composition, surface coating composition and amount, surface charge (zeta potential) if applicable, mass fraction in dispersion, and properties of the dispersion itself (chemical composition, pH, electrical conductivity, dissolved species). Then, even if a full parameter space investigation is not undertaken by the investigators, a subsequent study could then compare results by varying one or more of these properties while keeping the others consistent. Now, we realize that many factors may preclude such full disclosure, including intellectual property concerns, proprietary data, and so on. However, in general, more information is better, and this is certainly something to strive for.

Index